The Lighthouse and the Observatory

An observatory and a lighthouse form the nexus of this major new investigation of science, religion, and the state in late Ottoman Egypt. By linking astronomy, imperial bureaucrats, traditionally educated Muslim scholars, and reformist Islamic publications, such as *The Lighthouse*, Daniel Stolz reveals new connections between the making of knowledge, the performance of piety, and the operation of political power through scientific practice. Contrary to ideas of Islamic scientific decline, Muslim scholars in the nineteenth century used a dynamic tradition of knowledge to measure time, compute calendars, and predict planetary positions. The rise of a "new astronomy" owed much to projects of political and religious reform: from the strengthening of the multiple empires that exercised power over the Nile Valley; to the "modernization" of Islamic centers of learning; to the dream of a global Islamic community that would rely on scientific institutions to coordinate the timing of major religious duties.

Daniel A. Stolz is a visiting assistant professor in the Department of History and the Science in Human Culture Program at Northwestern University, USA.

SCIENCE IN HISTORY

Series Editors

Simon J. Schaffer, University of Cambridge

James A. Secord, University of Cambridge

Science in History is a major series of ambitious books on the history of the sciences from the mid eighteenth century through the mid twentieth century, highlighting work that interprets the sciences from perspectives drawn from across the discipline of history. The focus on the major epoch of global economic, industrial and social transformations is intended to encourage the use of sophisticated historical models to make sense of the ways in which the sciences have developed and changed. The series encourages the exploration of a wide range of scientific traditions and the interrelations between them. It particularly welcomes work that takes seriously the material practices of the sciences and is broad in geographical scope.

The Lighthouse and the Observatory

Islam, Science, and Empire in Late Ottoman Egypt

Daniel A. Stolz
Northwestern University

CAMBRIDGE
UNIVERSITY PRESS

CAMBRIDGE
UNIVERSITY PRESS

University Printing House, Cambridge CB2 8BS, United Kingdom

One Liberty Plaza, 20th Floor, New York, NY 10006, USA

477 Williamstown Road, Port Melbourne, VIC 3207, Australia

314-321, 3rd Floor, Plot 3, Splendor Forum, Jasola District Centre, New Delhi - 110025, India

79 Anson Road, #06-04/06, Singapore 079906

Cambridge University Press is part of the University of Cambridge.

It furthers the University's mission by disseminating knowledge in the pursuit of
education, learning and research at the highest international levels of excellence.

www.cambridge.org
Information on this title: www.cambridge.org/9781316647257
DOI: 10.1017/9781108164672

First published 2018
First paperback edition 2019

A catalogue record for this publication is available from the British Library

Library of Congress Cataloging in Publication data
Names: Stolz, Daniel A., 1985– author.
Title: The Lighthouse and the observatory : Islam, science, and empire in
late Ottoman Egypt / Daniel A. Stolz, Northwestern University.
Description: Cambridge : Cambridge University Press, 2018. |
Series: Science in history | Includes bibliographical references and index.
Identifiers: LCCN 2017038705 |
ISBN 9781107196339 (hardback : alk. paper)
Subjects: LCSH: Astronomy, Egyptian. | Astronomy – Egypt – History. |
Science and state – Egypt. | Islam and science – History.
Classification: LCC QB20 .S76 2018 | DDC 520.96209/034–dc23
LC record available at https://lccn.loc.gov/2017038705

ISBN 978-1-107-19633-9 Hardback
ISBN 978-1-316-64725-7 Paperback

For Judy

Contents

Figures

Tables

Acknowledgments

I owe this book, first, to my two institutional homes and intellectual communities of the last decade. At Princeton, Muhammad Qasim Zaman, Michael Gordin, M. Şükrü Hanioğlu, and Michael Cook were ideal readers; many pages still bear the imprint of their encouragement, criticism, and inspiration. At Northwestern, Ken Alder, Helen Tilley, and Steve Epstein, in addition to offering valuable criticism at several stages of the project, have gone to extraordinary lengths to help me see it to completion.

I am deeply indebted to Cambridge University Press: Lucy Rhymer, Simon Schaffer, and Jim Secord have all been generous with their interest and support. I would also like to acknowledge the press's anonymous readers, who made valuable suggestions. I thank Cassi Roberts and Helen Flitton for overseeing production, and Lawrence Osborn for copy-editing.

Parts of chapters 2 and 3 were previously published by Cambridge University Press in my article, "Positioning the Watch Hand: 'Ulama' and the Practice of Mechanical Timekeeping in Cairo, 1737–1874," *International Journal of Middle East Studies* 47 (2015): 489–510.

Many colleagues have taken time to read and comment on portions of the manuscript. For their gentle but perceptive criticisms, I thank On Barak, Elesha Coffman, Mariana Craciun, Alireza Doostdar, Jörg Matthias Determann, Edna Bonhomme, Stefanie Graeter, Sarit Kattan Gribetz, Camilo Leslie, Fredrik Meiton, Jaimie Morse, Projit Mukharji, Junaid Quadri, Paul Ramírez, Aaron Rock-Singer, Aviva Rothman, Tunç Şen, Ahmed El Shamsy, the members of the Religion and Culture seminar at the Center for the Study of Religion (Princeton), 2011–2012, and the members of the Program Seminar in the History of Science (Princeton), 2011–2013. I would also like to thank Andras Hamori for consulting on the translation in the appendix.

Several generous fellowships and grants enabled the research for this book. At Princeton, I benefited from the support of the Center for the Study of Religion, the Hyde Fellowship, and the Princeton

Institute for International and Regional Studies. At Northwestern, the Graduate School, the Science in Human Culture Program, and the History Department have all provided research funding. I would also like to acknowledge the support of the Whiting Foundation, as well as a Bernadotte E. Schmitt research grant from the American Historical Association.

I thank the staff of the Princeton University Firestone Library, the Northwestern University Library, the Egyptian National Archives and Library, the al-Azhar Library, the British Library, the British National Archives, the Royal Astronomical Society Library (London), the Science Museum's facility in Wroughton, the Cambridge University Special Collections Library, the Paris Observatory Library, the Bibliothèque nationale de France, and the University of Michigan Special Collections Library. For extraordinary help obtaining manuscripts, it is a pleasure to acknowledge Avner Ben-Zaken and Dr. Mustafa Budak.

Dr. Ashraf Shaker of the Egyptian National Research Institute of Astronomy and Geophysics welcomed me to the Helwan Observatory and shared his unrivaled knowledge of the institution's history. Although this book has strayed far from our conversations, I am grateful for his generosity, and for his selfless efforts to document and preserve the modern history of astronomy in Egypt.

Many of the people who made these pages possible cannot be acknowledged here: from friends and family, to fellow-travelers of the archive, their contributions are too numerous to describe. To be complete, however, this book must make space for Judy, and for our children, who have made space for it in their home.

Note on Chronology and Transliteration

One reason that people needed astronomers in late Ottoman Egypt was to make sense of the relationship between several different calendars that were commonly in use. For the convenience of today's readers, I have simplified this situation: in general, I use Gregorian dates. When discussing a date given in a text, however, I use the calendar employed by the text's author, which is the *hijrī* ("Islamic") system, unless otherwise indicated. The Gregorian date follows in parentheses. While some degree of confusion is inevitable, my hope is that it might deepen the reader's appreciation for the necessity of the chronological knowledge that was cultivated by many of the people whose stories are told in this book.

Transliteration of Arabic terms follows the style of the *International Journal of Middle East Studies*. Outside of technical terms (e.g., *mīqāt*), diacritical marks are generally omitted, excepting ʿ*ayn* and non-initial *hamza*.

Transliteration of Ottoman Turkish terms follows modern Turkish orthography where possible.

Abbreviations

Azhar	al-Maktaba al-Azhariyya, Cairo
BJHS	*British Journal for the History of Science*
BnF	Bibliothèque nationale de France
BSOAS	*Bulletin of the School of Oriental and African Studies*
CEDEJ	Centre d'études et de documentation économiques, juridiques et sociales (Cairo)
CSSH	*Comparative Studies in Society and History*
DWQ	Dar al-Watha'iq al-Qawmiyya (Egyptian National Archives)
ENL	Egyptian National Library
IJMES	*International Journal of Middle East Studies*
IRCICA	İslam Tarih, Sanat ve Kültür Araştırma Merkezi
JAOS	*Journal of the American Oriental Society*
JRAS	*Journal of the Royal Asiatic Society*
Michigan	Special Collections Library, University of Michigan
MNRAS	*Monthly Notices of the Royal Astronomical Society*
QNL	Qatar National Library
RAS	Royal Astronomical Society
RGO	Royal Observatory, Greenwich

For scientific manuscripts in the ENL, I have adopted the sub-collection and topical abbreviations used in King's *Survey*.

DM	Dar al-Kutub: Miqat
K	Astronomy and Mathematics (MSS acquired between 1930 and 1950)
ṬM	Talʿat: Miqat
TR	Taymur: Riyada
ṬR	Talʿat: Riyada
ZK	Zakiyya

Introduction: Astronomy, Empire, and Islamic Authority at the End of Days

In the town of Zagazig, in the eastern Nile Delta, a schoolteacher named Shaykh Mustafa Muhammad al-Shafiʿi brought good news to readers of his almanac for the year 1899: humanity would survive the middle of November. A need for existential reassurance had originated in Vienna, where the controversial geologist and science popularizer Rudolf Falb had predicted that a comet would demolish the earth on 13 November.[1] Journalists and astronomers from England to Russia debated Falb's prediction, their attitudes varying from frenzied alarm to dismissive condescension, as imminent doom became a matter of discussion in many parts of the world.[2] In Cairo, the daily newspaper *al-Muʾayyad* published a series of articles, beginning on 28 October, in which Shafiʿi elaborated on the errors underlying Falb's prediction.[3] The satirical journal *Himarat Munyati* chimed in the following week, also debunking Falb.[4] For many in Egypt, however, definitive relief came only on 15 November, when astronomers at the state observatory in Cairo gave the all-clear.[5]

---

[1] Falb later revised his prediction, claiming that the earth would merely pass through the comet's tail, which might flush the atmosphere with poisonous gases. See "The Menacing Comet," *Fortnightly Review* 66, November 1899, p. 769. On the debate over Falb's theory of subterranean lava tides in the history of seismology, see Deborah Coen, *The Earthquake Observers* (Chicago: University of Chicago Press, 2013), 53–55. For Falb's life, see H.G. Heller, *Rudolf Falb: Eine Lebens- und Charakterskizze nach persönlichen Erinnerungen* (Berlin: Friedrich Gottheiner, 1903).

[2] On Falb's prediction in Polish and Russian pamphlets, see Ekaterina Melnikova, "Eschatological Expectations at the Turn of the Nineteenth-Twentieth Centuries: The End of the World is [Not] Nigh?" *Forum for Anthropology and Culture* 1 (2004), 261. For reactions (mostly satirical) in the British press, see "Biela's Comet and I," *Fun*, 7 November 1899, p. 147; "When the End of the World Is Nigh," *Review of Reviews*, March 1894, p. 260; and "Old and New Astronomy," *Quarterly Review* 188, no. 375 (July 1898), 138.

[3] See *al-Muʾayyad* nos. 2901–2905, beginning on 23 Jumada II, 1317 (28 October 1899). Though I have not been able to locate an original copy of Shafiʿi's almanac, the relevant entry is preserved in a contemporaneous manuscript: see Mustafa Muhammad al-Shafiʿi, *Risala fi dhawat al-adhnab wa-takhtiʾat man zaʿm fanaʾ al-ʿalam bi-istidam al-ard bi-najm dhi dhanab*, MS TR 82, ENL, f. 2.

[4] "Al-Najma dhat al-dhanab," *Himarat Munyati* 2, no. 37 (3 Rajab 1317).

[5] See *al-Ahram*, 14 November 1899 and 15 November 1899.

1

Even as consensus emerged on the fallacy of Falb's prediction, the comet scare's broader significance for Egypt became a matter of debate. For Mustafa al-Shafi'i, refuting Falb offered an opportunity to promote a specifically Islamic understanding of the relationship between celestial motion and the end of days, which he argued would not occur for another 200,104 years.[6] By contrast, for the Anglophile publishers of another Cairo daily, al-Muqattam, Egyptians' alarm illustrated their society's need for a new type of "education and culture," which would uproot the tenacious hold of "fabulous tales" on unenlightened minds.[7] Meanwhile, shortly after Shafi'i's articles appeared in print, an anonymous writer transcribed them into a handwritten compilation of Arabic texts entitled *A Treatise on Comets, and on disproving whoever claims that the world will be destroyed by collision of the earth with a comet.*[8] In addition to Shafi'i's articles (now a single "treatise"), the seventy-folio manuscript includes excerpts from several other Arabic newspapers and a number of Islamic historical works.[9] The manuscript ends with "the verses regarding the destruction of the world" (*al-āyāt fī kharāb al-'ālam*), a series of Qur'anic statements that describe cataclysmic celestial events.

Comet Tempel did not strike the earth, but debate over Falb's prediction coincided with a period of upheaval in the relationship between science, the state, and Islam in Egypt. At the opening of the nineteenth century, astronomy had been deeply embedded in a broader culture of Islamic learning, practiced within certain circles of 'ulama': scholars of Islam who interested themselves in the workings of celestial motion alongside the study of Prophetic reports (hadith), jurisprudence (*fiqh*), and Qur'anic exegesis (*tafsīr*). Yet this astronomical knowledge held very limited authority in the ritual practices of most Muslims. By the end of World War I, after a century of Ottoman-Egyptian state-building, British occupation, and anticolonial struggle, astronomy took place largely outside of an Islamic scholarly context, among technically trained

---

[6] Shafi'i believed that the world would end when the plane of the earth's equator coincides with the plane of the earth's orbit around the sun (i.e., when the obliquity of the ecliptic has declined to zero), an idea with deep roots in Islamic eschatological writing. Shafi'i, *Risala fi dhawat al-adhnab*, f. 26. Compare Khalil ibn Aybak al-Safadi, *al-Ghayth al-musjam fi sharh lamiyyat al-'ajam* (Beirut: Dar al-Kutub al-'Ilmiyya, 1975), 414; and 'Ali ibn Abi al-Hazm, *The Theologus Autodidactus of Ibn al-Nafis*, trans. and ed. Max Meyerhof and Joseph Schacht (Oxford: Clarendon, 1968), 71–74.

[7] *Al-Muqattam*, 15 November 1899, p. 1.

[8] Shafi'i, *Risala fi dhawat al-adhnab*, f. 1. The title may be the work of a later collector or cataloguer.

[9] Shafi'i, *Risala fi dhawat al-adhnab*, ff. 44–59 and 60–69. Some, but not all, of the historical works were also quoted in Shafi'i's articles.

bureaucrats who had passed through a new system of civil and military schools, as well as European universities and observatories. At the same time, more and more Muslims turned to astronomy to define the timing of daily prayers, to determine the beginning and end of the sacred month of Ramadan, and even to interpret the Qur'an.

How did new, apparently secular sites of state science come to play an authoritative role in the religious practices of Muslims, while an older knowledge culture – a science produced in Islamic scholarly discourses – was refashioned as an object of historical memory? Astronomy in the nineteenth century was an unusual science in the degree to which it was both crucial, globally, for the governance of imperial states, and vital, in Islamic contexts, for new projects of religious reform.[10] While surveyors timed the passage of stars in order to measure longitude and draw maps, Muslim activists sought to create a more unified *umma* (Muslim community) by using astronomy to define the timing of ritual duties. The history of astronomy in late Ottoman Egypt therefore provides a window onto the relationship between the globalization of science, the building of empire, and the fashioning of Islam in a transformative period for all three. This is a story of practitioners and institutions that were decidedly new, yet which depended upon – and expanded – a tradition of knowledge that had been cultivated by Muslim scholars for a millennium. It is a story in which science and technology became mobile as a result of the global ambition and resources of an empire and a religious tradition rooted firmly outside of Europe. And it is a story that points to a tense but powerful alignment between state-building and religious reform in the late nineteenth and early twentieth centuries. The Ottoman Empire was politically bounded, while the Muslim *umma* knew no such limitations. But imperial authorities and transnational religious activists alike seized upon technical institutions as a means of forging knowledge that could travel. As a result, science came to play a new role in defining the possibilities of both political and religious belonging in the twentieth-century Middle East.

---

[10] On surveying and empire, see Kapil Raj, *Relocating Modern Science* (Basingstoke: Palgrave Macmillan, 2007), 181–222; Bernard Cohn, *Colonialism and Its Forms of Knowledge* (New York: Oxford University Press, 2004), 7–8, 80–82; D. Graham Burnett, *Masters of All They Surveyed: Exploration, Geography, and a British El Dorado* (Chicago: University of Chicago Press, 2000); Martina Schivaon, "Geodesy and Mapmaking in France and Algeria: between Army Officers and Observatory Scientists," trans. Charlotte Bigg and David Aubin, in *The Heavens on Earth: Observatories and Astronomy in Nineteenth-Century Science and Culture*, ed. David Aubin et al. (Durham, NC: Duke University Press, 2010), 199–224; and Simon Schaffer, "Keeping the Books at Paramatta Observatory," in *Heavens on Earth*, ed. Aubin et al., pp. 118–47.

Recasting Science and Modernity in the Middle East: A Role for 'Ulama'

Historians of the modern Middle East have long recognized a connection between the rise of new sciences and the great sociopolitical transformations of the region in the nineteenth century. In Egypt, new understandings of space, language, and education made the land and its inhabitants colonizable years before British warships opened fire on Alexandria in 1882.[11] Inoculation campaigns, birth and death registries, and the placement of medical officers in police stations enabled new forms of state power, even while breeding their own forms of resistance.[12] Practitioners of the human sciences defined new "subjects of knowledge" (like "population"), giving rise to newly conceived social problems and agendas for solving them.[13] Even the physical environment underwent a sea change, as increasingly ambitious approaches to water and labor management transformed the land.[14] By the beginning of the twentieth century, science animated new social and political elites: from the *vulgarmaterialismus* of the Young Turks, to the medicalizing discourse of the emergent Iranian middle class.[15]

However, perhaps one reason that science appears to have been so disruptive in the nineteenth-century Middle East is that our understanding of its history in this context has been rooted precisely among new social groups, such as French-trained technocrats, missionary-educated science popularizers, and graduates of the civil and military schools.[16] Of course, even these new kinds of actors drew deeply on the resources

---

[11] Timothy Mitchell, *Colonising Egypt* (Berkeley, CA: University of California Press, 1988).

[12] Khaled Fahmy, *All the Pasha's Men* (Cambridge: Cambridge University Press, 1997); Khaled Fahmy, "Medicine and Power: Towards a Social History of Medicine in Nineteenth-Century Egypt," *Cairo Papers in Social Science* 23, no. 2, ed. Enid Hill (Cairo: AUC Press, 2000), 16–62; see also Liat Kozma, *Policing Egyptian Women: Sex, Law, and Medicine in Khedival Egypt* (Syracuse, NY: Syracuse University Press, 2011).

[13] Omnia El Shakry, *The Great Social Laboratory: Subjects of Knowledge in Colonial and Postcolonial Egypt* (Stanford, CA: Stanford University Press, 2007).

[14] Alan Mikhail, *Nature and Empire in Ottoman Egypt* (Cambridge: Cambridge University Press, 2011). Mikhail argues that these changes began in the late eighteenth century.

[15] M. Şükrü Hanioğlu, "Blueprints for a Future Society: Late Ottoman Materialists on Science, Religion, and Art," in *Late Ottoman Society: The Intellectual Legacy*, ed. Elisabet Özdalga (London: RoutledgeCurzon, 2005), 28–116; Cyrus Schayegh, *Who is Knowledgeable, is Strong: Science, Class, and the Formation of Modern Iranian Society, 1900–1950* (Berkeley, CA: University of California Press, 2009). On technology and revolution, see On Barak, *On Time: Technology and Temporality in Modern Egypt* (Berkeley, CA: University of California Press, 2013), 175–204.

[16] See also M. Alper Yalçınkaya, *Learned Patriots: Debating Science, State, and Society in the Nineteenth-Century Ottoman Empire* (Chicago: University of Chicago Press, 2015); Pascal Crozet, *Les sciences modernes en Égypte: transfert et appropriation, 1805–1902* (Paris: Geuthner, 2008).

of their society's intellectual traditions. Arabic debates over Darwin, for example, reinvigorated Islamic philosophy and natural theology.[17] Yet we know little about the practice of science in the nineteenth century among the old learned elite of Ottoman-Islamic society, the 'ulama'.[18] Despite renewed attention to the role of 'ulama' in shaping modern articulations of law, politics, and piety, the significance of their venerable scientific traditions in this period remains largely unexamined.[19]

Astronomy recasts the narrative of science in the modern Middle East, bringing 'ulama' from the margins toward center stage. In part, astronomy's capacity to tell such a story derives from the distinctive place that astronomy long occupied in Islamic society. Beginning in the eighth century AD (more or less), scholars working under the patronage of the new Islamic Empire developed elaborate, written traditions of mathematical astronomy.[20] Such scholars translated, revised, and built upon Greek, Persian, and Indian sources in order to set the calendar, keep time, predict planetary positions, and make astrological judgments, but also to answer specifically Islamic questions such as the correct timing and direction of prayer and the visibility of the lunar crescent at the beginning and end of Ramadan.[21] Undermining an older narrative of scientific

---

[17] Marwa Elshakry, *Reading Darwin in Arabic* (Chicago: University of Chicago Press, 2013), especially pp. 131–59.

[18] An important exception is chapter 4 of Crozet, *Sciences modernes*. See also Ekmeleddin İhsanoğlu, "The Introduction of Western Science to the Ottoman World: A Case Study of Modern Astronomy (1660–1860)," in *Science, Technology, and learning in the Ottoman Empire* (Aldershot: Ashgate, 2004). By comparison, on the astronomical exchange between Europe and the Persianate world of the Qajar and late Mughal empires, see David Pingree, "An Astronomer's Progress," *Proceedings of the American Philosophical Society* 143, no. 1 (1999): 73–85; Simon Schaffer, "The Asiatic Enlightenments of British Astronomy," in *The Brokered World: Go-Betweens and Global Intelligence, 1770–1820*, ed. Simon Schaffer et al. (Sagamore Beach: Science History Publications, 2009), 49–104; S.M. Razaullah Ansari, "European Astronomy in Indo-Persian Writings," in *History of Oriental Astronomy*, ed. S.M. Razaullah Ansari (Dordrecht: Kluwer, 2002), 133–44; and Kamran Arjomand, "The Emergence of Scientific Modernity in Iran: Controversies Surrounding Astrology and Modern Astronomy in the Mid-Nineteenth Century," *Iranian Studies* 30, nos. 1–2 (Winter/Spring, 1997): 5–24.

[19] Muhammad Qasim Zaman, *The Ulama in Contemporary Islam: Custodians of Change* (Princeton, NJ: Princeton University Press, 2002); Malika Zeghal, *Gardiens de l'Islam: les oulémas d'Al Azhar dans l'Egypte contemporain* (Paris: Presses de la Fondation nationale des sciences politiques, 1996); Meir Hatina, *'Ulama', Politics, and the Public Sphere: An Egyptian Perspective* (Salt Lake City: University of Utah Press, 2010).

[20] For the debate over whether these developments originated under the 'Abbasids or late Umayyads, see George Saliba, *Islamic Science and the Making of the European Renaissance* (Cambridge, MA: MIT Press, 2007), 1–72; and Dimitri Gutas, *Greek Thought, Arabic Culture* (New York: Routledge, 1998).

[21] E.S. Kennedy, *Studies in the Islamic Exact Sciences*, ed. David King and Mary Helen Kennedy (Beirut: American University of Beirut Press, 1983); George Saliba, *A History of Arabic Astronomy: Planetary Theories during the Golden Age of Islam*

"decline," often attributed to post-twelfth-century religious "orthodoxy" and its putative animosity toward science, historians have shown that astronomy continued to thrive in Muslim societies well beyond its so-called "Golden Age."[22]

In fact, elements of this astronomy remained important and dynamic areas of knowledge for certain 'ulama' in the nineteenth-century Middle East. However, my goal in illuminating their practices, which I call "scholarly astronomy," is not to add one more exhibit to the catalogue of evidence against the decline thesis.[23] Rather, I am interested in the social context of scholarly astronomy in late Ottoman Egypt: the relationships between students, teachers, and patrons; between texts, instruments, and practices; between technical questions and social problems; between astronomy and other areas of scholarship. Such a contextual understanding of scholarly astronomy challenges conventional narratives of modernity in the Middle East in terms of both the degree of continuity and the manner of change. In the first place, the implications of new sciences did not necessarily resonate where older ways of knowing remained intellectually coherent and socially relevant. In the second place, where new techniques did appeal to late Ottoman practitioners of scholarly astronomy, 'ulama' made these techniques usable by adapting them to their own terminology and technical routines. They did so, moreover, with little fanfare. For such scholars, the rise of centers of astronomical knowledge in London and Paris was part of a long history, part classical and part Islamic, in which progress was rooted in continuities of genre (such as the $z\bar{\imath}j$, or astronomical handbook), and continuities of practice (especially tabular calculation). These acts of technical and historiographical translation were essential for the introduction of new sciences within centers of Islamic learning, like the al-Azhar mosque in Cairo. But if a necessary characteristic of the "modern" is a perception of rupture with the past, or "epochal change,"[24] even these innovative

---

(New York: New York University Press, 1994); David A. King, *Astronomy in the Service of Islam* (Aldershot: Ashgate, 1993); David A. King, *In Synchrony with the Heavens: Studies in Astronomical Timekeeping and Instrumentation in Medieval Islam*, 2 vols. (Leiden: Brill, 2004).

[22] For an overview, see Saliba, *Islamic Science*; and Jamil Ragep, "When did Islamic science die (and who cares)?" *Viewpoint: Newsletter of the BSHS* 85 (2008): 1–3. For more detailed studies, see Robert Morrison, *Islam and Science: The Intellectual Career of Nizam al-Din al-Nisaburi* (Abingdon: Routledge, 2007); Ahmad Dallal (ed.), *An Islamic Response to Greek Astronomy* (Leiden: Brill, 1995); and contributions to F. Jamil Ragep et al., eds., *Tradition, Transmission, Transformation* (Leiden: Brill, 1996).

[23] For a recent and powerful example, see Khaled El-Rouayheb, *Relational Syllogisms and the History of Arabic Logic* (Leiden: Brill, 2010).

[24] C.A. Bayly, *The Birth of the Modern World* (Oxford: Blackwell, 2004), 11.

practitioners of scholarly astronomy testify to the tenuousness of any connection between new scientific techniques and "modernity" in late Ottoman Egypt.

In this sense, the history of astronomy contributes to our understanding of the late Ottoman era of "reforms," the *Tanzimat*, which included the centralization of rule in the bureaucracy, the establishment of new schools and legal codes, the elimination of legal differences between Muslim and non-Muslim communities, and constraints on the exercise of arbitrary power. Rather than characterizing this era in terms of straightforward, top-down Westernization beginning with the Rose Chamber Edict of 1839, recent scholarship has pointed to the origins of late Ottoman reform in sociopolitical and even cultural transformations of the eighteenth century. Changes in urban space, literary production, and economic relations were a gradual process in which provincial elites and middling actors played crucial roles.[25] Even in the sciences, where late Ottoman reformers saw themselves quite self-consciously as having much to gain from European ways, Ottoman and Islamic knowledge traditions remained relevant, dynamic, and crucial to the formation of new political and pedagogical institutions.

Science and State in a "Colonized Colonizer"

Powerful as scholarly astronomy remained, its practitioners had to compete with new kinds of astronomers in late Ottoman Egypt. Some of these new astronomers were prolific and famous: men like Mahmud Hamdi and Isma'il Mustafa, who trained in the Observatoire de Paris and eventually served as cabinet ministers.[26] (Adjacent to Cairo's famous Tahrir Square lies Midan al-Falaki, "The Astronomer's Square," which is named for Mahmud Hamdi.) Others toiled in obscurity, but their work had very public ramifications: men like Mahmud Naji, a Survey Department official who produced the government's almanacs in the

---

[25] On Ottoman cultural change in the eighteenth century, see Dana Sajdi, ed., *Ottoman Tulips, Ottoman Coffee* (London: I.B. Tauris, 2007). For a reinterpretation of the *Tanzimat* and subsequent eras through the history of technology, see Avner Wishnitzer, *Reading Clocks, Alla Turca: Ottoman Temporality and Its Transformation during the Long Nineteenth Century* (Chicago: University of Chicago Press, 2015). On urban space, see Stefan Weber, *Damascus: Ottoman Modernity and Urban Transformation, 1808–1918* (Aarhus: Aarhus University Press, 2009); Jens Hanssen, *Fin de siècle Beirut: The Making of an Ottoman Provincial Capital* (New York: Oxford University Press, 2005); and Keith Watenpaugh, *Being Modern in the Middle East* (Princeton, NJ: Princeton University Press, 2006).

[26] On Mahmud Hamdi's career, see Pascal Crozet, "La Trajectoire d'un scientifique égyptien au XIXe siècle: Mahmud al-Falaki (1815–1885)," in *Entre réforme sociale et mouvement national: Identité et modernization en Égypte* (Cairo: CEDEJ, 1995), 285–309.

early twentieth century. These astronomers not only introduced certain techniques and knowledge into Egypt, but also cultivated a new relationship between science and the state. Their rise must therefore be understood, first, in relation to Egypt's unusual political dynamics in the late Ottoman period: modern science, like the modern state in Egypt, emerged in relation to simultaneous and overlapping projects of empire.

With the evacuation of Napoleon's ill-fated Armée d'Orient in 1801, Egypt fell into several years of political uncertainty, resolved between 1805 and 1811 with Mehmed Ali Pasha's consolidation of power as governor (vali) in Cairo. Whereas the occupants of this office in the eighteenth century had typically been in a weak position relative to the local military, economic, and learned elites, Mehmed Ali harbored – and made good on – his own imperial and dynastic ambitions. In the Sudan, the Pasha's conquests in the 1820s introduced a colonial rule that would endure, in various forms and with one substantial interruption, until 1954. In the 1830s, it seemed that a similar fate might be in store for much of the Ottoman lands, as the Pasha sent his forces on an astonishingly successful campaign through the Levant and into the Ottoman heartland. The 1833 Peace of Kütahya granted Mehmed Ali and his son Ibrahim Pasha the governorship of Egypt, Western Arabia (the Hijaz), the Syrian lands, and Crete. As was so often the case in the Ottoman wars of the eighteenth and nineteenth centuries, however, great power politics came to the empire's rescue: after winning new economic concessions from Istanbul, Great Britain forced the Pasha to abandon his conquests outside the Sudan in 1841.[27] In exchange, however, he received a dynastic hold on the governorship of Egypt. The ambitions and outlook of Mehmed Ali's successors varied, beginning with the cautious retrenchment of Abbas Pasha, but they consistently maintained their household's unique legal privileges within the Ottoman state.[28]

As this narrative implies, Egypt during the long nineteenth century is best understood as part of a larger Ottoman sphere, rather than as an independent nation-state. Although Egyptian nationalist historiography has long celebrated Mehmed Ali as "the founder of modern Egypt,"[29] the Pasha, his dynastic successors, and much of the governing

---

[27] On the Ottoman economic concessions under the treaty of Balta Limanı, see E. Roger Owen, The Middle East in the World Economy (London: I.B. Tauris, 1993), 74–75.

[28] Ehud Toledano, State and Society in Mid-Nineteenth-Century Egypt (Cambridge: Cambridge University Press, 1990), 12.

[29] Afaf Lutfi al-Sayyid Marsot, Egypt in the Reign of Muhammad Ali (Cambridge: Cambridge University Press, 1984); Amin Sami, Misr wa-l-Nil (Cairo: Matba'at Dar al-Kutub al-Misriyya, 1938); Jamal al-Din al-Shayyal, Tarikh al-tarjama wa-l-haraka al-thaqafiyya fi 'asr Muhammad 'Ali (Cairo: Dar al-Fikr al-'Arabi, 1951).

elite in the nineteenth century were Turcophones with strong cultural, familial, and political ties to the Ottoman Empire.[30] The expansion of the Ottoman-Egyptian state spurred the growth of indigenous, Arabic-speaking elites, but it was not until the early twentieth century that post-Ottoman national identity became a serious possibility.[31] Even the basic geographic borders of "Egypt" were in flux during this period, meaning we should understand the territory as a category to be contextualized, rather than a predetermined and immutable location.[32] In fact, one of the key contexts for the emergence of a new relationship between astronomy and the Ottoman-Egyptian state was an effort to lend new specificity to "Egypt" as both a geographic and historical entity.

If much was Ottoman about late Ottoman Egypt, however, much was also distinctively Egyptian. In the history of empire, what is particularly striking about Egypt in this period is its location within multiple, nestled imperialisms. The province's political status became further complicated in the 1870s, during the ambitious reign of the viceroy Ismail Pasha (r. 1863–79). Ismail, the first to legally use the title of "Khedive" (having purchased the privilege from the Sultan), revived the experiments in technical education begun under Mehmed Ali, and established a full-fledged system of civil and military schools. He also built up Egypt's railways and steam fleet, and sought to expand his empire in East Africa. New neighborhoods of Cairo were laid out with boulevards, plazas, an opera house, and theater.[33] The Suez Canal, the concession for which Said had granted in 1856, opened in 1869. While the cotton boom of the American Civil War financed some of these projects, by the 1870s the Khedive was deeply in debt to European bondholders, who forced Egypt to accept Anglo-French financial oversight in 1876. But the increasingly

---

[30] Toledano, *State and Society;* Khaled Fahmy, *Mehmed Ali: From Ottoman Governor to Ruler of Egypt* (Oxford: Oneworld, 2009). The "Ottoman turn" in the historiography of Egypt owes much to work on the eighteenth century, particularly Jane Hathaway, *The Politics of Households in Ottoman Egypt: The Rise of the Qazdağlis* (Cambridge: Cambridge University Press, 1997).

[31] On the emergence of sociopolitical elites outside the state, see Robert Hunter, *Egypt Under the Khedives 1805–1879: From Household Government to Modern Bureaucracy* (Cairo: AUC Press, 1999). For a less dichotomous account, emphasizing the consolidation of power within the household of Mehmed Ali, see Ehud Toledano, "Social and Economic Change in 'The Long Nineteenth Century,'" *Cambridge History of Egypt*, vol. 2, ed. M.W. Daly (Cambridge: Cambridge University Press, 1998), 252–84, especially 256–63. On the debates over post-Ottoman political community, see Israel Gershoni and James P. Jankowski, *Egypt, Islam, and the Arabs: the Search for Egyptian Nationhood, 1900–1930* (New York: Oxford University Press, 1986).

[32] Matthew Ellis, "Between Empire and Nation: the Emergence of Egypt's Libyan Borderland, 1841–1911" (Ph.D. Diss., Princeton University, 2012).

[33] Janet Abu-Lughod, "Tale of Two Cities: The Origins of Modern Cairo," *CSSH* 7 (1965): 429–57; Mitchell, *Colonising Egypt*, pp. 64–68.

close relationship between the viceregal household and European interests stoked resentment among Egypt's other elites, including newly ascendant rural landlords and mid-ranking military officers, as well as certain ʿulamaʾ. These groups came together during the ʿUrabi Revolution (1879–82), which imposed a consultative government on the Khedive Tawfiq Pasha (r. 1879–92).[34] The ʿUrabist project only came to an end with British military intervention in 1882. Justified originally as a limited effort to restore the friendly Khedive and guarantee Egypt's financial obligations to European creditors, the British invasion grew into indefinite occupation, and became a formal protectorate in 1914.

In sum, until the outbreak of World War I, the sovereignty of Istanbul, the Nilotic quasi-empire of the Ottoman-Egyptian dynasty, and the "Veiled Protectorate" of Great Britain existed in one space. Astronomy was entwined in these overlapping histories of empire, or what Eve Troutt Powell has termed the phenomenon of the "colonized colonizer."[35] Thus, late Ottoman Cairo had multiple state observatories, each the project of distinct imperial ambitions. The first, which Mehmed Ali ordered built near the school of engineering and government press in Bulaq in the 1840s, was itself erected on the remains of a French observatory long abandoned by Napoleon's troops.[36] Two decades later, Ismail Pasha had a new observatory built in the tower of an old barracks, inside an enclave of military education at ʿAbbasiyya. Dissatisfied with this facility, British surveyors eventually built a new observatory at Helwan in 1903. It would be a mistake, however, to draw a line between "Egyptian" observatories at Bulaq and ʿAbbasiyya and a "colonial" observatory at Helwan. At each of these sites, astronomers contributed to a common set of empire-building projects – especially surveying, cartography, and time regulation – which sought to link Cairo with other points in both Egypt and Europe.

While state astronomy flourished within the nurturing culture of overlapping empires, it cannot be reduced to the political interests it served. Men like Mahmud Hamdi, Ismaʿil Mustafa, and Mahmud Naji utilized their knowledge, prestige, and access to resources like print, draftsmen, and (literally) armies of assistants, in order to articulate – publicly – new understandings of history, political community, and the performance of religious duties. In concert with European Orientalists, they helped articulate the notion that science in Muslim society, though an important

---

[34] Juan Cole, *Colonialism and Revolution in the Middle East* (Princeton, NJ: Princeton University Press, 1993).

[35] Eve Troutt Powell, *A Different Shade of Colonialism* (Berkeley, CA: University of California Press, 2003), 6.

[36] On the Bulaq Observatory, see Crozet, *Sciences modernes*, pp. 194–99.

historical phenomenon, had long since declined into insignificance. And yet, at the same time, state astronomers adopted certain practices long cultivated by Muslim scholars of astronomy – especially the mathematical determination of the times of prayer – and popularized them as normative for all Muslims. In other words, state astronomers translated their power into cultural authority.

Islam, Science, and Authority

Muslims in Egypt – and beyond – increasingly defined the correct performance of religious duties according to the knowledge of new astronomical practitioners in the late nineteenth and early twentieth centuries. By the middle of the 1920s, mosques in Cairo were legally obligated to issue the five daily calls to prayer according to times that were defined and published by Survey Department bureaucrats. Meanwhile, an increasing number of readers of the Qur'an understood God's revelation to contain a new cosmology – not only heliocentricity, but asteroids, galaxies, and the proper motion of stars. Many who fasted during Ramadan now saw the state observatory as the best arbiter of the holy month's beginning and end.

The religious authority of new scientific practitioners arose in part from the interest of others in adopting new scientific standards for their own purposes – not least, the cultivation of piety. In other words, the disciplinary power of new scientific practices did not flow on a one-way street, subjugating population to an abstract and monolithic state. It was also a resource that diverse actors used to fashion new social identities and movements.[37] Islamic reformists, in particular, helped to forge new articulations of science, religion, and politics toward the end of the late Ottoman period. Such reformists – activists and journalists like Muhammad Rashid Rida, publisher of *al-Manar* ("The Lighthouse") – were harshly critical of the Islamic scholarly establishment of their day.[38]

---

[37] On the need for more attention to the "productive" powers of discipline in colonial history, see Projit Mukharji, *Nationalizing the Body: the medical market, print, and daktari medicine* (London: Anthem Press, 2009), 10.

[38] On Rashid Rida, see Amal Ghazal, *Islamic Reform and Arab Nationalism: Expanding the Crescent from the Mediterranean to the Indian Ocean* (London: Routledge, 2010); Ahmad Dallal, "Appropriating the Past: Twentieth-century Reconstruction of Pre-Modern Islamic Thought," *Islamic Law and Society* 7, no. 3 (2000): 325–58; Umar Ryad, *Islamic Reformism and Christianity* (Leiden: Brill, 2009); Mahmoud Haddad, "Arab Religious Nationalism in the Colonial Era," *JAOS* 117 (1997): 253–77; Malcolm Kerr, *Islamic Reform: the Political and Legal Theories of Muhammad 'Abduh and Rashid Rida* (Berkeley, CA: University of California Press, 1966); and Albert Hourani, *Arabic Thought in the Liberal Age* (Cambridge: Cambridge University Press, 1962), ch. 9.

Rida, along with likeminded colleagues and disciples who helped give rise to the Salafi movement in the early twentieth century, saw most 'ulama' as hidebound, incapable of meeting the challenges of European (including Russian) imperialism. Reformists like Rida propagated a view of Islamic history in which the degenerative forces of local custom and scholastic ignorance had left the Muslim community (*umma*) fragmented and weak.[39] The notion that Islam's scientific traditions had disappeared into oblivion fits a larger narrative in which *all* Islamic scholarship (law, theology, exegesis) had long since deviated from the truth.

A dim view of Islamic intellectual history was not all that religious reformists and state astronomers shared. Just as state astronomers labored to make clocks in distant parts of the Nile Valley show the same time, religious reformists labored to make Muslims in distant communities observe the same practice. These projects intersected most prominently where astronomy had intersected with Islamic duties for over a thousand years: the timing of prayer, and the timing of Ramadan. Now, however, an odd but convenient alliance – a technical bureaucracy and a nascent religious movement – popularized the performance of these practices according to astronomical definitions. For the orderly functioning of a state, as well as for a global community of believers newly connected with each other through the press, such standardizing work was essential.

Understanding the assimilation of new scientific institutions into the agenda of Islamic reformism sheds light on the nature of authority in modern Islam, a subject that scholars in religious studies have vigorously debated. On one view, the rise of radicalism in Islamic politics may be traced to a "fragmentation of religious authority": whereas the interpretation of the Qur'an and the parsing of legal texts were once narrowly held skills of 'ulama', mass literacy and communication have allowed a much broader array of actors to "speak for Islam."[40] By contrast, while not denying the increased presence of lay voices, others have shown that the 'ulama' have also gained new kinds of authority through the use of mass media, as well as from the creation of a national religious establishment in certain countries (including Egypt).[41] Across this literature,

---

[39] In Rida's lifetime, appeals to the "*salaf*," or forebears of Islam, authorized a variety of discourses. The exclusive identification of Salafism with an "ultra-conservative" or "fundamentalist" movement emerged later. See Henri Lauzière, *The Making of Salafism* (New York: Columbia University Press, 2016).

[40] Dale F. Eickelman and James Piscatori, *Muslim Politics* (Princeton, NJ: Princeton University Press, 1996), 131.

[41] Malika Zeghal, "Religion and Politics in Egypt: The Ulema of al-Azhar, Radical Islam, and the State (1952–1994)," *IJMES* 31, no. 3 (1999): 371–99; Jakob Skovgaard-Petersen,

however, authority has largely been understood in terms of the relations among laypeople, religious scholars, and the state.[42] But the science that was newly understood as necessary to the correct performance of prayers or the correct interpretation of the Qur'an did not simply become lay knowledge or state knowledge rather than scholarly knowledge in the late Ottoman period. Rather, laypeople decided to pray and read in new ways as part of the passage of science from 'ulama' to a new kind of technically learned elite – a shift that was both contested and contingent. 'Ulama' fought hard to claim competence over new technologies and areas of knowledge. The triumph of new actors must be partially ascribed to the tremendous material resources they derived from service to the state. But it was also a function of the degree to which they could be enlisted in the agenda of certain religious activists who wanted to cultivate a more globally uniform community.

Critical analysis of the relationship between Islam and modern technologies and institutions has sometimes borne a problematic implication, which the anthropologist Hussein Agrama has identified as "the notion of contemporary Islamic religiosity as essentially a form of modern falsehood."[43] The fact that Muslims have incorporated new ways of knowing into their religious practices does not make modern Islam a peculiar form of religiosity, unique among modern religions in the kind of explanation that it requires. (It would be far more peculiar had Islam been immune to the scientific and technological developments of the long nineteenth century.) What becomes remarkable, in light of such deep connections between modern science and religiosity, is the putative association of science with the various forms of diminishment or differentiation of religion that secularization theory has sought to describe. In an influential work, José Casanova proposed that, instead of understanding secularization as a lessening of religion (including public religion), we should instead view secularization as the emergence of new distinctions between religion and other spheres, particularly science and

Defining Islam for the Egyptian State (Leiden: Brill, 1997); Zaman, *The Ulama in Contemporary Islam*.

[42] On authority as relational, see Gudrun Krämer and Sabine Schmidtke, "Introduction," in *Speaking for Islam: Religious Authorities in Muslim societies*, ed. Krämer and Schmidtke (Leiden: Brill, 2006); and Muhammad Qasim Zaman, *Modern Islamic Thought in a Radical Age: Religious Authority and Internal Criticism* (Cambridge: Cambridge University Press, 2012), 29–34. For a critique of logocentric bias in the study of authority, see Devin DeWeese, "Authority," in *Key Themes for the Study of Islam*, ed. Jamal Elias (Oxford: Oneworld, 2010), 26–52.

[43] Hussein Ali Agrama, *Questioning Secularism: Islam, Sovereignty, and the Rule of Law in Modern Egypt* (Chicago: University of Chicago Press, 2012), 15.

politics.[44] "Differentiation" itself, however, entails a variety of processes, not all of which necessarily move in tandem with one another. In the case of Islam in late Ottoman Egypt, while expertise in natural science and technology became the province of social groups newly differentiated from the old learned elite of 'ulama', it was precisely through this social differentiation that scientific knowledge – as well as the state – became more intimately involved in religious practice.

Another Geography: Empire, Religion, and the Global History of Modern Science

If science was key to the making of empire and religion in late Ottoman Egypt, the reverse is true as well: late Ottoman history offers insights into the role of empire and religion in the making of science. For historians of science, the question of how to incorporate the "non-West" into the history of the modern sciences has begotten a number of approaches.[45] In recent years, a number of scholars have shown that European imperialism, rather than being the vehicle that brought modern sciences to other parts of the world, provided a framework within which modern sciences emerged from new circulations of people, knowledge, and objects.[46] This approach has incorporated diverse regions, peoples, and knowledge traditions into the history of the modern sciences, and indeed the history of modernity. Despite the attention that non-European knowledge has started to receive in this literature, however, *empire* remains typically understood as a practice of Europeans.[47] Late Ottoman Egypt's position as both agent and object of empire – a position shared, notably, with

---

[44] José Casanova, *Public Religions in the Modern World* (Chicago: University of Chicago Press, 1994), 24. For a critique of Casanova's reasoning, see Talal Asad, *Formations of the Secular* (Stanford, CA: Stanford University Press, 2003), 182.

[45] For an overview, see Sujit Sivasundaram, "Focus: Global Histories of Science," *Isis* 101 (2010): 95–97.

[46] Raj, *Relocating Modern Science*; Schaffer et al., eds., *The Brokered World*; Sujit Sivasundaram, "Science," in *Pacific Histories*, ed. David Armitage and Alison Bashford (Basingstoke: Palgrave Macmillan, 2014), 237–60. This literature revises earlier models of scientific "diffusion" from the West, and the closely related "tools of empire" interpretation of technology and imperialism. Cf. George Basalla, "The Spread of Western Science," *Science* 156, no. 3775 (1967): 611–22; Daniel Headrick, *The Tools of Empire: Technology and European Imperialism in the Nineteenth Century* (New York: Oxford, 1981).

[47] But see the growing literature on early modern Chinese science: Laura Hostetler, *Qing Colonial Enterprise: Ethnography and Cartography in Early Modern China* (Chicago: University of Chicago Press, 2001); Carla Nappi, *The Monkey and the Inkpot: Natural History and Its Transformations in Early Modern China* (Cambridge, MA: Harvard University Press, 2009); Dagmar Schaefer, *The Crafting of the 10,000 Things* (Chicago: University of Chicago Press, 2011).

the late Qing Empire – enabled its own, distinctive geography of knowledge circulation.[48] When Ottoman-Egyptian astronomers went abroad to gain advanced training in France, England, and even the United States, they not only acquired knowledge and skills that they would bring back to Egypt; they also became staff members – sometimes for years – at the European and American institutions where they studied. The government of Egypt lent an important surveying instrument to the government of France, whose interests in cartography were not so different from those of the viceroys. Egyptians who had trained to translate French and English science into Arabic did the reverse as well, contributing to European Orientalism alongside astronomy. Thus, the history of astronomy in the late Ottoman period helps expand our conception of "contact zones" to include not only extra-European sites of scientific exchange, but also the Observatoire de Paris and the United States Naval Observatory.[49]

Just as late Ottoman Egypt's political history can reorient our understanding of science and empire, the role of scholars of Islam in translating astronomy in late Ottoman Egypt brings to light a different set of pathways than are usually understood in the global history of science. Scholarly astronomy was a deeply sophisticated tradition that developed over the course of a thousand years across a number of regions, evincing a substantial degree of continuity and development across time and space; its practitioners were not guides, translators, or healers, whose knowledge of locally specific terrains, languages, or flora was assimilated into European "centers of calculation" or collection.[50] Neither was their work rendered legible to European colonial administrators as a kind of "vernacular knowledge"[51] – but the language in which they generally wrote, Arabic, was read by Muslim scholars from the western shores of Africa to the Strait of Malacca.[52] Islamic history offers much more than

---

[48] On the "hyperimperial" context of treaty-port China, see Ruth Rogaski, *Hygienic Modernity* (Berkeley, CA: University of California Press, 2004).

[49] Cf. Kapil Raj, "The Historical Anatomy of a Contact Zone: Calcutta in the Eighteenth Century," *Indian Economic and Social History Review* 18 (2011): 55–82, drawing on the work of Mary Louise Pratt.

[50] Cf. Bruno Latour, *Science in Action* (Cambridge, MA: Harvard University Press, 1987), 232.

[51] Cf. Helen Tilley, "Global Histories, Vernacular Science, and African Genealogies," *Isis* 101 (2010): 117.

[52] In this respect, scholarly astronomy as a literary tradition resembled the "cosmopolitan" Sanskrit of the first millennium, more than the later "vernacularization" of South Asian literary culture. See Sheldon Pollock, *The Language of the Gods in the World of Men* (Berkeley, CA: University of California Press, 2006), 23.

a "non-Western" counterpoint in the study of science and religion.[53] The language and networks of Islamic scholarship were crucial for rendering science and technology mobile in the nineteenth century.

Put another way, what late Ottoman-Egyptian history points to is not the existence of "another reason" in non-Western cultures of science, as Gyan Prakash has characterized Indian science in the same period.[54] Rather, it points to the role of other geographies: pathways and trajectories that were orthogonal to European empire and knowledge-production, but which were nevertheless, and in their own way, trans-regional. Whether it was the use of Egyptian survey instruments in Paris, the study of English-made watches among 'ulama' in Cairo, or the translation of Islamic justifications of Copernicanism from Persian into Turkish and then Arabic, such pathways were as "global" as any, even if they trace a pattern of circulation that passed only intermittently through Europe.

The following pages explore these pathways in three parts. Part I explains and compares astronomical practice within two distinct social contexts of late Ottoman Egypt. Chapter 1 introduces the practice of scholarly astronomy, focusing on the tradition of *mīqāt* (astronomical timekeeping) as it was pursued by 'ulama'. The chapter takes as its focal point Muhammad al-Khudari (d. 1870/71), a deaf scholar from the port of Damietta whose commentary on a fifteenth-century astronomical handbook (*zīj*) gained him wide repute in the middle of the nineteenth century. For scholars such as Khudari, scientific practice derived many of its conventions, and much of its significance, from a broader training in Islamic discursive traditions, especially law, the study of Prophetic reports, and Qur'an commentary. However, the specific ritual uses of astronomy – the determination of prayer times, for example – played a relatively minor role in Khudari's astronomy. For most of his readers and students, the primary significance of his work lay in the need for astronomical knowledge to navigate the multiple calendars in use in late Ottoman society, and to make astrological judgments.

Chapter 2 centers on the emergence of a different kind of astronomer in the middle of the nineteenth century. The lives of Mahmud Hamdi "al-Falaki" (the astronomer) and Isma'il Mustafa "al-Falaki" illuminate the new institutions, powers, and purposes that constituted the pursuit of astronomy for servants of the Ottoman viceroys of Egypt. In contrast to

---

[53] Such a counterpoint is, of course, badly needed: see Thomas Dixon et al., eds., *Science and Religion: New Historical Perspectives* (Cambridge: Cambridge University Press, 2010).

[54] Gyan Prakash, *Another Reason: Science and the Imagination of Modern India* (Princeton, NJ: Princeton University Press, 1999).

scholars like Muhammad al-Khudari, these men were conscripted into state academies and ordered to undergo years of training in state observatories in Cairo and Paris. They used their skills to oversee and execute technical projects that contributed to the growth of a more centralized, more powerful bureaucracy in Cairo that not only defended its autonomy from Istanbul, but also sought to perpetuate and extend its rule in parts of the Sudan and East Africa. Moreover, through the labor of their training in Paris, Mahmud and Isma'il also contributed to French astronomical projects, including survey work very similar to the surveying of the Nile Valley that they later oversaw. In other words, Paris and Cairo were connected in the middle of the nineteenth century through a web of mutually supportive imperial and scientific ambitions. This "viceregal astronomy," practiced in a new set of spaces – state observatories, government ministries, instrument workshops, eclipse expeditions – helped to produce new geographical conceptions of the larger space to which people in Egypt belonged.

The figure of the viceregal astronomer was a crucial way in which late Ottoman Egypt adopted new scientific techniques, but it was not the only way. Part II explores the world of scholars who did not participate in the new social groups, institutions, and linkages between science and political power that characterized viceregal astronomy, yet who played leading roles in the emergence of certain technologies in late Ottoman-Egyptian society. Thus, Chapter 3 uncovers the history of a genre of manuals that 'ulama' authored on the use of mechanical timepieces in eighteenth- and nineteenth-century Cairo. These manuals, which describe the correct manipulation of "the position of the watch hand," enabled their users to employ mechanical timepieces in the practice of specifically Ottoman and Islamic temporal routines – notably, the adjustment of clocks and watches to tell time in relation to local sunset. New technology did not render scholarly astronomy irrelevant; rather, the spread of mechanical clocks and watches seems to have broadened the audience for its practitioners. The case of mechanical timekeeping points to the dynamic role of 'ulama' in the changing material and technological culture of the Middle East in this period. When the viceregal astronomers began to establish themselves as authorities in public timekeeping, they co-opted and built upon practices already established within this older genre on "the position of the watch hand."

Chapter 4 extends the discussion of the role of 'ulama' in the translation of technology, shifting the focus from mechanical timepieces to tables for the prediction of planetary motion. From Cairo, to Aleppo, to Istanbul, a number of Ottoman scholars became particularly interested in the work of Jérôme de Lalande (d. 1807), whose astronomical textbook and tables they believed enabled them to predict the positions of the stars and planets with unprecedented precision and accuracy. Yet the appeal of Lalande's work did not spur its late Ottoman users to break radically with their own astronomical practices. Instead, they translated Lalande as a *zīj*, the basic genre of scholarly astronomy. In both Arabic and Turkish, Lalande was remade according to conventions of the *zīj*'s format, framed within a classical and Islamic history of the *zīj*'s development, and sometimes even rendered into the *zīj*'s geocentric cosmology. Just as mechanical timekeeping did not render *mīqāt* irrelevant, but rather extended its powers and relevance, knowledge of French astronomical textbooks did not necessarily lead scholars to abandon older genres, techniques, and worldviews, but rather led to a redeployment of those resources. During a period of intense debate over the future of centers of Islamic learning, particularly at Cairo's al-Azhar mosque, such acts of translation enabled the teaching of new sciences.

Part III tracks the rising authority of the new, state-centered culture of science in the late nineteenth and early twentieth centuries, and the simultaneous transformation of scholarly astronomy into an object of historical memory. Chapter 5 shows how discussions of astronomy in the emergent Arabic periodical press helped to define new terms of debate over science and religion. The increasing social distinction between religious scholars and scientific experts went hand in hand with an increasing involvement of science in the normative interpretation of religious texts, including the Qur'an. This particular arrangement of authority was promoted through new representations and histories of science, which transcended typical divisions within the Arabic press. *Al-Muqtataf* ("The Digest"), the preeminent venue in which readers of Arabic engaged with European and American science in this period, was founded, published, and edited by students of American Protestant missionaries; it would seem to have had little in common with Rashid Rida's *al-Manar*, the preeminent venue in which a new Islamic discourse of Salafism – reviving the example of *al-salaf al-ṣāliḥ*, the "pious ancestors" – emerged in the early twentieth century. Yet, both journals propagated a view of modern science as a product of new technologies and institutions, and of science in Islamic society as the memory of a bygone age. This dichotomy served agendas of "reform" common to both journals.

In 1898, the government of Egypt created a new Survey Department under the directorship of a British officer. In 1923, a year after Egypt's formal independence, the Ministry of Religious Endowments ordered that every mosque in Cairo issue the call to prayer at the times published in the Survey Department's almanac. Chapter 6 illuminates the contested work of synchronization that led from one event to the other, as the way in which Muslims measured their religious duties became enmeshed in the efforts of British surveyors and astronomers to measure land, synchronize clocks, regulate commerce, and facilitate maritime navigation. Although the peculiar power of the colonial state is important to this story, the linking of observatory and mosque was not strictly a matter of coercion. The standardization of prayer times under the Survey Department worked to the extent that many British and Egyptians – from those regulating clocks in the observatory, to those praying in mosques – came to share a common set of assumptions (and frustrations) regarding the practice of measurement. The achievement of uniform prayer time – like the achievement of uniform mean time in the same period – was not *only* a consolidation of power at the center at the expense of a periphery; it was also a fragile system, continually produced through the labor of a new middle class as well as a technical elite.

In the early twentieth century, new practices of measurement made their way not only into the daily performance of Muslim prayer, but also into the annual observance of Ramadan. Chapter 7 examines debates in this period over the role of astronomical knowledge in determining the beginning and end of Islam's holiest month. In most accounts of these debates, advocates of a "scientific" method, based on astronomical prediction, failed to overcome the partisans of a "traditional" method, based on actual sighting of the lunar crescent. However, the privileging of lunar crescent observation as the exclusive Islamic norm was itself the work of early twentieth-century reformists who sought to replace locally variable traditions of calculation and observation with a single practice. Moreover, these reformists promoted lunar sighting not out of opposition to science, but out of a desire to create an observance of Ramadan that would be unified both nationally and transnationally. Indeed, Rashid Rida's vision of an Islamic practice that would be both more global and more uniform was explicitly tied to notions of synchrony and standardization that he saw modeled at the new Helwan Observatory.

By the end of the late Ottoman period in the 1920s, Egyptians debated the terms of a new nation-state and a modern Islam in ways that granted the authority of technically trained bureaucrats to draw borders, tell time, interpret the meaning of texts, and even to write history. While new social groups, new institutions, and a new relationship between science, religion, and the state may lie at the end of this story, however, they do not dominate the story itself. Too often, writing the history of modern sciences has marginalized people whose interests and practices happened not to contribute directly to these highly particular ways of knowing. We would misunderstand the role of scholarly astronomy as an intellectually viable and socially important practice in late Ottoman Egypt if we acknowledged only as much of it as participated in the making of the modern sciences. And we would misunderstand the modern sciences themselves if we assumed that their emergence was synonymous with the displacement of older forms of knowledge. In his refutation of Falb, Mustafa al-Shafi'i recounted the history of predicting comets in Europe from Halley to Laplace,[55] but he also drew on occult Islamic sources that existed only in manuscript.[56] He argued that Falb's calculations were wrong according to post-Newtonian celestial mechanics, while also mistaken from the perspective of Aristotelian physics.[57] Even as he explained the discovery of the asteroid belt, he warned that an unusual display of meteors in mid-November might arise from a conjunction of four planets in the "fiery house" of Sagittarius. For Shafi'i and his readers, understanding astronomy meant navigating amid observatory data and scriptural portents, Aristotelian categories and astrological signs, printed newspapers and manuscript folios. As late Ottomans looked to the stars from a landscape in which the borders of science and religion were shifting, astronomy could not predict the end of days, but it helped redraw the world in which they lived.

---

[55] Shafi'i, *Risala fi dhawat al-adhnab*, ff. 15, 27–29. On comets in early modern European culture, see Sara Schechner Genuth, *Comets, Popular Culture, and the Birth of Modern Cosmology* (Princeton, NJ: Princeton University Press, 1997).

[56] In some instances, Shafi'i may have been drawing on printed discussions of manuscript work. On the discussion of al-Suyuti's occult astronomy in İbrahim Hakkı's eighteenth-century Ottoman encyclopedia, for example, see İhsanoğlu, "Introduction of Western Science," 25.

[57] The reference to Aristotle is explicit: Shafi'i, *Risala fi dhawat al-adhnab*, f. 8.

Part I

Geographies of Knowledge

1   The Deaf Shaykh: Scholarly Astronomy in Late Ottoman-Egyptian Society

On the fourteenth day of Sha'ban, 1254 (2 November 1838), Shaykh Shihab al-Din Ibrahim ibn Hasan put the finishing touches to a seventy-folio manuscript that he called *Sharh al-Lum'a fi Hall al-Kawakib al-Sab'a* (*The Commentary on the Brilliancy of the Solution of the Seven Planets*).[1] His writing was cramped. Hundreds of overlapping lines of ink, many of them flowing almost to the paper's edge, had sprung from their author's continuing preoccupation with commentary (*sharh*) on astronomical scholarship. Several years earlier, Ibrahim ibn Hasan had completed a twenty-eight-folio commentary on a seventeenth-century introduction to the science of astronomical timekeeping (*'ilm al-mīqāt*).[2] Now, with the *Commentary on the Brilliancy*, he had demonstrated his mastery of texts both older and more recent. The *Brilliancy* on which he commented was an astronomical handbook (*zīj*) composed by the fifteenth-century scholar Ibn Ghulam Allah.[3] However, Ibrahim ibn Hasan explicitly based his interpretation of the *Brilliancy* on a commentary on the same handbook that his teacher Muhammad al-Khudari had completed in 1239 (1823–24).[4]

---

[1] Ibrahim ibn Hasan "Shihab al-Din," *Sharh al-Lum'a fi Hall al-Kawakib al-Sab'a*, MS DM 638, ENL. For the dating of this manuscript, see fols. 13r and 41r in addition to the colophon.

[2] Ibrahim ibn Hasan "Shihab al-Din," *Sharh Muqaddimat Mahmud Qutb al-Mahalli*, MS TM 225, ENL. King gives this commentary the approximate date of 1300 (1882–83), but the colophon of this manuscript states that it was written in 1250 (1834–35); a later copy, MS DM 889.3, ENL, was finished in 1284 (1867).

[3] *Al-Lum'a fi Hall al-Kawakib al-Sab'a* (*The Brilliancy of the Solution of the Seven Planets*), by Abu al-'Abbas Ahmad ibn Ghulam Allah ibn Ahmad al-Kawm al-Rishi. Despite the centuries of use it enjoyed in Eastern Mediterranean lands, the *Brilliancy* has been the subject of very limited study, all of it contained in surveys. See Benno van Dalen's forthcoming *zīj* survey; David A. King, *A Survey of the Scientific Manuscripts in the Egyptian National Library* (Winona Lake, Ind.: Eisenbrauns, 1986), 65; B.A. Rosenfeld and Ekmeleddin İhsanoğlu, *Mathematicians, Astronomers, and Other Scholars of Islamic Civilization and Their Works (7th-19th c.)* (Istanbul: IRCICA, 2003), 269 (#800). See also David A. King, "The Astronomy of the Mamluks," *Isis* 74 (1983): 536.

[4] See Ibrahim ibn Hasan, *Sharh al-Lum'a*, MS DM 638, ENL, fol. 1v.

Ibrahim ibn Hasan and Muhammad al-Khudari labored over their astronomical scholarship during eventful years. Under the long rule of Mehmed Ali Pasha (r. 1805–48), Egypt saw the introduction of military conscription, cotton monoculture, new systems of military and civil education, and Arabic printing presses – all part of the emergence of an Egyptian empire within the Ottoman state. By 1832, the Pasha's armies had conquered much of the Eastern Mediterranean, Northern Sudan, and Western Arabia. Not three months before Ibrahim ibn Hasan reached the end of his *Commentary on the Brilliancy* in 1838, the Ottoman Empire and the United Kingdom agreed to the Treaty of Balta Limanı, opening a new and decisive chapter in the conflict between Istanbul and its renegade governor in Cairo. By 1841, the Pasha was forced to abandon most of his conquests outside the Sudan, but he achieved his longstanding aspiration to permanent and hereditary rule over the province of Egypt.

Ibrahim ibn Hasan was aware that he lived in unusual times. His *Commentary on the Brilliancy* drew attention to two moments in particular: Napoleon's invasion in 1213 (1798), "when the French entered Egypt"; and the founding of one of Mehmed Ali Pasha's new technical academies, the Engineering School (Tur. *mühendishane*, Ar. *muhandiskhāna*), in 1235 (1820), "when education appeared in Egypt" (*hīna zahara al-ta'līm bihā*).[5] With these allusions, Ibrahim ibn Hasan invoked two processes that are often placed at the center of the cultural and intellectual history of Egypt, of the Ottoman Empire, and to some extent even of Islam, in the long nineteenth century: European encroachment, and efforts to establish new educational institutions modeled on European counterparts.[6]

And yet, in ninety-eight dense folios of astronomical scholarship that he wrote in the 1830s, Ibrahim ibn Hasan left little evidence of these processes. Like the texts on which they serve as commentaries, his works teach the use of cosmographical models, mathematical techniques, observed values, and observational instruments that Muslim scholars had used since the fifteenth century or earlier. Indeed, compared with his teacher Muhammad al-Khudari's *Commentary on the Brilliancy*, Ibrahim ibn Hasan's explication of the geocentric cosmos was even expanded with diagrams.[7] Of what meaning, then, were the French invasion and

---

[5] Ibrahim ibn Hasan, *Sharh al-Lum'a*, MS DM 638, ENL, fol. 62v. On the founding of the Engineering School, see Crozet, *Sciences modernes*, 92n15.
[6] Benjamin Fortna, *Imperial Classroom: Islam, the State, and Education in the Late Ottoman Empire* (Oxford: Oxford University Press, 2002); Indira Falk Gesink, *Islamic Reform and Conservatism: al-Azhar and the Evolution of Modern Sunni Islam* (London: I.B. Tauris, 2014).
[7] Ibrahim ibn Hasan, *Sharh al-Lum'a*, MS DM 638, ENL, fol. 18v.

Mehmed Ali Pasha's new technical academies? For Ibrahim ibn Hasan, they served to illustrate the significance of the conjunction of Jupiter and Saturn. This periodic celestial phenomenon, he explained, "denotes a special event" (*ḥādith khāṣṣ*). The two planets were conjunct in "the last of the fire triplicity in 1213 [1798–99]," and again in "the beginning of the earth triplicity in 1235 [1819–20]."[8] Thus, although Ibrahim ibn Hasan attributed significance to these events, they were not relevant to him in the way we might expect. He did not mention new educational institutions or the French because they contributed to his understanding of astronomy. They certainly did not diminish his appreciation for the centuries-old astronomical texts that he devoted himself to explicating. To the contrary, they only deserved mention because they worked to illustrate the knowledge contained in these texts. From Ibrahim ibn Hasan's perspective, the French and their sciences were present, but only in his peripheral vision. Had the French invaded when Saturn and Jupiter bore a different relationship to each other in the zodiac, or had Mehmed Ali not founded a new kind of school specifically in 1820, Ibrahim ibn Hasan might well have followed Muhammad al-Khudari's example and not mentioned them at all.

For the time being, let us adopt a perspective that is closer to Ibrahim ibn Hasan's than to more common ways of looking at Ottoman-Egyptian society in the nineteenth century. Rather than assuming that the powers of European science and technology posed an obvious problem to which Ottoman intellectuals had to respond, let us take seriously the work of scientific practitioners for whom this dynamic was marginal. Only by shifting our perspective toward the vantage point of scholars like Ibrahim ibn Hasan can we understand the significance of a way of making knowledge in the nineteenth century that remained confidently rooted in venerable locations, texts, techniques, and epistemologies.

Ibrahim ibn Hasan was one of many late Ottoman practitioners of what I call "scholarly astronomy," a kind of science that shared practices of patronage, pedagogy, and writing with diverse kinds of knowledge that Muslim scholars ('ulama') made in late Ottoman Egypt. Muhammad al-Khudari and Ibrahim ibn Hasan moved within physical spaces and personal networks, and according to material constraints, particular to a geography of Islamic learning. When they explicated an astronomical

---

[8] The four triplicities each comprise three signs of the zodiac separated by 120°. Successive conjunctions of Jupiter and Saturn typically occur in the same triplicity (a "small conjunction," *al-qirān al-aṣghar*). The rare shift between triplicities was sometimes called a "middle conjunction" or "shift of transit" (*intiqāl al-mamarr*), but Ibrahim ibn Hasan did not use this terminology. See E.S. Kennedy, "The Sasanian Astronomical Handbook of Zij-i Shah," in *Studies in the Islamic Exact Sciences*, 259.

text like the *Brilliancy*, they understood the norms and objectives of "commentary" in relation to Islamic exegetical, juristic, and linguistic knowledge traditions in which they were also productive participants. Even as they continued to learn and teach geocentric cosmology, such scholars were not opposed to novelty, or to modifying the centuries-old texts that they devoted so much effort to understanding. But they had specific agendas and ways of communicating that led them to make sophisticated mathematical knowledge differently from their contemporaries at the Engineering School and other new institutions.

Understanding the perspective of scholars like Muhammad al-Khudari and Ibrahim ibn Hasan can inform our view of a crucial period in the relationship between the Middle East and other parts of the world. Historians of the Middle East have drawn attention to the nineteenth century, especially its latter half, as a period of radical intensification in the region's connection to global markets, movements, and networks. Ibrahim ibn Hasan and Muhammad al-Khudari wrote at the dawn of an era when the telegraph, printing press, steamer, and railroad began to move goods, people, and information between the Ottoman Empire, Europe, Asia, and even the Americas with astonishing speed.[9] The very geography of the Eastern Mediterranean changed, as new cities emerged in concert with new water routes (like the Mahmudiyya and Suez Canals), new institutions of governance (like sanitary commissions and Red Sea quarantine), and new, transnational forms of organization and expression (such as labor unions and the theater).[10] By placing the nineteenth-century Middle East in such global contexts, historians have drawn attention to the ways in which globalization was far from a universal experience or homogenizing force.[11]

Such nuanced narratives capture much about Egypt in the nineteenth century, but the enduring relevance of scholarly astronomy in the same context points to certain phenomena that fit uneasily – or disappear entirely – within a framework of globalization. "The age of steam and print" was also an age of manuscripts and sundials. Knowledge that moved through telegraphy, print, and new schools was not persuasive

---

[9] James Gelvin and Nile Green, eds., *Global Muslims in the Age of Steam and Print* (Berkeley, CA: University of California Press, 2014).

[10] Mikhail, *Nature and Empire*; Valeska Huber, *Channelling Mobilities: Migration and Globalisation in the Suez Canal Region and Beyond, 1869–1914* (Cambridge: Cambridge University Press, 2013); Michael Reimer, *Colonial Bridgehead: Government and Society in Alexandria, 1807–1882* (Boulder, CO: Westview Press, 1997); Ilham Khuri-Makdisi, *The Eastern Mediterranean and the Making of Global Radicalism, 1860–1914* (Berkeley, CA: University of California Press, 2010).

[11] Barak, *On Time*, 239–40.

or even interesting to all literate people. New global networks were not the only ways of moving knowledge across significant distances. In other words, scholarly astronomy points to the highly contingent and bounded reach of the modernization projects typically associated with the late Ottoman period. Even as rising generations of Ottoman-Egyptian technical officials were schooled in Paris and new neighborhoods of Cairo, scholars like Ibrahim ibn Hasan moved along different, intersecting, trajectories, animated by different objectives and norms. Their work represented an intellectually viable, socially important, much-more-than-local practice in its own right. And it was not displaced, or even necessarily disrupted, by the emergence of new sites of knowledge.

Scholarship on the relationship between astronomy and Islam has tended to focus either on the use of astronomy to define the correct performance of Islamic rituals, or on the role of Islamic theology in shaping astronomers' epistemology.[12] Ibrahim ibn Hasan's exercise in historical astrology points to the limitations of both approaches. In the first place, scholarly astronomy contained more than its intersections with Islamic ritual; when scholars used a *zīj* like the *Brilliancy* to calculate prayer times or the beginning of Ramadan, they did so as part of a larger set of practices that measured and interpreted time in late Ottoman society. In the second place, among the most popular of these practices was astrology (*aḥkām al-nujūm*, literally "judgments of the stars"), which stood in tension with prevailing Islamic theology. Of course, no single answer will ever suffice to characterize the relationship between a changing set of mathematical, observational, legal, and textual practices ("astronomy"), and a vast and contingent assemblage of discursive traditions ("Islam").[13] But if we consider the relationship between astronomy and Islam only in terms of the latter's rituals or tenets, we focus on the relationship between a particular subset of astronomical practices and similarly specific interpretations of piety.

Instead, following a discussion of the sources we can use to understand scholarly astronomy's extensive activity in late Ottoman Egypt, I situate this activity in the context of Islamic scholarship through a

---

[12] King, *Astronomy in the Service of Islam*; George Saliba, "The Development of Astronomy in Medieval Islamic Society," *Arab Studies Quarterly* 4 (1982): 211–25; F. J. Ragep, "Freeing Astronomy from Philosophy: An Aspect of Islamic Influence on Science," *Osiris*, 2nd. Ser., 16 (2001): 49–71; A.I. Sabra, "The Appropriation and Subsequent Naturalization of Greek Science in Medieval Islam: A Preliminary Statement," *History of Science* 25 (1987): 223–43.

[13] On Islam as discursive tradition, see Talal Asad, "The Idea of an Anthropology of Islam," 14–15, Georgetown University Center for Contemporary Arab Studies Occasional Paper Series, 1986.

series of fundamental questions about practice: *Where* did scholars of astronomy learn and do their work in late Ottoman Egypt? *What* did they aim to know? *How* did they try to know it? And finally, how did they *not* know about certain major developments, such as heliocentric planetary theory, that might seem pertinent to their concerns? In exploring the locations, objectives, and methods of scholarly astronomy, my goal is not to assess its technical development, much less to provide a comprehensive description. Rather, I aim to show that locations, material resources, literary genres, normative virtues, and textual techniques particular to Islamic scholarship shaped the practice of a kind of astronomy that transcended Islamic ritual needs, and sometimes contradicted Islamic tenets.

The Scope of Scholarly Astronomy in Late Ottoman Egypt: Some Remarks on Sources

A reasonable way to gain perspective on scholarly astronomical practice in late Ottoman Egypt is to take notice of the quantity of its textual production – or rather, the fraction of its textual production that happens to be preserved in well-catalogued library collections. The Egyptian National Library (ENL), al-Azhar Library, and the Municipal Library in Alexandria together contain (or contained) more than 300 manuscripts of scholarly astronomy produced *after* the formal introduction of new forms of astronomical training in Egypt in 1820 ("when education appeared in Egypt," in Ibrahim ibn Hasan's words) and before the close of the late Ottoman era at the end of World War I.[14] Since many of these manuscripts contain more than one work, the total number of titles is higher, at over 350.[15] Of course, residing in an Egyptian library in the twentieth century does not mean that a text was produced in Egypt in the nineteenth century. However, for the ENL (the largest of the three collections), the cataloguing work of David A. King et al. allows us to ascribe Egyptian provenance to a majority of the library's 263 relevant astronomical manuscripts, while 68 are of non-Egyptian origins, and 30 of unknown provenance. There is little reason to think that these proportions would differ greatly in the al-Azhar or Alexandria collection.

---

[14] David A. King, *Fihris al-Makhtutat al-'Ilmiyya al-Mahfuza bi-Dar al-Kutub al-Misriyya*, 2 vols. (Cairo: al-Hay'a al-'Amma al-Misriyya li-l-Kitab, 1986); Jami' al-Azhar, *Fihris al-Kutub al-Mawjuda bi-l-Maktaba al-Azhariyya* (Cairo, 1946–52); and Ahmad Abu 'Ali and Amin al-Watani, *al-Maktaba al-Baladiyya: Faharis al-Tabi'iyyat wa-l-Riyadiyyat wa-l-Qawanin wa-l-shara'i'* (Alexandria: Sharikat al-Matbu'at al-Misriyya, 1347 [1928]).

[15] For a title that is present in multiple collections, I only count the title once. However, such duplicates do contribute to the count of manuscripts (as opposed to titles), since the number of copies of a work may be significant.

Furthermore, the presence of non-Egyptian manuscripts in Egyptian collections is not necessarily a confounding variable. Late Ottoman Egypt's intellectual and political elite were deeply connected with other parts of the region, especially the Ottoman lands in Syria (provenance of twenty-two manuscripts) and Anatolia and the Balkans (twenty-eight manuscripts). It would be misleading to characterize Turkish or Syrian manuscripts as "foreign" to the Egyptian milieu.

While the collections that I surveyed are the most significant repositories of scientific manuscripts in Egypt, they only begin to suggest what a broader but unobtainable perspective might reveal. "Global" catalogue surveys, though helpful, tend to reflect the alignment of manuscript collecting and cataloguing with historiographical paradigms that render the practice of scholarly astronomy in the nineteenth century almost invisible. For example, the only American collections included in Ekmeleddin İhsanoğlu's *History of Ottoman Astronomical Literature* (*Osmanlı Astronomi Literatürü Tarihi*, hereafter *OALT*) are those of Princeton University, the University of Chicago, the Newberry Library, the Library of Congress, and the Metropolitan Museum of Art.[16] These are exceptionally valuable collections by the usual standards: they contain old copies of classically "important" texts. But if one is looking either for relatively late copies of such texts, or for relatively new texts that few historians have previously considered important, then many prestigious collections contain almost nothing of value. The most "important" collection that I have identified outside of Egypt belongs to the University of Michigan, which holds a number of manuscripts from the collection of Max Meyerhof, an ophthalmologist and pioneering scholar of Arabic medicine who lived in Egypt in the early twentieth century.[17] Surveys like *OALT* overlook this collection, obscuring a significant part of the documentary record of nineteenth-century Egyptian science. It is partly to overcome such blind spots in our historical perspective that scholars have undertaken new digital projects, such as the Islamic Scientific Manuscripts Initiative (ISMI), which promises to enable more-robust quantitative as well as qualitative assessments of Islamic scientific production from the eighth through nineteenth centuries.[18]

[16] Ekmeleddin İhsanoğlu et al., eds., *Osmanlı Astronomi Literatürü Tarihi* (Istanbul: IRCICA, 1997), II: 960–61.
[17] Joseph Schacht, "Max Meyerhof," *Osiris* 9 (1950): 7–32; Evyn Kropf, "Islamic Manuscripts Collection: About the Collection," University of Michigan Library Research Guides, http://guides.lib.umich.edu/islamicmss/about, accessed 26 June 2015.
[18] As of early 2017, although the ISMI database included 2,200 individuals and 600,000 images, only 123 codices were available through open access. https://ismi.mpiwg-berlin.mpg.de/drupal-ismi/, accessed 26 April 2017.

If one must be cautious when generalizing from sources that have been preserved, collected, and catalogued according to contingent standards, the available evidence nevertheless yields a remarkably different portrait of astronomical practice than previous scholarship has described. In the most significant study of the scientific activity of nineteenth-century Egyptian 'ulama', Pascal Crozet has argued that scholars "perpetuated the corpus of scientific writing in usage during the previous century."[19] Yet, the four authors whom Crozet places at the center of this "corpus" (Ibn al-Ha'im, Ibn al-Majdi, Sibt al-Mardini, and Ridwan Effendi) account for only sixteen of the titles I identified in the ENL, al-Azhar, and Alexandria collections.[20] By contrast, at least eighty titles are works entirely new to the nineteenth century, while the remainder include not only copies of, but also new commentaries on, earlier work. It is worth bearing in mind, moreover, that even a nineteenth-century copy of a fourteenth-century handbook (for example) is a document of the copyist's period, reflecting the decision of at least one person to spend time and resources writing out one specific text rather than another. This decision is best understood in the context of the copyist and, possibly, his patron and readers. Thus, the notion of "perpetuation" fails to capture the scope of nineteenth-century scholarly astronomy on at least two levels: it overlooks a significant degree of original composition, and it occludes the potential for new uses or understandings in the reproduction of older texts.

The reasoning that has led historians to take such a narrow view has little to do with their ability to count texts, of course, and more to do with their perspective on which texts count. Crozet's study focuses on works that he can show were widely read, or at least disseminated, generally because they were printed. Print sources, however, grant access to very particular kinds of Arabic and Turkish writing in the nineteenth century. Until the 1860s in Turkish and 1870s in Arabic, printing was almost entirely a government enterprise, and subsidies from political elites remained indispensable for many periodicals into the twentieth century.[21] Technical publishing was especially dominated by a small number of government presses, such as Egypt's Matba'at Bulaq, the mission of

---

[19] Crozet, *Sciences modernes*, 211.

[20] This number increases only slightly if we consider mathematical works not directly related to astronomy, which Crozet includes in his study. I only included mathematical texts if their typical use would lie in astronomy: treatises on sexagesimal arithmetic, for example.

[21] Ami Ayalon, *The Press in the Arab Middle East* (New York: Oxford University Press, 1995); M. Şükrü Hanioğlu, *A Brief History of the Late Ottoman Empire* (Princeton, NJ: Princeton University Press, 2008), 94–95.

which was to serve the new viceregal schools.[22] In other words, while the printing of scholarly astronomy says something about the extent to which scholarly astronomy interested the educational projects of the viceregal government, it tells us less about the practice of scholarly astronomy on its own terms. Moreover, the size of a text's readership, always a limited way of evaluating its social significance, may be particularly misleading in the case of technical manuals that were never intended to be widely read. If Ibrahim ibn Hasan's commentary on al-Mahalli's *al-Muqaddima fi 'Ilm al-Miqat (Introduction to the Science of Timekeeping)* was studied by only a few students, those students were able to perform calculations and provide guidance for others in ways that would have left ephemeral (if any) documentation.[23] Manuscripts of scholarly astronomy provide evidence for practices that were embedded in a variety of social relations, not only the relation of authors and readers.

While drawing on a range of manuscript and print sources from nineteenth-century Egypt, I will make thematic reference to the life of Muhammad al-Khudari, and especially to his *Commentary on the Brilliancy*, a text produced only in manuscript. Khudari makes for a fitting case study, rather than his student Ibrahim ibn Hasan or other practitioners, for two related reasons: the relative availability of information about his life, and the fact that his work was widely copied and well-regarded throughout the nineteenth century, and not only in Egypt. At least twenty-two complete copies by at least ten distinct scribes are extant or were catalogued in the twentieth century.[24] Some

[22] J. Heyworth-Dunne, "Printing and Translations under Muhammad 'Ali of Egypt: the Foundation of Modern Arabic," *JRAS* 3 (1940): 333. Science-popularizing journals such as *al-Muqtataf* (est. 1876) and *Mecmua-i Fünûn* (est. 1862) were privately published, but explicitly committed to cultivating a modern outlook among readers by translating science from European languages.

[23] The almanac marginalia of the Anatolian scholar Sadullah al-Ankaravi (d. 1855) constitute a remarkable exception to this observation. See Gülçin Tunalı Koç, "An Ottoman Astrologer at Work: Sadullah El-Ankaravi and the Everyday Practice of *İlm-ı Nücûm*," in *Les Ottomans et le temps*, ed. François Georgeon and Frédéric Hitzel (Leiden: Brill, 2012), 39–59.

[24] The 1239 (1823–24) autograph that King et al. cataloged (MS DM 94, ENL) could not be located at the Egyptian National Library (Bab al-Khalq) in 2010–11. Later copies include MSS KhM 3, DM 270, ṬM 125.2, DM 976, TR 200, DM 208, and DM 200, ENL; MSS [292] 14478, [374 'Arusi] 42753; [336 Halim] 34492; [379 'Arusi] 42758; [396 Bakhit] 45605; and [397 Bakhit] 45606, Azhar; MSS 4831 and 3100, Zahiriyya; MS Bağdadlı Vehbi Efendi 909, Süleymaniye; MS 871, AUB; Mecmu-i Antaki 52/1, Aleppo; MSS 523.4089927 and 520.89927, QNL; MS 520.89927, JMI; MS 520.89927, JMA; and Isl. Ms. 722, Michigan. Note that the geographic distribution of these manuscripts is not a reliable guide to the text's travel in the nineteenth century, since many of these collections (particularly the ones in North America and the Gulf states) were assembled in the twentieth century.

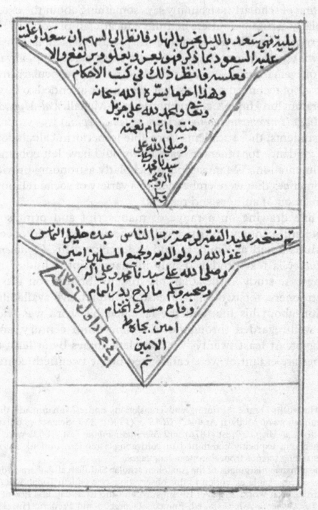

Figure 1.1 Final page of an 1889 copy of Muhammad al-Khudari's
Commentary on the Brilliancy of the Solution of the Seven Planets. The top
section reproduces the author's colophon. The bottom is the scribe's
colophon, attesting that this copy was completed by "the one in need
of the mercy of the Lord of people, his slave Khalil al-Nahhas," who
completed it in I Jumada 1306 (January 1889). The writing in the top
left corner is the scribe's notation that he has reached the end of the
corresponding section in the manuscript on which he based this copy.
Muhammad al-Khudari, *Sharh al-Lum'a fi Hall al-Kawakib al-Sab'a*,
p. 227, Isl. Ms. 722, University of Michigan Library (Special Collections
Library), Ann Arbor.

were done within the author's lifetime by his close relations or associ-
ates, such as his half-brother (his father's son), who was also named
Muhammad al-Khudari, and his student Ahmad al-Wafa'i.[25] Other
copies date to after Khudari's death, including a 1306 (1889) copy that
shows considerable investment in its production, including extensive
correction, a two-color border, and multiple copyists (Figure 1.1).[26]
One copy, on which the date is obscured, belonged to a judge in
Fayyum.[27] Illustrating the text's renown beyond Egypt, an anonymous
scholar in Aleppo in the 1860s referred to Khudari's commentary
as "well known," and included it on a list of landmark astronomical
texts.[28] In the early twentieth century, the legal reformist Ahmad al-
Husayni, who drew on European anatomical and chemical science in
his own work, referred to Khudari's *Commentary on the Brilliancy* as
"unparalleled in this science."[29] While the extensive copying and reputation of his work make Khudari an atypical figure in certain ways, they
demonstrate a key point about scholarly astronomy in the nineteenth
century: far from an antiquarian pursuit, it was, for many, a dynamic
way of making socially valued knowledge.

Damietta and Cairo: A Geography of Scholarly Astronomy

To understand the social relations within which the practice of scholarly astronomy occurred, we must locate these relations in space.[30]
Practitioners moved between Cairo and other, sometimes quite distant locations; between the different spaces of learning found within
Cairo; and through the network of personal relationships that linked
these spaces and sustained the transmission of knowledge. The "map"
of scholarly astronomy that emerges when we trace these dimensions
illustrates intimate connections as well as important distinctions between
scholarly astronomy and a broader context of Islamic learning. It also
suggests that, despite the Ottoman-Egyptian government's assault on

---

[25] MS [397 Bakhit] 45606, Azhar; MS 520.89927, Qatar.
[26] Isl. Ms. 722, Michigan. The manuscripts that I consulted, which also include MS 397
Bakhit 45606, Azhar, and MSS 523.4089927 and 520.89927, Qatar, do not contain significant textual variations, except for the Azhar manuscript's inclusion of certain tables
not present in the others.
[27] MS 523.4089927, Qatar.
[28] MS ṬR 182, ENL.
[29] Ahmad al-Husayni, *Tabaqat al-Shafi'iyya*, MS Taymur: Tarikh 1411, ENL, p. 58.
[30] On geographic approaches in science studies, see Adi Ophir and Steven Shapin, "The
Place of Knowledge: A Methodological Survey," *Science in Context* 4 (1991): 3–21;
Thomas Gieryn, "Three Truth Spots," *Journal of the History of the Behavioural Sciences* 38
(2002): 113–32; and David Livingstone, *Putting Science in Its Place* (Chicago: University
of Chicago Press, 2003).

the economic independence of 'ulama' in the nineteenth century, and its increasing investment in new institutions of scientific education, scholarly astronomy was surprisingly successful in maintaining access to material support.

Muhammad al-Khudari was born in Damietta in 1798–99.[31] Lying on the Mediterranean at the eastern mouth of the Nile, Damietta was one of Egypt's principal cities and among the key ports of the Ottoman Empire at the beginning of the nineteenth century, linking the "bread-basket" of the Nile Valley with Syria and Anatolia.[32] It is not surprising to find scholars of astronomy like Khudari and Ibrahim ibn Hasan associated with it. In the 1830s, Lane observed that Damietta was a center for the production of the *qibliyya*, a kind of compass that Egyptians commonly used to determine the direction of prayer (*qibla*) in this period.[33] A Mediterranean city, Damietta was home to consuls, merchants, and a circle of Greek Christians whose transnational networks and multilingualism produced some of the first translations of French and English literature into Arabic in the nineteenth century – including Jérôme de Lalande's *Abrégé d'Astronomie*, translated in 1808.[34] At this point, however, translations were the exception. Lalande's Arabic *Abrégé* (*al-Mukhtasar fi 'Ilm al-Falak*) appears not to have circulated beyond the original manuscript, whereas Khudari's *Commentary on the Brilliancy* was widely copied and studied.

What is typical about Muhammad al-Khudari, however, is not his link to Damietta specifically, but rather the move he made, around the age of seventeen, from Damietta to Cairo. He had completed his memorization of the Qur'an under the tutelage of a local teacher (a certain Shaykh

---

[31] Biographical information about Muhammad al-Khudari is based primarily on Ahmad al-Husayni, *Tabaqat al-Shafi'iyya*, MS Taymur: Tarikh 1411, ENL, p. 58. Husayni's account is credible because its author was a student of Khudari's half-brother, and the only source of biographical information to be so directly connected to our author. See al-Husayni, *Tabaqat*, p. 62. By contrast, the relevant entries in the standard biobibliographical dictionaries are incomplete, contradictory, and lacking citations. Yusuf Sarkis, for example, seems to have been unaware that there were two Khudari brothers; his entry on "al-Dimyati al-Khudari" lists the works of "our" Khudari but describes the life of his brother. Yusuf Sarkis, *Mu'jam al-Matbu'at al-'Arabiyya wa-l-mu'arraba* (Cairo: Maktabat Yusuf Ilyan Sarkis, 1928), s.v. "al-Dimyati al-Khudari."

[32] Mikhail, *Nature and Empire*; Jamal al-Din al-Shayyal, *Mujmal Tarikh Dimyat* (Alexandria: Matba'at Don Bosco, 1949).

[33] E.W. Lane, *An Account of the Manners and Customs of the Modern Egyptians: The Definitive 1860 Edition* (Cairo: AUC Press, 2003), 218.

[34] J.J. de Lalande, *al-Mukhtasar fi 'ilm al-falak*, trans. Basili Fakhr et al., MSS Arabe 2554 and 2555, BnF. On the Damietta translators, see Peter Hill, "The First Arabic Translations of Enlightenment Literature: The Damietta Circle of the 1800s and 1810s," *Intellectual History Review* 25 (2015): 209–33; and Peter Hill, "Early Translations of English Fiction into Arabic," *Journal of Semitic Studies* 60 (2015): 177–212.

al-Sharnubi),[35] and he was ready to join the circles of learning (*ḥalaqāt*) at the al-Azhar mosque. In this path, he was following a common itinerary for Sunni Muslim scholars in search of knowledge. The architectural and administrative division of al-Azhar into numerous regionally defined "pavilions" (sg. *riwāq*), including Syrian, East African, and even – by the 1860s – a Southeast Asian pavilion (*riwāq al-jāwā*), was a formal, albeit approximate, reflection of the transnational Islamic geography within which the mosque occupied one of the oldest and most distinguished centers.[36]

Gathered in the courtyards of al-Azhar and its adjoining madrasas, well-traveled scholars focused on the sciences of the Arabic language (grammar, semantics, and rhetoric), traditions of the Prophet (hadith), jurisprudence (*fiqh*), Qur'anic exegesis (*tafsīr*), creed (*'aqīda*), and logic (*manṭiq*).[37] Khudari formed impressive connections in several of these areas, studying under shaykhs including the future Shaykh al-Azhar Hasan al-Quwaysini (d. 1838–39). He also acquired the epithet *al-Shafi'i*, indicating his study and adoption of the Shafi'i legal tradition (*madhhab*). But the best indication of the range of his learning are the texts he ultimately wrote over the course of his life: a treatise in verse (*manẓūma*) on Qur'anic vocabulary, a treatise on the principles of Qur'anic exegesis, supercommentaries (sg. *ḥāshiya*) on major works of rhetoric, grammar, and inheritance law, a treatise on the principles of jurisprudence (*uṣūl al-fiqh*), and an important work on logic, in addition to his *Commentary on the Brilliancy* and a text on the sundial. For 'ulama', astronomy belonged to a constellation of sciences (*'ulūm*) that constituted what it meant to be a scholar (*'ālim*).[38]

But if astronomy was always one among a variety of scholarly pursuits, it stood apart from most areas of knowledge in the location where this pursuit occurred in the early nineteenth century: not inside al-Azhar, for

[35] Possibly related to 'Abd al-Majid al-Sharnubi al-Azhari, who was active in 1889. See 'Abd al-Majid al-Sharnubi al-Azhari, *Sharh dala'il al-khayrat li-l-Juzuli* (Cairo: Maktabat al-Adab, 1994), 79.

[36] For a description of al-Azhar in the middle of the nineteenth century, see 'Ali Mubarak, *al-Khitat al-tawfiqiyya al-jadida* (Bulaq: al-Matba'a al-Kubra al-Amiriyya, 1304–06 [1886–89]), 4:22. For an overview of the mosque's architectural history, see Nasser Rabbat, "Al-Azhar Mosque: An Architectural Chronicle of Cairo's History," *Muqarnas* 13 (1996): 45–67. On the *riwāq al-jāwā*, see Michael Laffan, "An Indonesian Community in Cairo: Continuity and Change in a Cosmopolitan Islamic Milieu," *Indonesia* 77 (2004): 2–5.

[37] On the books commonly read at al-Azhar in the middle of the nineteenth century, see Mubarak, *al-Khitat*, 4: 27–28.

[38] Ahmad Dallal, *Islam, Science, and the Challenge of History* (New Haven, CT: Yale University Press, 2010), 19–22. For another example from nineteenth-century Egypt, see the biography of Muhammad b. Ahmad al-Dasuqi in Mubarak, *al-Khitat*, 11:9.

the most part, but rather in people's houses. This distinction arose partly from theological and legal considerations. Certain astronomical practices, astrology in particular, were the subject of venerable theological critiques. According to most schools of jurisprudence (*madhahib*), moreover, even those aspects of astronomical knowledge that facilitated the fulfillment of religious duties – the science of determining prayer times, for example – were not necessary for the education of every scholar, but only for a single scholar in every community (*farḍ kifāya*). In commonly studied books of *fiqh*, discussion of prayer times was typically limited to basic shadow schemes. As David A. King has noted, the sophisticated methods of tabulating prayer times developed by Muslim astronomers since the ninth century, and with particular energy in Cairo since the tenth century, were rarely considered normative for fulfilling the duty of prayer.[39]

The absence of astronomy, among other "rational" and mathematical sciences, from madrasas has often been produced as evidence that such sciences became marginal, or even illegitimate, among pious 'ulama' in post-classical Muslim societies.[40] But this is a flawed interpretation, for several reasons. First, the history of one science does not represent the fate of all: while astronomy was not well-supported inside al-Azhar in the early nineteenth century, logic, for example, was de rigueur. Second, even the weakness of astronomy inside the mosque should be understood as specific to the period in question. Accounts of the dearth of astronomical studies at al-Azhar in the late eighteenth and early nineteenth centuries describe contingent circumstances – not a tale of uninterrupted decline since the twelfth century. As Sonja Brentjes has remarked, the notion that post-classical madrasas simply excluded the rational sciences as a matter of course contradicts a remarkable amount of evidence, from madrasa libraries stocked with mathematical texts, to the endowing of medical madrasas under the Ottomans.[41]

Moreover, even when certain sciences did lose their standing inside the madrasa, and were pursued in other spaces, the location of study should not necessarily be understood in terms of the legitimacy of the subject. In the eighteenth and nineteenth centuries, fully fledged, even eminent Azhari scholars regularly studied astronomy outside the mosque

---

[39] David A. King, *The Call of the Muezzin* (Leiden: Brill, 2004), 549–50, ProQuest ebrary, Web, accessed 11 January 2015.

[40] For this thesis as articulated by an eminent orientalist, see G.E. von Grunebaum, "Muslim World View and Muslim Science," *Dialectica* 17 (1963): 353–67.

[41] Sonja Brentjes, "The Prison of Categories – 'Decline' and Its Company," in *Islamic Philosophy, Science, Culture, and Religion*, ed. Felicitas Opwis and David Reisman (Leiden: Brill, 2012), 139–43.

without incurring ill repute. Jabarti evocatively describes his father's practice of teaching astronomy to students on the rooftop of the family's house.[42] Several of the astronomical manuscripts cited in this book belong to collections left to al-Azhar by prominent nineteenth-century shaykhs, including Mustafa al-'Arusi and Ibrahim al-Saqqa.[43] Khudari's teacher in astronomy was probably the former Shaykh al-Azhar and famous polymath Hasan al-'Attar, who was reputed to have taught him "the science of wisdom and nature."[44] 'Attar was a controversial figure, because of his association with Napoleon's savants and, later, with the government of Mehmed Ali. But if Khudari and others studied astronomy with Hasan al-'Attar in the latter's house – or on top of it – rather than in a circle (ḥalqa) at al-Azhar, there was nothing untoward or even unusual about such study. As Jonathan Berkey has shown, the transmission of knowledge among 'ulama' was "fundamentally and persistently an informal affair," in which prestige and authority attached to individual teachers, not institutions.[45]

If the location of learning did not determine the legitimacy of knowledge, however, it bore directly on the kinds of material support available for the pursuit of knowledge. Mosques and madrasas benefited from waqfs (pious endowments) dedicated to teaching positions in specific fields, with fiqh often occupying a privileged position.[46] Although many prominent astronomers in the fourteenth through sixteenth centuries had benefited from endowed muwaqqit (timekeeper) posts at mosques in Cairo and Damascus, these positions were generally modest and frequently defunct by the eighteenth century.[47] Thus, 'Abd al-Rahman al-Jabarti's

[42] Jane Holt Murphy, "Improving the Mind and Delighting the Spirit: Jabarti and the Sciences in Eighteenth-Century Ottoman Cairo" (Ph.D. Diss., Princeton University, 2006), 131.

[43] MSS 317 al-Saqqa 28898, 379 'Arusi 42758, 391 'Arusi 42802, 386 'Arusi 42765, and 374 'Arusi 42753, Azhar.

[44] "Fann al-ḥikma wa-l-tabī'a." Al-Husayni, Tabaqat, p. 42. Ḥikma in this context could denote philosophy or even medicine, a specialty of 'Attar's. However, since Khudari is not known to have written on philosophy or medicine, I translate the term more inclusively as "wisdom." On ḥikma, see A.M. Goichon, "Ḥikma," Encyclopaedia of Islam, 2nd edition, ed. P. Bearman et al. (Brill Online.)

[45] Jonathan Berkey, The Transmission of Knowledge in Medieval Cairo: A Social History of Islamic Education (Princeton, NJ: Princeton University Press, 1992), 17–18.

[46] On waqfs and Islamic scholarship, see George Makdisi, The Rise of Colleges: Institutions of Learning in Islam and the West (Edinburgh: Edinburgh University Press, 1981). From a different perspective, Chamberlain emphasizes the importance of endowed teaching positions as an object of economic competition. Michael Chamberlain, Knowledge and Social Practice in Medieval Damascus, 1190–1350 (Cambridge: Cambridge University Press, 1994).

[47] King, "The Astronomy of the Mamluks," 534–35; Murphy, "Improving the Mind," 147. Exceptions, however, include Muhammad al-Banhawi (d. 1886–87), "an eminent

chronicles relate that most people at al-Azhar in the late eighteenth century could not afford astronomical instruments, but that experts in astronomy "are found in their houses" (*mawjūdūn fī buyūtihim*).[48] The location of astronomy outside the mosque was understood primarily in terms of material resources, rather than legal or theological norms.

And in material terms, having roots outside the mosque may have worked to the advantage of astronomy over other kinds of scholarly pursuits in nineteenth-century Egypt. Mehmed Ali Pasha's policy of bringing *waqf*s under state control struck a powerful blow against the financial autonomy of the spaces of Islamic scholarship.[49] Scholars of astronomy, however, being accustomed to securing support outside of endowed teaching positions, were relatively well-positioned to withstand this shock. The Pasha's destruction of Cairo's old political-military elite may have posed a more serious challenge to astronomers than his appropriation of *waqf*s. As Jane Murphy has shown, astronomy and related sciences were an important means by which Arabophone scholars had forged connections with elite households in eighteenth-century Cairo.[50] But the classic connection between scholarly astronomy and political patronage survived the destruction of the old elite households, as Egypt's new political elite – the Pasha's household and, eventually, new classes of land magnates and bureaucrats – continued to patronize scholars of astronomy.[51] Indeed, the very date on which the Pasha carried out the infamous Citadel massacre in 1811 was chosen in consultation with astrologers.[52] Another example is Sulayman Pasha Abaza, a provincial governor under the Khedive Ismail, who employed the scholar of astronomy Khalil al-'Azzazi to build a sundial for one of his mosques.[53] The viceregal government's enormous investment in new kinds of astronomical practitioners and institutions did not preclude individual officials from patronizing scholars like 'Azzazi, who continued to fulfill a variety of needs.

scholar and *muwaqqit*" at al-Azhar, and Muhammad al-Sannar, author of a 1225 (1810–11) treatise on geometry, which identifies him as "the *muwaqqit* at the noble Ahmadi shrine," i.e., the gravesite of al-Sayyid al-Badawi in Tanta. Husayni, *Tabaqat*, 77; Muhammad al-Sannar, *Kitab Rawdat al-Fasaha fi 'Ilmay al-Handasa wa-l-Misaha*, MS TR 95, ENL.

[48] Mubarak, *al-Khitat*, 4:17.
[49] Gesink, *Islamic Reform and Conservatism*, 13–16.
[50] Murphy, *Improving the Mind*, 114.
[51] On the "court" as the typical source of scientific patronage, see A.I. Sabra, "Situating Arabic Science: Locality versus Essence," *Isis* 87 (1996): 662.
[52] Fahmy, *Mehmed Ali*, 36.
[53] Crozet, *Sciences modernes*, 214. See also a text on the prediction of eclipses that 'Ali al-Khashshab wrote "in the house of... Hanafi ibn Shahin" in 1293 (1876–77). *Risala mukhtasira fi ta'dil al-zaman wa-l-kusufat wa-l-khusuf*, MS DM 261, ENL.

The nature of such relationships between scholars of astronomy and new elites, including new scientific elites, was mediated by the geography of learning in nineteenth-century Cairo. Consider the daily routine of Mustafa al-Bulaqi al-Burullusi. Bulaqi was born around 1780 and died in 1847, making him about a generation older than Muhammad al-Khudari; the two shared certain teachers, and it is not unlikely that they knew each other.[54] Bulaqi himself began to teach and give legal responsa (*iftā'*) at al-Azhar beginning in 1808. At the same time, however, he continued to reside (as his name suggests) in the area of Bulaq, where Mehmed Ali placed many of the new institutions of his rule – notably the Engineering School, and later the School of Languages and the Bulaq Press. 'Ali Mubarak, writing about twenty years after Bulaqi's death, emphasized his interaction with these new institutions:

[Bulaqi] had a great inclination toward the sciences of mathematics, engineering, arithmetic, and astronomy. He greatly loved to meet the people qualified in these sciences, such as Mahmud Bey al-Falaki... and others among the masters of the Engineering School that was in Bulaq, until he became competent in those sciences and composed a treatise in the science of timekeeping on the sine quadrant, and composed many treatises on algebra and trigonometry. He came to al-Azhar every day and gave the sermon at the mosque of the Sultan Abu al-'Ala', and he had a regular lesson there between the sunset and evening prayers.[55]

As Bulaqi's daily rounds suggest, old spaces of learning like al-Azhar and nearby mosques were separate, but not walled off, from new spaces of learning. It was in the former spaces that Bulaqi had received his "complete education" (*taḥṣīl tāmm*), and in which he gave sermons and lessons. Yet his interests and the location of his home brought him into the orbit of the new technical institutions in Bulaq, where no barrier prevented him from engaging in regular conversation with Mahmud al-Falaki, an instructor who had recently learned French and begun teaching astronomy from the works of Delambre and Francoeur.[56] Mahmud al-Falaki did not teach Delambre at al-Azhar, nor did Bulaqi, whose written astronomical work concerned topics long familiar to scholars of astronomy (e.g., the sine quadrant).[57] If astronomy meant different

---

[54] Mubarak, *al-Khitat*, 9:33.
[55] Mubarak, *al-Khitat*, 9:33. Crozet also discusses Bulaqi as an example of connections between old and new milieus of learning in *Sciences modernes*, 204.
[56] Ismail-Bey Moustapha, "Notice nécrologique de S.E. Mahmoud Pacha El Falaki," in *Notices biographiques sur S.E. Mahmoud-Pacha El Falaki* (Cairo: Impremerie Nationale, 1886), 9. Crozet, *Sciences modernes*, 195.
[57] Bulaqi's works, which do not appear to be extant, are described in Mubarak, *al-Khitat*, 9:33.

practices in different spaces, however, practitioners were connected socially.

Bulaqi's relationship with Mahmud al-Falaki reminds us that astronomical knowledge in late Ottoman Egypt was constituted not only in specific places, but also by the networks of people who moved through these spaces. Sometimes it is possible to trace an intellectual genealogy across several generations. 'Abd al-Hamid Mursi Ghayth, author of an early twentieth-century manual on the calculation of almanacs, was a student of Husayn Zayid, who authored a "new *zīj*" specifically for use at al-Azhar when astronomy was introduced in the 1880s.[58] Zayid was a student of Khalil al-'Azzazi, a mid nineteenth-century author of texts on timekeeping and arithmetic who was also the grandson of a Sufi shaykh whose *mawlid* was still celebrated in the Eastern Delta in the 1880s.[59] If we dig beneath the surface of such pedagogical lineages, moreover, additional kinds of connections emerge. 'Azzazi's introduction to arithmetic was lithographed in 1881. The copyist, Husayn Yahya, was also the author of a 1285 (1868) manuscript almanac that includes a calculation of the *qibla* of London according to the method of one 'Ali al-Khashshab, whom Yahya acknowledged as his teacher.[60] Khashshab himself wrote a manuscript on the prediction of eclipses in 1293 (1876), but must have been active earlier.[61] As I will argue further in Chapter 4, these kinds of connections point to an oft-forgotten context of the "reform" of al-Azhar in the late nineteenth century: notwithstanding the resistance among certain 'ulama' to introducing new sciences inside the mosque, such sciences could only be introduced successfully because 'ulama' had long been studying them elsewhere.

The network of scholars that produced astronomical texts extended beyond those with a particular interest in astronomy. Muhammad al-Khudari's half-brother, who was the scribe for one of the early copies of the *Commentary on the Brilliancy*, had no other connection to astronomy. He was, however, both a student of his older half-brother and later, in his own right, one of the most renowned teachers of his generation; it

---

[58] 'Abd al-Hamid Mursi Ghayth, *Kitab al-Manahij al-Hamidiyya fi Hisabat al-Nata'ij al-Sanawiyya* (Cairo, 1923); Husayn Zayid, *al-Matla' al-sa'id fi hisabat al-kawakib al-sab'a 'ala al-rasd al-jadid* (Cairo: al-Matba'a al-Baruniyya, 1304 [1887]), 2–3.

[59] Mubarak, *al-Khitat*, 15:10; Khalil al-'Azzazi, *Tashil al-raqa'iq fi hisab al-daraj wa-l-daqa'iq* (n.p.., al-Matba'a al-Bahiyya, 1299 [1881–82]); Khalil ibn Ibrahim al-'Azzazi, *Muqaddima fi 'amal mawaqi' 'aqarib al-sa'at 'ala qadr al-hisas al-shar'iyya li-kull 'ard*, MS TR 204, ENL. See also Crozet, *Sciences modernes*, 214–16.

[60] Husayn Yahya, *Jadwal al-tawqi'at al-yawmiyya wa-fihi samt al-qibla*, MS K 4011.2, ENL.

[61] Khashshab, *Risala mukhtasira fi ta'dil al-zaman*, MS DM 261, ENL.

was said in the early twentieth century that every scholar at al-Azhar had studied either directly with him or with one of his students.[62] Ahmad al-Wafaʾi, another of Khudari's scribes, similarly left no greater record of astronomical scholarship – but he did carry the epithets *al-Ashʿarī al-Shādhilī*, denoting his allegiance to a specific theological school and Sufi order, respectively. Since Muhammad al-Khudari composed his *Commentary on the Brilliancy* after leaving Cairo and returning to Damietta, where he passed the rest of his life, such connections suggest the routes by which the *Commentary* quickly became known in places like Fayyum and Aleppo.

The individual relationships that produced astronomical texts, like the spaces in which astronomy was taught, studied, and practiced, show the depth of scholarly astronomy's connection to a broader context of Islamic scholarship, as well as some of the factors that distinguished astronomy from other kinds of knowledge. Scholars of astronomy, their students, and the scribes who copied their texts were all, for the most part, like any other ʿulama'. From towns and villages near and far, they converged on Cairo in order to study with masters of hadith, *fiqh*, and other Islamic knowledge traditions. While the study of astronomy took them beyond the mosques and madrasas in which they spent much of their time, this shift in location was not indicative of a marginal or dubious activity, but rather of the different kinds of material support allocated to different kinds of knowledge in different spaces. In short, scholarly astronomy was a kind of knowledge that people who were profoundly at home at al-Azhar pursued outside of al-Azhar. Far from weakening their social standing, such mobility enabled astronomers to sustain themselves in a difficult period for ʿulama' by forging relationships with patrons and interlocutors in the new elite, even as they remained rooted in venerable spaces of Islamic learning.

What Did Muhammad al-Khudari Know?

In the introduction to his *Commentary on the Brilliancy*, Muhammad al-Khudari explained his choice of a fifteenth-century text in terms of the interests that he expected of his nineteenth-century readers. He wrote, "The (most commonly) used of (this science's) books in our time is *The Brilliancy of the Solution of the Seven Planets*, because of the slackening of interest in other tomes, and people's contentment with the apparent phenomena (*ẓawāhir al-aʿmāl*) rather than the examination of aspects of

[62] Al-Husayni, *Tabaqat*, p. 62.

proofs."[63] Khudari's sweeping evaluation of his contemporaries' interests was less damning than it might appear. The commentary that follows it is replete with mathematical demonstrations, alternative procedures, and historical, legal, and etymological arguments; Khudari could hardly have thought that his readers were incurious. They were curious, however, about specific things. Ibn Ghulam Allah's *Brilliancy* is a typical *zīj*: a set of tables and instructions that allow the user to set the calendar, tell time, and predict planetary positions and eclipses.[64] Except in a very cursory manner, it does not concern itself with the geometry of planetary models, which was the domain of *'ilm al-hay'a* (literally, "the science of structure [of the spheres]").[65] Although late Ottoman scribes still produced copies of a few popular *hay'a* texts, they appear to have been more rarely copied than *zīj*s or related *mīqāt* texts.[66]

If Khudari's readers were not particularly interested in *hay'a*, however, their interest in the *zīj* still suggests a great variety of possible practices. In fact, Khudari's expectation that his readers would use the *zīj* broadly to study "the apparent phenomena" hints at the difficulty of reducing scholarly astronomy to a single activity *par excellence* such as ritual timekeeping. In fact, scholars in nineteenth-century Egypt used a *zīj* like the *Brilliancy* with an eye toward three interrelated objects of knowledge: chronology (the study of dates), horology (the study of measuring time), and astrology, which gave meaning to the divisions of time. Chronology, horology, and astrology did not constitute all the possible uses of the *zīj*, much less all the practices of scholarly astronomy. Focusing on these practices and their place in nineteenth-century Egyptian society, however, will contextualize the well-known intersection of astronomical timekeeping (*mīqāt*) with Islamic ritual duties as a

[63] Khudari was playing with language from Ibn Ghulam Allah's own introduction to the *Brilliancy*. However, where the fifteenth-century scholar had written of a *decline* in learning, the nineteenth-century scholar substituted a comment on the *specificity* of his readers' interests. Compare Isl. Ms. 722, Michigan, p. 6, with pp. 7–8.

[64] The classic history of the *zīj* is E.S. Kennedy, "A Survey of Astronomical Tables," *Transactions of the American Philosophical Society* 46 (1956): 123–77.

[65] On the development of *hay'a*, see F.J. Ragep, *Nasir al-Din al-Tusi's Memoir on Astronomy*, vol. 1 (New York: Springer-Verlag, 1993), 29–35. Although *'ilm al-hay'a* eventually came to supplant *'ilm al-nujūm* as a general term for astronomy in many classifications, Khudari's usage invokes the specific sense of *hay'a* as that part of astronomy concerned with understanding the arrangement of the orbs.

[66] For nineteenth-century copying of classic *hay'a* texts, see MS DH 88, ENL, a 1243 (1827–28) copy of Nizam al-Din al-Nisaburi's *Tawdih al-tadhkira*; and MSS TH 32, TH 42, DH 103, DH 91, and DH 86, ENL, all late copies of 'Ali ibn Muhammad al-Sharif al-Jurjani's *Sharh al-tadhkira*. Both are based on Tusi's *Tadhkira fi 'ilm al-hay'a*. (See King, *Survey*, 1/2/10.1.)

relatively small piece of a larger relationship between 'ulama', astronomical knowledge, and temporality.

Converting Time

Let us begin where readers of Muhammad al-Khudari's *Commentary on the Brilliancy* began. The first three of the *Brilliancy*'s twelve sections introduced the mathematical principles of the solar and lunar cycles, the relationship between these cycles and the Arab, Coptic, Hebrew, Seleucid (*rūmī*), and Yazdigird (*fārisī*) calendars, and major dates of significance in each of these calendars. Muhammad al-Khudari's commentary on these sections typically occupies slightly more than a quarter of the total manuscript text, suggesting that he thought chronology merited his full attention.[67]

Late Ottoman Egypt was a chronologically plural society. Coptic, Muslim, Jewish, and Melkite communities each used distinct methods for measuring and expressing the passage of time: their calendars each had a distinct epoch ("zero" year), number of days in the year, method of determining the beginnings of the months, and approach to intercalation (leap days or months). Texts such as Khudari's *Commentary* were not simply introductions to the principles of these systems. Perhaps more importantly for readers, they offered extensive instructions on the use of tables, which were included prominently in the text, for calculating the beginnings of months according to different calendars, knowing when their leap years will occur, and, ultimately, being able to convert dates between calendars (*istikhrāj al-tawārīkh*) (see Figure 1.2).

Such procedures recognized the fact that, despite the link between specific calendars and particular religious communities, every community – and certainly the Muslim community of Egypt – lived with multiple calendars simultaneously.[68] Thus, what we now call the "Islamic" (*hijrī*) calendar was used in Muslim scholarly culture to date historical events and texts, to record the death dates of scholars (especially the transmitters of hadith), and to observe Muslim holidays. To track the agricultural seasons, however, even Muslims used the Coptic calendar, the liturgical calendar of the country's largest Christian community, since it was the oldest and most commonly used solar calendar in Egypt.[69] As

---

[67] See, e.g., Isl. Ms. 722, Michigan, and MSS 523.4089927 and 520.89927, QNL.

[68] On the multiconfessional experience of time in late Ottoman cities, see Bernard Lory and Hervé Georgelin, "Les temps entrelacés de deux villes pluricommunautaires: Smyrne et Monastir," in *Les Ottomans et le temps*, ed. François Georgeon and Frédéric Hitzel (Leiden: Brill, 2012), 173–201.

[69] Lane, *Manners and Customs*, 219.

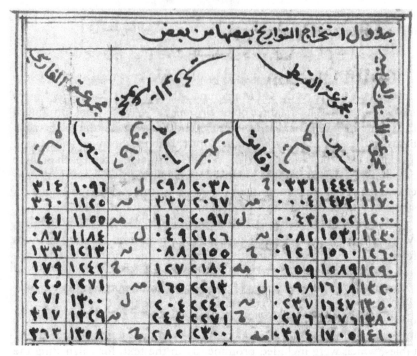

Figure 1.2 Table for the conversion of dates between the Arab, Coptic, Seleucid (*rūmī*), and Yazdigird (*fārisī*) calendars. Hebrew dates were calculated using a separate set of tables. Muhammad al-Khudari, *Sharh al-Lum'a fi Hall al-Kawakib al-Sab'a*, p. 53 (detail), Isl. Ms. 722, University of Michigan Library (Special Collections Library), Ann Arbor.

On Barak has shown, the notion that the *hijrī* calendar embodied a specifically "Muslim" temporal domain emerged only in contrast with the identification of the Gregorian calendar, which was officially adopted in Egypt in 1875, as a marker of "secular" temporality.[70] In fact, scholars of astronomy often referred to the *hijrī* system as the "Arab" (rather than "Islamic") calendar, denoting its association with the Arabian origins of the early Muslim community, rather than with the religious duties of the global *umma*.[71]

Given the link between specific calendars and particular communities, chronology was an admixture of mathematics with historical and legal

---

[70] Barak, *On Time*, 122–23.
[71] See e.g., Khudari, *Sharh al-Lum'a*, Isl. Ms. 722, Michigan, p. 9.

knowledge. Khudari's readers knew that the Coptic calendar reckoned years from "the beginning of the rule of Diocletian, killer of the martyrs," and that some say its first day was a Thursday, but "correctly reckoned" it was a Friday.[72] It also fell within the purview of scholarly astronomy to maintain the complex knowledge required to navigate the web of fasts and festivals that structured the passage of time in a multiconfessional society. Indeed, the pages of tables necessary to calculate when the Jewish and Christian holidays would fall in a given year must have made the purely lunar *hijrī* calendar seem like simplicity itself. The procedures for calculating the period of the Great Lent (*al-ṣawm al-kabīr*), the timing of which differs among the Eastern rites as well as between the Eastern and Western churches, was an area of particular concern. Of course, interconfessional knowledge was not necessarily presented in the spirit of ecumenism. "God disgrace them and disgrace their festivals," was Khudari's final comment on this section of the *zīj* (*qabbaḥahum allāh wa-qabbaḥa a'yādahum*).[73] Nevertheless, his facility with the Jewish and various Christian calendars suggests that such knowledge served to facilitate public commerce and, perhaps, keep the peace.[74]

It was within such larger historical, legal, and mathematical discussions that scholars' well-known interest in the problem of determining the beginning and end of Ramadan was typically located. Scholars of astronomy debated the nuanced *fiqh* positions that had evolved around the difference between the "calculated" (*bi-l-ḥisāb*) month, based on the passage of the moment of conjunction, and the "observed" (*bi-l-ru'ya*) or "legal" (*shar'ī*) month, based on sighting of the new crescent. Khudari, for example, displayed his training in the Shafi'i school (*madhhab*) when he distinguished between "the general legal ruling for everyone," which is to determine the month according to crescent sighting, and the ruling for the individual who is able to calculate (*al-ḥāsib*), who is obligated to use mathematics.[75] Within this distinctively Shafi'i framework, Khudari even adopted the strong position that "calculation" (*ḥisāb*) meant actually determining the month by its mathematical beginning, rather than merely using astronomical knowledge to evaluate the credibility of witnesses who claim to have seen the new crescent. Whereas anyone reading the book would qualify as "one

---

[72] Khudari, *Sharh al-Lum'a*, Isl. Ms. 722, Michigan, p. 23.

[73] Khudari, *Sharh al-Lum'a*, Isl. Ms. 722, Michigan, p. 69.

[74] On the importance of calendrical knowledge to public safety in Jewish communities in early modern Europe, see Elisheva Carlebach, *Palaces of Time: Jewish Calendar and Culture in Early Modern Europe* (Cambridge, MA: Belknap Press, 2011).

[75] Khudari, *Sharh al-Lum'a*, Isl. Ms. 722, Michigan, pp. 197–99.

able to calculate" (*ḥāsib*), the lengthy instructions that it provides in this area – including guidance on the use of the astrolabic quadrant – would have been necessary, in the author's view, for the reader to fulfill a religious duty. Yet such engagement in *fiqh* was only one province of a much larger chronological terrain that scholarly astronomy navigated.

Telling Time

Scholarly astronomy offered diverse ways of organizing time not only at the scale of days, months, and years, but also at the level of hours and minutes. In the *Brilliancy*, for example, the sixth and seventh sections treat topics such as observing the solar altitude, determining right ascensions of stars, calculating the semidiurnal arc, and understanding the difference between equal and seasonal hours. Outside of the *zīj* literature, shorter manuals and tables devoted to timekeeping were also quite common.[76] I will address the practice of scholarly timekeeping in depth in the context of the spread of mechanical timepieces in Cairo. Here, I would like to clarify how we should – and should not – think of Muslim scholarly timekeeping as "Islamic."

Doubtless, the times of prayer constituted an important area of interest for Muslim scholars of astronomy. Such scholars tabulated the times of prayer in Egypt in the late Ottoman period, as they had since the tenth century. And Muhammad al-Khudari often directed his comments on timekeeping to such specifically ritual contexts, or what he called "Sharia judgments" (*al-aḥkām al-sharʿiyya*).[77] As in the case of Ramadan and chronology, however, the ritual uses of timekeeping were one element of a larger picture. Manuals devoted to determining the times of prayer were rare; more often, prayer timing appeared within a broader set of practices in introductions to *mīqāt*. Moreover, even the calculation of prayer times in late Ottoman Egypt was "Islamic" in a different sense than would hold true today. Because the prayer intervals vary according to the length of daylight, they are substantially easier to tabulate according to a solar, rather than lunar, calendar. For this reason, until the twentieth century, Islamic prayer tables in Egypt frequently used the Coptic calendar.[78]

---

[76] Ibrahim ibn Hasan's commentary on al-Mahalli's *Introduction to the Science of Timekeeping* is a good example. See also the mid nineteenth-century works of Ahmad ibn Qasim, *Wasilat al-mubtadiʾ in li-ʿilm ghurrat al-shuhur wa-l-sinin* and *Tuhfat al-Ikhwan*, MS DM 1016, ENL.

[77] Khudari, *Sharh al-Lumʿa*, Isl. Ms. 722, Michigan, p. 139. For an example, see Isl. Ms. 722, Michigan, p. 142.

[78] E.g., Muhammad ibn ʿAbd al-Rahman al-Nabuli, *Natijat mawqiʿ ʿaqrab al-saʿat ʿala qadr hisas awaʾil awqat al-salawat fi al-shuhur al-qibtiyya*, MS 317 al-Saqqa 28898, Azhar.

This procedure suited the convenience of the scholar who made the tables, and it may also have reflected an expectation that even the Muslim scholarly audience of the tables was as likely to know the Coptic date as the *hijrī* date. To tabulate the times of prayer according to the latter, as is typically done today, would have struck nineteenth-century ʿulamaʾ as a bizarrely inefficient procedure. The modern conception of specifically "Islamic," "Christian," and "secular" calendars had yet to materialize.

Judging Time

Seasons, holidays, and prayer intervals were not the only ways in which late Ottomans understood the meaning of time. In 1895, the well-known scientific monthly *al-Muqtataf* (published in Cairo) deemed it necessary to warn its readers of "the fraudulence of magic and astrology" (*fasād al-siḥr wa-l-tanjīm*).[79] Such warnings only underscore the degree to which astrology was a pervasive aspect of late Ottoman society. Notwithstanding the modernizing discourse of certain elites, celestial considerations weighed on an array of decisions, and not only among the less educated. As Gülçin Tunalı Koç has shown, in a remarkable essay based on the almanac notes of the Ankara mufti and astrologer (*müneccim*) Sadullah El-Ankaravi (d. 1855), horoscopes helped ambitious bureaucrats decide whether to accept a new position, assuaged the anxiety of grooms, offered sartorial guidance, and provided a means for the astrologer to evaluate his household help – or even to locate a lost pipe.[80]

Astrological considerations carried as much weight in Cairo as they did in Ankara or Istanbul. E.W. Lane observed that astrology was "studied by many persons" in the 1820s and 1830s, particularly to determine the suitability of prospective marriages.[81] One such student of astrology, Ibrahim ibn Hasan, offered a more expansive view of its utility, articulating a common connection between astrology and a panoply of occult sciences.[82] Knowing the ascendant house of the zodiac, he wrote, is essential for nativities, as well as for "the science of letters, talismans, times,

---

[79] "Fasad al-sihr wa-l-tanjim," *al-Muqtataf* 19 (1895): 470. The warning came in response to a reader's question, and referred to the journal's having previously published numerous items on the fallacy of astrology.

[80] Koç, "An Ottoman Astrologer at Work," 53–55. As Koç remarks, choice of clothing was fraught with political and professional significance for Tanzimat-era bureaucrats. Consulting the stars before getting dressed underscored the authority, not the triviality, of astrological knowledge.

[81] Lane, *Manners and Customs*, 264.

[82] On the history of this connection, see Matthew Melvin-Koushki, "Powers of One: The Mathematicalization of the Occult Sciences in the High Persianate Tradition," *Intellectual History of the Islamicate World* 5 (2017): 127–99.

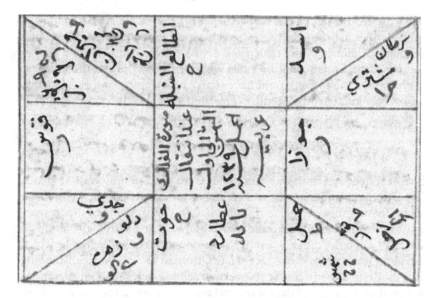

Figure 1.3 Chart depicting the position of the planets in the zodiac at the vernal equinox of 1239 (1824). Muhammad al-Khudari, *Sharh al-Lum'a fi Hall al-Kawakib al-Sab'a*, p. 148 (detail), Isl. Ms. 722, University of Michigan Library (Special Collections Library), Ann Arbor.

spiritual beings, extraction of precious metals, answering questions, conditions of the ill, fulfillment of wishes, circumstances of officials, and other purposes."[83] Thus, through astrology, celestial phenomena such as the ascendant (*ṭāli'*) entered a variety of other lofty spheres, including alchemy, medicine, and politics.

While many treatises and commentaries from the late Ottoman period (indeed, a plurality of the manuscripts that I surveyed) are devoted explicitly to astrology, focusing on such textual categories obscures the broader place of astrology in scholarly practice. At several points in the *Commentary on the Brilliancy*, Muhammad al-Khudari refers his reader to "the discourse of the astrologers" (*kalām al-munajjimīn*) for information on the astrological significance of certain phenomena (see Figure 1.3).[84] On the one hand, such comments emphasized that Khudari's own work was *not* astrology, echoing a distinction that Muslim astronomers since

---

[83] Ibrahim ibn Hasan, *Sharh Muqaddimat Mahmud Qutb al-Mahalli*, MS ṬM 225, ENL, fol. 18v.

[84] Khudari, *Sharh al-Lum'a*, Isl. Ms. 722, Michigan, pp. 147, 148, 225, 227.

the tenth century, responding in part to theological criticism, had drawn between calculating the expected positions of celestial bodies and ascribing causal significance to such positions.[85] On the other hand, Khudari's reference to "the discourse of the astrologers," far from dismissing the value of astrology, likely suggests how the author expected students to use his work: master the procedures and tables of the *zīj* in order to calculate celestial phenomena, the astrological significance of which they would then determine using other books. For example, Khudari explains how to calculate the ascendant for a variety of scenarios (e.g., eclipses and conjunctions), at one point stating that he expected this information to be used for "the judgments of astrology" (*aḥkām al-tanjīm*).[86] Even where Khudari did not make this application explicit, however, it remained a likely context for his readers' interest, especially in the lengthy discussions on the prediction of planetary motion. "Astronomical" texts were used for "astrological" purposes, while, by the same token, "astrology" depended upon astronomical knowledge.

Historians have often placed timekeeping (*mīqāt*) and astrology at opposite poles of astronomy's relationship to Islam. Timekeeping, with its links to prayer and fasting, was "astronomy in the service of Islam"; astrology, by compromising God's omnipotence, earned astronomers the enduring suspicion of theologians and jurists. This analysis has a basis in polemical stances taken by many astronomers and their critics in more than one context of Islamic history. It is also an appealing analysis for historians who wish to rebut the hoary notion that Islam and science are inherently in conflict, since it points to ways in which "good" science was promoted (and "bad" science demoted) by Islam *as a religion* – not simply by rulers who happened to be Muslim, or by "marginal" scholars in Islamic society. Thus, according to George Saliba, Islamic pietism deserves at least some of the credit for leading astronomers to articulate a firm distinction between astrology and other astronomical practices.[87]

By focusing on polemics and the epistemological developments that emerged from them, however, we risk obscuring a larger picture. Timekeeping and astrology were pursued not only by many of the same people, but through a significantly overlapping set of mathematical practices and shared texts like the *zīj*. Ritual timekeeping was no more the

---

[85] Robert Morrison, "The Islamic aspects of cosmology, astronomy and astrology," in *The New Cambridge History of Islam, Volume 4*, ed. Robert Irwin and William Blair (Cambridge: Cambridge University Press, 2010), 603.

[86] Khudari, *Sharh al-Lumʿa*, Isl. Ms. 722, Michigan, p. 175.

[87] Saliba, "Development of Astronomy."

"essence" of an "Islamic science"[88] than astrology was its antithesis. Both were part of a capacious relationship between scholarly astronomy and the measurement and interpretation of time. It is impossible to characterize the uses of this astronomy as "Islamic" or "un-Islamic" without simultaneously distorting the nature of astronomical practice and reducing "Islam" to specific kinds of theology and piety that were frequently honored in the breach. Yet the fact that 'ulama' moved through a specifically Islamic geography of learning and mastered a variety of Islamic knowledge traditions was hardly incidental to the way they did astronomy. What is necessary is a way of understanding scholarly astronomy's Islamic context that moves beyond ritual and theology.

How Did Muhammad al-Khudari Know?

At the age of about twenty-five, while attending Shaykh Hasan al-Quwaysini's lessons on creed ('aqīda), Muhammad al-Khudari fell ill. The sickness, apparently typhoid fever, rendered him permanently deaf. He cut short his studies at al-Azhar and returned to Damietta by 1823–24, presumably to enjoy the support of his family.[89] He seems not to have left his native city again.[90] Yet, the loss of his hearing did not deter Khudari from pursuing the life of a scholar. According to one of his half-brother's students, he developed his own way to communicate by sign language, and he "worked by himself in the sciences," including the "Sharia sciences" as well as the philosophical, theological, natural, and mathematical sciences.[91] Indeed, it is to this second period in Damietta that we can trace Khudari's prodigious scholarly output – starting, significantly, with the Commentary on the Brilliancy and a work on the sundial, both of which he completed between 1823 and 1825.[92]

---

[88] Cf. David A. King, "Science in the Service of Religion: the Case of Islam," in Astronomy in the Service of Islam, 245.

[89] The worked examples in the Commentary on the Brilliancy, calculated for the year in which it was written (1239 [1823–24]), are for the latitude of Damietta.

[90] İhsanoğlu claims that Khudari later returned to Cairo and taught at al-Madrasa al-Taybarsiyya, but he appears to base this on the biography of Khudari in Sarkis' Mu'jam, which, as noted above, actually describes the life of Khudari's half-brother. Cf. OALT, 611.

[91] Husayni, Tabaqat, p. 58.

[92] The earliest copy of Khudari's work on the sundial is from 1240 (1824–25). See Ibrahim Khuri, Fihris Makhtutat Dar al-Kutub al-Zahiriyya: 'Ilm al-Hay'a wa-Mulhaqatuhu (Damascus: Majma' al-Lugha al-'Arabiyya, 1389 [1969]). The Azhar catalogue is mistaken in claiming that his work on grammar, the Hashiya 'ala Sharh ibn 'Aqil 'ala al-Alfiyya li-Ibn Malik, was composed in 1225 (1810–11). The first printed edition of this book, a 1272 (1856) lithograph, states that Khudari finished it in 1250 (1834), which is

While Khudari never reflected in writing on his reasons for focusing on astronomy in the immediate aftermath of his illness, the timing seems more than coincidental – especially since astronomy was not to become an enduring interest for him. (The two works of the early 1820s were not only his first, but also his last in the field.) As a recent history of disability in the Ottoman Empire has observed, the privileged authority of aural transmission in Islamic scholarly culture made deaf scholars exceedingly rare, especially when compared with the abundance of blind scholars.[93] With astronomy, however, Muhammad al-Khudari had hit upon one of the few areas of learning in which seeing mattered more than hearing. Obviously, astronomers had to observe the movements of celestial bodies, as well as the position and length of shadows. Just as importantly, however, scholarly astronomical knowledge was irreducibly *textual* knowledge, knowledge that depended not only – like all Islamic scholarship – on skills of language, but more specifically and unusually on the skilled manipulation of *writing*. Of course, most scholars of astronomy were not deaf. But Khudari's decision to pursue astronomy specifically as a deaf scholar is good warrant for thinking that the apparent centrality of textuality to scholarly astronomical practice is more than an artifact of the evidence that happens to remain of this practice. Like Khudari's deafness, the overwhelmingly textual nature of this evidence points to the need to understand how particular textual practices shaped the ways in which scholars of astronomy made knowledge – more specifically, how they read, wrote, and calculated.

Reading Piously

According to Muhammad al-Khudari, his study of the *Brilliancy* that preceded the composition of a commentary on the text was a grueling experience. He wrote, "my shortcomings impeded me, along with lack of information, and my preoccupation with many troubles and controversies."[94] While Khudari often began his works with a similar apologia,

---

more likely than his having written it at the age of twelve. See Muhammad al-Khudari al-Dimyati, *Hashiya 'ala Sharh ibn 'Aqil 'ala Matn al-Alfiyya* (Cairo: Matba'at al-Hujar al-Nayyira, 1272 [1856]).

[93] Sara Scalenghe, "Being Different: Intersexuality, Blindness, Deafness, and Madness in Ottoman Syria" (Ph.D. Diss., Georgetown University, 2006), 152. On the trope of the blind scholar and Enlightenment empiricism, see Jessica Riskin, *Science in the Age of Sensibility: The Sentimental Empiricists of the French Enlightenment* (Chicago: University of Chicago Press, 2002), ch. 1.

[94] Khudari, *Sharh al-Lum'a*, Isl. Ms. 722, Michigan, p. 7.

nowhere else did he mention "lack of information" as an obstacle that he had to overcome.[95] That he had special difficulty reading the text is confirmed by the extraordinary solution that came to him:

I was still beginning and collecting (material) when help came to me, and I began (the work) full of hope for guidance from God. For I saw, while asleep, that I drank of the water that gushed from between the fingers of (the Prophet Muhammad) peace be upon him, so that I drew extraordinary knowledge from it, after being quite enlivened by drinking it. People were hastening towards it, crowding around it, and I rejoiced in my vision, saying, "Oh, my heralds of glad tidings!" So (the book) drew strength from resolve, and earlier decisions helped me by the power of God, and by the grace of God it became a commentary fit to open hearts, and a decisive word upon whose grove sings the nightingale of correct inquiry.[96]

This moment frames the entire commentary that follows. Khudari is not merely acknowledging the divine favor necessary to the acquisition of knowledge. He maintains that this commentary is the fruit of a specific encounter with the Prophet Muhammad, without which he would never have been able to finish the work.

Such a claim is not unheard of – the great Sufi figure Ibn 'Arabi wrote that one of his major books was handed to him by the Prophet in a dream – but it calls for explanation.[97] If the chief obstacle that Muhammad al-Khudari's dream helped him to overcome was his "lack of information," the most obvious way for a scholar to have overcome such a difficulty would have been to study the text with someone more knowledgeable. Yet he never mentioned such a person, nor do any of the other sources on Khudari's life. He had already completed his studies with Hasan al-'Attar when he returned Damietta, where he "worked *by himself* in the sciences" (*ishtaghala li-nafsihi*, my emphasis). I suggest that Khudari overcame his "lack of information" autodidactically, and that the dream was critical to his completion of the text because it authorized the knowledge he had acquired in this unusual fashion.

Autodidacticism was a dubious practice in a culture that prized the personal transmission of knowledge from one authority to the next. As Jonathan Berkey has written, "respected shaykhs imparted to their pupils more than mere knowledge; they also imparted authority, an authority over texts and over a body of learning that was intensely personal, and

---

[95] For comparison, see the pro forma apology in Khudari, *Hashiya 'ala Sharh ibn 'Aqil*, 2.

[96] Khudari, *Sharh al-Lum'a*, Isl. Ms. 722, Michigan, p. 7.

[97] 'Abd al-Ghani ibn Isma'il al-Nabulusi, *Jawahir al-Nusus fi Hall Kalimat al-Fusus*, ed. 'Asim Ibrahim al-Kayyani al-Husayni al-Shadhili al-Darqawi (Beirut: Dar al-Kutub al-'Ilmiyya, 2008), 45.

that could be transmitted only through some form of direct personal contact."[98] The study of texts outside of such a master–student relationship required special legitimation. Khudari's dream plays this role, testifying that, despite the author's largely untutored study of the text, the commentary is still of trustworthy origin.[99]

Muhammad al-Khudari's readers would have understood the details of his dream to frame the commentary not only as authoritative knowledge, but also as an exercise in piety. Khudari's vision of "the water that gushed from between the hands" of the Prophet evokes a number of widely narrated hadiths that recall a time when the early Muslim community found itself without enough water to perform ritual ablutions before the afternoon prayer. The Prophet placed his hands in a vessel of water, and, said one witness, "I watched the water gush from under his fingers until the last one of [those present] had performed ablution."[100] Thus, Khudari was comparing the solution to his predicament (lack of information) with a well-known episode when the Prophet's intervention miraculously enabled the community to overcome an obstacle (lack of water) to the performance of a religious duty. In other words, the language of the dream articulates an analogy between two devotional acts: writing a commentary on an astronomical handbook, and praying.

As this analogy suggests, Muhammad al-Khudari approached the attainment of astronomical knowledge as an exercise in piety in ways that transcended the specifically ritual uses of this knowledge.[101] Thus, he introduced the science of astronomical timekeeping (*fann al-mīqāt*) as

the most noble of the sciences after the religious sciences, for by means of it one arrives at the contemplation of the heavenly bodies... and their precise ordering according to perfect system and variation, so that the observer is baffled by the subtleties of wisdom and marvels of creation they comprise, and he submits to

---

[98] Berkey, *Transmission of Knowledge*, 24. For a cross-cultural history of autodidacticism, see Avner Ben-Zaken, *Reading Hayy Ibn Yaqzan* (Baltimore, MD: Johns Hopkins University Press, 2011).

[99] On dreams of the Prophet as signs of "personal distinction" and God's favor, see Jonathan Katz, *Dreams, Sufism and Sainthood: The Visionary Career of Muhammad al-Zawawi* (Leiden: Brill, 1996), 221–22.

[100] Muhammad ibn Isma'il al-Bukhari, *Sahih al-Bukhari* (Cairo: Wizarat al-Awqaf, 1990), Kitab al-Wudu', no. 32; see also Bukhari, *Sahih*, Kitab al-Wudu', no. 46. A similar tradition, pertaining to the battle of Hudaybiyya, relates that the Prophet "caused the water to rise like springs between his fingers, and we drank and performed ablution." Bukhari, *Sahih*, Kitab al-Manaqib, no. 25.

[101] On Islamic justifications for astronomy in the fourteenth century, see Morrison, *Islam and Science*, 28. It is noteworthy that Khudari did not invoke the ritual uses of astronomy at all in his introduction, despite the fact that they alone were classified by jurists as the "obligatory" aspect of astronomy, whereas practicing astronomy in order to contemplate God's power was only "recommended" (*mandūb*).

the greatness of their creator and the majesty of their inventor, saying, "*Our Lord! not for naught hast Thou created (all) this! Glory to Thee!*" (Qur'an 3:191)[102]

Even in the specific kind of astronomy that concerned timekeeping (*mīqāt*), Khudari did not frame the religious significance of this science in terms of determining the times of prayer or the beginning of Ramadan. Rather, he invoked much broader (and older) tropes such as contemplation of the Creator's greatness.[103] These tropes linked the study of astronomy with a kind of theology of nature, the end of which was to intensify one's sense of subordination to God.

Writing Pedagogically

The link between attaining knowledge and cultivating piety was not the only trait that scholarly astronomy shared with other kinds of Islamic learning. 'Ulama' also employed many of the same genres to record astronomical knowledge that they used in their other studies. Forms such as the commentary (*sharḥ*), supercommentary (*ḥāshiya*), versification (*manzūma*), and epistle (*risāla*) carried different expectations about the role of the author in relation to another text (or texts), the competency of the reader, and the purpose of reading. In this discussion, I will focus on the *sharḥ*, which was among the most important genres through which 'ulama' studied and taught all kinds of knowledge, astronomy included. Through the norms of the *sharḥ*, the practice of astronomy was thereby linked not only with specific texts, but also with a particular conception of what it meant to master such texts.

Although the long-held view of the *sharḥ* as little more than a record of intellectual stasis continues to resonate, a number of historians in recent decades have developed approaches to understanding the kinds of change for which commentaries provide evidence.[104] Studies have shown that commentary was a vital site for debate and innovation in diverse fields, enabling scholars to preserve the integrity of a discursive tradition while adopting new stances on important issues.[105] As Brinkley Messick

---

[102] For a full translation of Khudari's introduction to the *Commentary on the Brilliancy*, see the Appendix. Throughout this book, Qur'anic quotations are given according to the translation of 'Abdullah Yusuf 'Ali, *The Meaning of the Holy Qur'an*, 11th edition (Beltsville, MD: Amana Publications, 2008).

[103] Conceiving of astronomy as an exercise in devotion to the divine is much older than Islam. On Ptolemy's view of celestial bodies as gods, see Liba Taub, *Ptolemy's Universe: The Natural Philosophical and Ethical Foundations* (Chicago: Open Court, 1993).

[104] For an overview of the historiography of the *sharḥ*, see Zaman, *Ulama in Contemporary Islam*, 38–9.

[105] See, e.g., Muhammad Qasim Zaman, "Commentaries, Print, and Patronage: 'Hadith' and the Madrasas in Modern South Asia," *BSOAS* 62 (1999): 60–81; El-Rouayheb,

has noted, the *sharḥ* inserts its words directly into the relevant passage of the parent text (*matn*), so that "interpretations literally become part of the text interpreted."[106] Thus, in the *Commentary on the Brilliancy*, Khudari's interjections are grammatically contiguous with the words or phrases of Ibn Ghulam Allah that they frame.

Muhammad al-Khudari articulated a specific approach to the genre of *sharḥ* in the introduction to the other astronomical commentary that he wrote, which explicates a classic manual on the sundial:

This is the commentary they require who attempt to understand *The Provisions of the Traveler for Knowledge of the Hour-Angle* (*Zad al-Musafir fi ma'rifat wad' fadl al-da'ir*) by the Imam, the eminent scholar Shihab al-Din Ahmad ibn al-Majdi, which unlocks its words and breaks its defenses, making apparent what is hidden in it and completing that which is missing from it. For, despite its small size and little bulk, it contains principles of the science (*maqāṣid al-fann*) that are scattered among tomes and lacking in many abbreviated works. Therefore, I wanted to write (the commentary) in an easy way (*wajh sahl*) that would be a help at the beginning (of one's study) and an aid to memory (*tadhkira*) at the end.[107]

As a commentator, Khudari's concerns centered on the needs of students. He tried to write in an "easy way," to explain difficulties in the *matn*, and to provide presupposed or supplementary information from other texts. In a similar vein, Khudari decided to write on the *Brilliancy* because the only extant commentary on it that he could find was a "fragment" that was "insufficient for one who needs it to clarify ambiguities and obscurities."[108] This conception of the author's task as the explanation of complex master-works to students is also a theme of Khudari's introduction to his work on grammar, which expresses a

---

Relational Syllogisms, ch. 3; Dimitri Gutas, "Aspects of Literary Form and Genre in Arabic Logical Works," in *Glosses and Commentaries on Aristotelian Logical Texts: The Syriac, Arabic and Medieval Latin Traditions*, ed. Charles Burnett (London: Warburg Institute, 1993), 28–76; Reviel Netz, *The Transformation of Mathematics in the Early Mediterranean World: From Problems to Equations* (Cambridge: Cambridge University Press, 2004); and Asad Q. Ahmed, "Systematic Growth in Sustained Error: A Case Study in the Dynamism of Post-Classical Islamic Scholasticism," in *The Islamic Scholarly Tradition: Studies in History, Law, and Thought in Honor of Professor Michael Allan Cook*, ed. Asad Q. Ahmed, Behnam Sadeghi, and Michael Bonner (Leiden: Brill, 2011), 343–77.

[106] Brinkley Messick, *The Calligraphic State: Textual Domination and History in a Muslim Society* (Berkeley, CA: University of California Press, 1993), 31.

[107] Muhammad al-Khudari al-Dimyati, *Sharh Zad al-Musafir fi ma'rifat wad' fadl al-da'ir*, MS ṬM 124, ENL, fol. 2r.

[108] The reference is to the work of the early seventeenth-century scholar Ibn Abu al-Khayr al-Husayni al-Armayuni al-Tahhan (OALT #127). See MS K 4009, ENL, which contains Tahhan's commentary on seven of the *zīj*'s twelve sections; perhaps the "fragment" that Khudari possessed was all that his predecessor ever completed.

similar emphasis on "ease" and "accessibility."[109] If pedagogical necessity was Khudari's consistent motive for writing, however, notice that
he conceived of his students as composed of two kinds of readers whom
modern scientific writing typically addresses through separate genres: those "at the beginning" of their studies, and those "at the end." In
this sense, a text like the *Commentary on the Brilliancy* can be read as
two texts, each of them equally but differently concerned with communicating a comprehensive understanding of a science through explication of a particular text.

As a way for novices to develop a comprehensive understanding of
mīqāt, the *sharḥ* proceeded largely through syntactic and lexical exegesis.
In the *Commentary on the Brilliancy*, Muhammad al-Khudari's most frequent comment is an interjection that supplies a word or phrase, such as
an ambiguous antecedent, that is implied in the *Brilliancy's* abbreviated
language. In lengthy mathematical discussions, he similarly interjects to
remind readers of their place in the procedure. He also provides extensive
definitions, treating no term as too basic to explain: "even" and "odd"
numbers are defined in the same text as concepts such as the "node"
and "daily motion" of planets.[110] Some of this work mustered skills and
knowledge developed specifically in Islamic traditions of Arabic exegesis
and lexicography. For example, commenting on the five- or six-day period
at the end of the Coptic year called "the days of *nasī*'," Khudari explains:

> It is called this because it is later than the month, for *nasa'a* means to delay. ...
> The meaning of *al-nasī'* in the [Qur'anic] verse is, as al-Baydawi says, "to delay
> the prohibition of the prohibited month. The idolaters, if the prohibited month
> came while they were fighting, would make it permitted and prohibit another
> month in its place, until they rejected the particular characteristics of the months
> and paid attention only to the number."[111]

The technical term *nasī'* linked an understanding of the Coptic calendar
with an old debate in Qur'anic exegesis over the nature of a pre-Islamic
Arabian practice, also called *nasī'*, that the Qur'an prohibited. Khudari,
following most Qur'an commentators (like Baydawi), understood the
practice to have been an arbitrary transposition of the months when
fighting was prohibited, rather than a technically developed system of
intercalation.[112]

---

[109] Khudari, *Hashiyya 'ala sharh Ibn 'Aqil*, 1:2.
[110] Khudari, *Sharh al-Lum'a*, Isl. Ms. 722, Michigan, pp. 16, 81, 101.
[111] Khudari, *Sharh al-Lum'a*, Isl. Ms. 722, Michigan, p. 25.
[112] On the opposing view of some Muslim astronomers (e.g., al-Biruni), see Caussin de
Perceval, "Mémoire sur le calendrier arabe avant l'islamisme," *Journal Asiatique* 4th
Ser., 1 (1843): 342–79.

For more advanced readers, Khudari's commentary systematically corrected, updated, and compared the *Brilliancy* with more recent sources, while adhering to its structure and defending its authority. Khudari had access to more than one manuscript of the *Brilliancy*, and his frequent editorial remarks suggest a keen eye for textual corruption.[113] In some cases, however, he appears to have introduced novel changes in this way. For example, Khudari noted that all of the copies he had of the *Brilliancy* gave sixty-three minutes as the maximum value for the moon's distance from the ecliptic in order for a lunar eclipse to occur, whereas he believed it should read sixty-eight minutes.[114] Given that Khudari had no manuscript evidence to support this assertion, other than the fact that three (ز) and eight (ث) are easily confused in transcription, he must have been relying on his own astronomical judgment, or on that of another, unmentioned text or scholar.[115] (Sixty-three minutes is the modern value as well as the value found in older copies of the *Brilliancy*.) Thus, Khudari understood his task to include using his own expertise to try to improve the text, but he sometimes employed conventions like the suggestion of manuscript error rather than contradict the *matn* directly.[116]

Such conventions expressed an understanding of scholarship as the cultivation not only of pious devotion toward God, but also of a specific disposition toward other people. Ibn Ghulam Allah had begged his reader's forgiveness, asking "the great ones of virtue to look upon [this text] with the eye of favor" (*yanẓurū bi-'ayn al-riḍā*), a concept on which Muhammad al-Khudari elaborated:

> That is, they should see its excellences and ignore its faults, for nothing created lacks one. Who is there who has never erred, as it is said,
> I could not see the entirety of the beloved's fault, nor part of it when I was pleased
> For the eye of favor is too dim to find fault, but the eye of wrath makes wrongs apparent.

These lines of poetry, attributed to the eighth-century 'Alid figure 'Abd Allah ibn Mu'awiyah, are in one sense an odd choice to illustrate the

---

[113] For example, "ascensions" (*maṭāli'*), where the correct reading is "ascendant" (*ṭāli'*). Khudari, *Sharh al-Lum'a*, Isl. Ms. 722, Michigan, p. 171.

[114] Khudari, *Sharh al-Lum'a*, Isl. Ms. 722, Michigan, p. 161.

[115] See MS Arabe 2526, BnF, fol. 8r, for an 1146 (1733) copy that gives the parameter as 63'. For a modern discussion, see *Nautical Astronomy* (Scranton: International Textbook Company, 1902), 35.

[116] On the tendency of commentators to mask their innovations in conventionality, see El-Rouayheb, *Relational Syllogisms*, 95.

virtue of an "eye of favor," since their author was mourning his betrayal by the friend whose faults he had previously been unable to see.[117] The point seems to be that another's flaws may be real and even grievous, yet the virtuous scholar – like the virtuous companion – will view them in the most favorable light. This explication of the "eye of favor" strongly echoes language in the seventeenth-century "fragment" of commentary on the *zīj* that Khudari had consulted, suggesting that the concept was a trope.[118]

A generous disposition toward other scholars did not limit the commentary to expressing only as much information as could be attributed to the author of the *matn*. Khudari consistently used the medium of commenting on the *Brilliancy* to explain and evaluate more recent sources. At times, this engagement was critical, as when he alerted his readers that the seventeenth-century Egyptian scholar al-Qalyubi misstated the age of the Coptic calendar relative to that of the Arabic calendar.[119] At other times, Khudari drew on recent scholars to buttress his own position, such as when he cited the eighteenth-century Ottoman jurist Khayr al-Din al-Ramli in support of his argument that those able to calculate are obligated to determine the new month mathematically, rather than by observation.[120] The reference to Ramli's work is noteworthy since, as Haim Gerber has shown, Ramli was a distinctive jurist who relied on his own *ijtihād* (usually defined as "independent legal reasoning").[121] Yet Khudari's attention to recent and dynamic scholarship is unsurprising, so long as we recognize that 'ulama' were often as concerned with the work of later scholars (*al-muta'akhkhirūn*) as they were with mastering the early authorities (*al-mutaqaddimūn*).[122]

In fact, to the extent that Khudari's *Commentary on the Brilliancy* can be said to have a systematic agenda for its more advanced readers, it is the comparison of the *Brilliancy*'s procedures (which are based on those

---

[117] Abu al-Hasan 'Ali bin Muhammad ibn Habib al-Basri al-Mawardi, *Adab al-Dunya wa-l-din* (Beirut: Dar al-Kutub al-'Ilmiyya, 1987), 20n3.

[118] See MS K 4009, ENL, p. 5.

[119] Khudari, *Sharh al-Lum'a*, Isl. Ms. 722, Michigan, p. 27. Probably Ahmad ibn Ahmad al-Qalyubi al-Shafi'i (d. 1659), author of a *Risala fi 'ilm al-Miqat*. A twelfth/eighteenth-century copy (produced by a Dimyati, no less) is extant in the ENL. See OALT #157.

[120] More precisely, Khudari cites Ramli's opinion according to "Ibn Qasim on al-*Tuhfa*." See 'Abd al-Hamid Shirwani, *Hawashi 'Abd al-Hamid al-Shirwani wa-Ahmad ibn Qasim al-'Abbadi 'ala Tuhfat al-muhtaj bi-sharh al-Minhaj*, vol. 3 (Beirut: Dar Sadir, n.d.), 373. 'Ali Mubarak states that Ibn Qasim's gloss was one of the commonly used works on Shafi'i law at al-Azhar in the nineteenth century. Mubarak, *al-Khitat*, 4:27–28.

[121] Haim Gerber, *Islamic Law and Culture 1600–1840* (Leiden: Brill, 1999), 72.

[122] Wael B. Hallaq, "Model *Shurut* Works and the Dialectic of Doctrine and Practice," *Islamic Law and Society* 2, no. 2 (1995): 126.

of Ibn al-Shatir, d. 1375) with those of the later *Zij-i Sultani* of Ulugh Beg (d. 1449). Khudari compares the two authorities on questions including the debate over the day of the week on which the *hijrī* calendar began, the value of the mean month (lunation), the equation of time, and the calculation of the moon's true position.[123] Much like commentaries on *fiqh* handbooks or hadith collections, astronomical commentary facilitated the development of new positions through the exposition of older texts. However, while this teaching of new procedures alongside older alternatives was related to the form of the *sharḥ* and the norms of authority in Islamic scholarship, it also enabled a kind of knowledge-making particular to astronomy. Where Muhammad al-Khudari was not certain whether the *Zij-i Sultani* was better than the *Brilliancy*, he enrolled his readers in an observational program to find out. Thus, regarding the two *zījs'* disagreement on determining the time of eclipses, Khudari wrote, "We do not understand which of them is correct, this disagreement having occurred several times, and we have seen each of their opinions hold true (*yuṣādif al-ṣiḥḥa*) at times and err at other times."[124] Looking forward, Khudari provided the respective predictions yielded by each procedure for the possible time, in Damietta, of a lunar eclipse on 15 Rabiʿ al-Thani 1242 (15 November 1826) and a solar eclipse on 29 Rabiʿ al-Thani 1242 (29 November 1826). While the use of such worked examples was a standard pedagogical tactic of scholarly commentary, here, the example appears to have had an additional purpose: Khudari and his readers, whether in 1826 or thereafter, would have been able to use these calculations, along with the observed phenomena from that year, to add another piece of data to the comparison of two competing procedures.

The norms of a pedagogical commentary shaped an approach to writing astronomy that one might describe as "expansively narrow," an approach committed to comprehensiveness through specificity. On the one hand, the basic task of the commentator was the exposition of a particular text, the structure and authority of which demanded respect. On the other hand, inasmuch as *all* information relevant to either a novice's or an adept's reading of this text could – indeed ought – to be included, exposition of a particular text could serve as an introduction to much of scholarly astronomy. The same commentary explains basic arithmetical techniques as well as the difference between procedures for predicting planetary positions. Through the study of a fifteenth-century adaptation of a fourteenth-century handbook, readers were also introduced to

[123] Khudari, *Sharh al-Lumʿa*, Isl. Ms. 722, Michigan, pp. 10–11, 14, and 117–18.
[124] Khudari, *Sharh al-Lumʿa*, Isl. Ms. 722, Michigan, p. 167.

more recent sources, some of which contradicted the *matn*. Such ways of communicating astronomical knowledge were deeply connected with, yet cannot be reduced to, norms of learning and authority shared with the broader practice of Islamic scholarship in which 'ulama' participated.

Calculating Textually

Perhaps it may be remarked that Muhammad al-Khudari's devotion to comparing fourteenth-century procedures with fifteenth-century counterparts seems not to contradict, but rather to reinforce, the idea that nineteenth-century scholars of astronomy were stuck in the past. In a limited sense, this is true. Khudari's reliance on tables ultimately based on the observations of Ibn al-Shatir or Ulugh Beg reflects the fact that more recent *zījs*, in particular the Mughal *Zīj-i Muhammad Shahi*, do not seem to have been widely known among Egyptian scholars.[125] But working with 400-year-old texts was not a matter of privileging the authority of antique knowledge per se. Rather, it reflected a contingent judgment regarding the relative unreliability of available observational instruments compared with tabular techniques for preserving and manipulating previously observed values. To elaborate on this assertion, it is necessary to discuss practices of calculation and observation among the 'ulama' in some detail.

While some aspects of scholarly astronomy were rooted in sophisticated mathematics, the *zīj* served to reduce the measurement of time and the prediction of celestial motion to relatively simple arithmetic procedures. Toward the middle of his *Introduction to the Science of Timekeeping*, the late seventeenth-century Damiettan scholar Mahmud Qutb al-Mahalli commented: "In these practices, it is necessary to master the multiplication and division of degrees and minutes, [for] therein lies the heart of this art" (*fī dhālika lubb hādhā al-fann*). Even these basic skills were open to negotiation, however. Commenting on a later passage in Mahalli's *Introduction*, Ibrahim ibn Hasan provided two procedures for calculating the position of a star at a time other than noon: one involving division, and one involving only addition, subtraction, and multiplication, "if division is difficult for you."[126] Division was not difficult for

---

[125] When Khudari cited "Ulugh Beg," he may have had in mind the work of scholars active in seventeenth- and eighteenth-century Egypt who based their tables on Ulugh Beg's observations and procedures. Examples include Ramadan al-Khawaniki, Mustafa al-Khayyat, and 'Abd al-Rahman al-Jabarti. See King, *Survey*, 2/2/3.6.

[126] Ibrahim ibn Hasan, *Sharh Muqaddimat Mahmud Qutb al-Mahalli*, MS ṬM 225, ENL, fol. 18r.

Ibrahim ibn Hasan, of course. But the culture of learning that produced the same text for those "at the beginning" and those "at the end" of their study placed a premium on accessibility. Just as Khudari did not assume that readers knew the difference between an even and odd number, his student did not assume that readers were comfortable with division.

If the ability to tell time, determine dates, or predict planetary positions was to depend only on addition, subtraction, and multiplication, however, even a novice had to learn to add, subtract, and multiply the correct terms, which in practice meant developing facility with a variety of tables. For efficiency's sake, many of these tables were densely constructed: for example, by combining the degrees of zodiacal houses like Aries and Virgo (which are equidistant from the solstices) into a single column, which must be read in different directions corresponding to each house. Cultivating such skills was a major focus of scholarly astronomical pedagogy, whether in zīj commentaries like Muhammad al-Khudari's or more specific manuals. Phrases such as, "enter the table... and take," and "enter the table... and find," are refrains in these texts.[127] Mahalli's *Introduction to the Science of Timekeeping* even instructed its readers on the use of their fingers to navigate such tables: "... enter the table by the number on the right, according to the number of days that have passed in the month, and pass your finger through the row, until you come to the cells of the aforementioned signs [of the first day of the month] on the left."[128] The description of tabular calculation in such embodied terms is consistent with the fact that scholars of astronomy explicitly conceived of tables as instruments, in the same category as the astrolabe, quadrant, or clock.[129] Indeed, the ability to manipulate the values contained in tables – manipulation understood literally, as the work of hands – was arguably the key instrumental skill of scholarly astronomy.

Of course, scholars of astronomy used observational instruments as well. The abundance of nineteenth-century manuals on the design and use of sundials, quadrants, and astrolabic quadrants suggests that such objects were reasonably common in the material culture of the 'ulama'.[130]

---

[127] E.g., on finding the "true" position of the sun (*muqawwam al-shams*), see Khudari, *Sharh al-Lum'a*, Isl. Ms. 722, p. 87.

[128] Ibrahim ibn Hasan, *Sharh Muqaddimat Mahmud Qutb al-Mahalli*, MS ṬM 225, ENL, fol. 6r.

[129] Ibrahim ibn Hasan, *Sharh Muqaddimat Mahmud Qutb al-Mahalli*, MS ṬM 225, ENL, fol. 10v.

[130] On the use of astrolabic quadrants, see for example, Khalil al-'Azzazi, *al-Kawkab al-Azhar fi al-'aml bi-l-rub' al-muqantar*, MS K 3990, ENL; Muhammad ibn 'Ali ibn Abi al-Fadl, *al-Riyad al-Zahirat fi al-'Aml bi-rub' al-muqantarat* (Cairo, 1322 [1904–05]). On the construction of the astrolabe, see a nineteenth-century copy of Abu Rayhan al-Biruni, *al-Isti'ab bi-l-wujuh al-mumkina fi sina'at al-asturlab*, MS K 8528, ENL. On

In this context, Jabarti's claim that such instruments were unaffordable for most scholars should be understood in light of certain distinctions regarding astronomical instrumentation. In his commentary on Mahalli's *Introduction to Timekeeping*, Ibrahim ibn Hasan references a number of instruments that he expected his readers to have access to, including the quadrant, sundial, sand clock, and *musātira* (a tool for measuring meridian transits).[131] But Ibrahim lamented the lack of a star catalogue based on observations more recent than those of Mustafa Effendi Suhrab for 1061 (1651–52).[132] This lack was due, he specified, not to a dearth of skilled observers, but to the scarcity of patrons for such a project, which only rulers (*mulūk*) could afford.[133] Ibrahim's implication was that different types of astronomical work demanded different degrees of precision, corresponding with instruments of different quality. Ibrahim associated the production of star catalogues with princes like Ulugh Bey, the patron of the Samarkand observatory and the *Zij-i Sultani*. It was this specific kind of observatory work that lay beyond the means of Egyptian scholars of astronomy in the late eighteenth and early nineteenth centuries. More common tasks like timekeeping and prognostication relied on a host of workaday instruments, to which Khudari and his peers evidently had access.

To put the matter bluntly, while scholarly astronomical practice was replete with instruments, in Ottoman and late Ottoman Egypt these instruments were relatively cheap. Tables were not the only ones made of paper. In an eighteenth-century manual on the correction of mechanical timepieces (which was regularly copied until the middle of the nineteenth century), 'Abd al-Latif al-Dimashqi suggests that the reader fashion a quadrant out of the back of the book itself.[134] Dimashqi did not discuss the degree of precision, let alone durability, that he expected of such an instrument. However, his assumption that the reader will take the suggestion as unproblematic offers an intriguing glimpse into a highly

inclined sundials, see Khalil al-'Azzazi, *Nukhbat qawl al-sadat fi ma'rifat ma yata'allaq bi-l-munharifat* (Cairo, 1315 [1897–98]).

[131] Ibrahim b. Hasan, *Sharh Muqaddimat Mahmud Qutb al-Mahalli*, MS ṬM 225, ENL, fols. 10v and 11r.

[132] Because Suhrab's star catalogue is described by the term *Satīḥiyāt*, King supposed that it was a recension of a thirteenth-century catalogue based on the observations of al-Satihi. From Ibrahim ibn Hasan's discussion, however, it is clear that *Satīḥiyāt* had become a general term for star catalogues, and that Suhrab's was based on original observations. Ibrahim ibn Hasan, *Sharh Muqaddimat Mahmud Qutb al-Mahalli*, MS ṬM 225, ENL, fols. 25v and 26v. Cf. King, *Survey*, 103, 200, and 239.

[133] Ibrahim ibn Hasan, *Sharh Muqaddimat Mahmud Qutb al-Mahalli*, MS ṬM 225, ENL, fol. 26v.

[134] 'Abd al-Latif al-Dimashqi, *al-Manhaj al-Aqrab li-Tashih Mawqi' al-'Aqrab*, Isl Ms. 808, Michigan, p. 24; see also MSS DM 1104 and TR 286, ENL.

improvisational culture of instrumentation, in which a text could be transformed into the observational instrument that it called for.

While employing such instruments, scholars conceived of them as substantial sources of error. In the seventeenth century, Mahalli had written: "One who treats knowledge of prayer times should produce the calculation of degrees and minutes that he can attain, then render them conservatively so that the prayer intervals have begun with certainty."[135] This conception of "certainty" is interestingly composite, relying not only on mathematics but also on the discretion of the scholar, who is advised to add a few minutes to the calculated beginning of a prayer interval in order to be sure that the proper time has arrived. Ibrahim ibn Hasan, however, critiqued even this qualified assertion that one could know the times "with certainty," commenting skeptically, "Or probabilistically (*zannan*), because this craft errs, whether because of variation in the thread of the *musātira*, or flaws in the centers and lines of quadrants, or the corruption of tables."[136] This description of astronomical knowledge as "probabilistic" rather than "certain" did not necessarily diminish its authority. In fact, such terms invoked the epistemology of Sunni jurisprudence, which held that jurists could generally reach only probable knowledge of God's will, but that such knowledge could – indeed must – serve as a basis for action. Ibrahim ibn Hasan used this principle to articulate the limitations of his instruments, but also to vindicate his reliance on them.[137]

Given that scholars of astronomy in nineteenth-century Egypt understood their observational tools as limited in comparison with the more expensive instruments of their predecessors, tables became key instruments not only because they could be used to turn complex calculations into simple arithmetic procedures, but also because – if preserved and manipulated correctly – they contained observational results better than what nineteenth-century scholars of astronomy thought they could achieve themselves. After all, whereas Ibrahim ibn Hasan regarded the imprecision built into the material of his quadrant or *musātira* as inevitable, the kind of error that crept into tables – textual corruption – was one that the scribal culture of 'ulama' was well

---

[135] Ibrahim ibn Hasan, *Sharh Muqaddimat Mahmud Qutb al-Mahalli*, MS ṬM 225, ENL, fol. 10v.

[136] Ibrahim ibn Hasan, *Sharh Muqaddimat Mahmud Qutb al-Mahalli*, MS ṬM 225, ENL, fol. 10v.

[137] On the probabilism of Sunni jurisprudence, see Bernard Weiss, *The Spirit of Islamic Law* (Athens, GA: University of Georgia Press, 1998), 110–12. For a comparable use of juristic epistemology to articulate the status of astronomical knowledge, see Morrison, *Islam and Science*, 70.

accustomed to preventing and correcting. According to Khudari, distinct norms had emerged around the copying of tables. Whereas most kinds of text should be corrected with marginalia, rather than erasure (which would raise doubts about other copies), "a qualified person" (*man huwa ahluhu*) could erase and overwrite scribal error in a table.[138] The basis for this distinction seems to have been the mathematical relationship of tabular content, which enabled competent scholars – "not [just] anyone" (*lā kulla aḥad*) – to make corrections with certainty.[139] In fact, astronomical tables were an unusual form of textual knowledge inasmuch as their "preservation" actually depended upon modification: many of the *zīj*'s tables had to be periodically recalculated to account for the precession of the equinoxes. Since this process was a matter of mathematical manipulation, however, it was not understood necessarily to introduce further uncertainty.

Reliance on fourteenth- and fifteenth-century tables was not a form of "blind imitation," as adherence to older authorities in Islamic jurisprudence has often been (mis)characterized, but rather a reasoned judgment that such tables provided the best observations available. This judgment was contingent upon the instruments available to scholars, who placed less trust in their everyday quadrants than in the ability of their textual and mathematical techniques to maintain the knowledge that the Samarkand Observatory had produced. If the net result of this epistemological calculus privileged the authority of observers and instruments that were several hundred years old, such a judgment was little different in kind from the judgment of nineteenth-century French astronomers who relied on Jérôme de Lalande's eighteenth-century observations. A star catalogue is a major undertaking, demanding, as Ibrahim ibn Hasan noted, uncommon resources (in time as well as money). Reliance on Ibn al-Shatir or Ulugh Beg did not privilege antique knowledge per se. To the contrary, scholars in the nineteenth century continued to refer to Ulugh Beg's *zīj* as *al-raṣd al-jadīd*, "the new observation."

How Muhammad al-Khudari Did Not Know: The Case of Heliocentricity

Late Ottoman scholars of astronomy had a variety of powerful tools for making socially valued knowledge. To a modern reader, however,

---

[138] Khudari, *Sharh al-Lum'a*, Isl. Ms. 722, Michigan, p. 9.
[139] Khudari, *Sharh al-Lum'a*, Isl. Ms. 722, Michigan, p. 9.

Figure 1.4 Illustration of an eccentric model of the sun's orbit in an 1889 copy of Khudari's *Commentary on the Brilliancy*. Muhammad al-Khudari, *Sharh al-Lum'a fi Hall al-Kawakib al-Sab'a*, p. 86 (detail), Isl. Ms. 722, University of Michigan Library (Special Collections Library), Ann Arbor.

perhaps the most striking characteristic of a work such as Muhammad al-Khudari's *Commentary on the Brilliancy* is something that it does not know, which is that the earth moves in an orbit around the sun. Khudari introduces the section of the *zīj* on planetary motion with a cosmographical sketch that not only is geocentric, but makes no reference to any alternative (Figure 1.4).[140] We are in the classic Greco-Islamic cosmos of nine spheres centered on a stationary earth, with the outermost sphere corresponding to the Qur'anic "throne" (see *inter alia* Qur'an 7:54).[141] Khudari explains how the varying longitudinal motion of the planets within this system requires astronomers to posit multiple spheres for each planet.[142] As for Copernicus, heliocentricity, Kepler, or Laplace, Khudari says not a word. Nor does he discuss other potentially interesting developments, such as the telescope or the recently discovered

---

[140] Khudari, *Sharh al-Lum'a*, Isl. Ms. 722, Michigan, pp. 82–3.

[141] On Islamic cosmology, see Angelika Neuwirth, "Cosmology," in *Encyclopaedia of the Qur'an*, ed. Jane Dammen McAuliffe (Brill Online).

[142] Khudari, *Sharh al-Lum'a*, Isl. Ms. 722, Michigan, p. 86. For an introduction to Ptolemaic astronomy, see Michael Crowe, *Theories of the World from Antiquity to the Copernican Revolution* (Mineola, NY: Dover Publications, 1990).

planet Uranus. Even the Gregorian calendar goes unmentioned, despite the great attention that Khudari devoted to chronology.[143] In this sense, *The Commentary on the Brilliancy* is characteristic of scholarly astronomical works from the nineteenth century, including many from later in the century. While Khudari wrote in the 1820s, scholars of astronomy working in the second half of the century, such as 'Ali al-Khashshab or Muhammad al-Nabuli, generally also did not write of any of the great changes that astronomy had undergone over the last several centuries in Europe.[144] Whereas a number of historians have examined discussions of heliocentricity among the 'ulama' in different Islamic societies between the eighteenth and early twentieth centuries, the *absence* of such discussions also reveals significant features of the community of scholars to which Muhammad al-Khudari belonged.[145]

The silence of scholars like Khudari regarding heliocentricity is best understood as an expression of active indifference. They knew *about* heliocentricity, in the sense that they must have encountered the idea, whether in writing or conversation; their silence was not the hermetic silence of unawareness. Nor, however, was it an expression of critique. We may presume that Khudari and others who taught geocentric cosmography harbored significant doubts regarding the heliocentric alternative; otherwise, we must attribute to them an implausibly radical commitment to preserving knowledge that they thought was not knowledge at all. Possibly their doubts were based in religious conviction, although it is worth recalling the serious technical objections that may also be raised regarding a heliocentric model, depending on the information available to an astronomer.[146] But the questions of whether and why scholars of astronomy were or were not persuaded by heliocentricity divert our attention from their principal response to it – not critique, but silence.

---

[143] The Gregorian calendar was not widely used in Egypt until the end of the nineteenth century. Barak, *On Time*, 121–23.

[144] Khashshab, *Risala mukhtasira fi ta'dil al-zaman*, MS DM 261, ENL. Muhammad al-Nabuli, *Fath al-Mannan 'ala al-Manzuma al-Musamma Tuhfat al-Ikhwan* (Cairo: Matba'at Mustafa al-Babi al-Halabi, 1325 [1907]), which the author completed in 1862 (as stated on p. 46 of this 1907 print edition). See also MS ṬM 166, Cairo; and Isl. Ms. 778, Michigan.

[145] Cf. Martin Riexinger, *Sanā'ullāh Amritsarī (1868–1948) und die Ahl-i-Ḥadīs im Punjab under britischer Herrschaft* (Würzburg: Ergon, 2004), 392–410; Arjomand, "The Emergence of Scientific Modernity in Iran"; George Saliba, "Copernican Astronomy in the Arab East: Theories of the Earth's Motion in the Nineteenth Century," in *Transfer of Modern Science and Technology to the Islamic World*, ed. Ekmeleddin İhsanoğlu (Turkey: IRCICA, 1992), 145–55; İhsanoğlu, "Introduction of Western Science."

[146] On the technical objections to Copernican astronomy, especially the problem of stellar parallax, see Crowe, *Theories of the World*.

Regardless of what Muhammad al-Khudari thought about heliocentricity, the fact that he never mentioned it in writing suggests something more significant than that he doubted it, which is that he did not think it mattered.

Heliocentric planetary models had been circulating in the Eastern Mediterranean since the seventeenth century. Muhammad al-Khudari had a direct link to this phenomenon through his teacher in "the science of wisdom and nature," Hasan al-'Attar.[147] A charismatic and controversial figure at al-Azhar in the late eighteenth and early nineteenth centuries, 'Attar associated not only with prominent 'ulama' of his generation such as 'Abd al-Rahman al-Jabarti and Isma'il al-Khashshab,[148] but also, during the French invasion, with the *savants* of the Institut d'Égypte. In an apologetic account of these encounters, 'Attar wrote that the French allowed him to use their "astronomical and engineering equipment" and "discussed with me various matters in these fields."[149] After the French evacuation, 'Attar left Egypt for other parts of the Ottoman Empire, possibly motivated by his desire to pursue learning in natural sciences.[150] Between 1803 and 1815 he spent several years in Istanbul, at one point living in the house of the chief imperial physician (*hekimbaşı*) and associating with European doctors.[151] In such cosmopolitan circles, heliocentric astronomy had long been a topic of translation and discussion. A fairly detailed treatment appeared in the introduction to Ibrahim Müteferrika's printed edition of Katip Çelebi's *Cihannüma* in 1732; a more abbreviated account could be found in the *Marifetname* (encyclopedia) of Ibrahim Hakkı (d. 1780), which was shortly to be printed as well (first in 1825).[152] The imperial vice-astronomer (*müneccim-i sani*)

[147] On 'Attar, see Gilbert Delanoue, *Moralistes et politiques musulmans dans l'Egypte du XIXe siècle*, vol. 2 (Cairo: Institut Français d'archéologie orientale du Caire, 1982), 344–57; and Peter Gran, *Islamic Roots of Capitalism* (Cairo: AUC Press, 1999). 'Attar's full bibliography appears in Gran, *Islamic Roots*, 200.

[148] Gran, *Islamic Roots*, 79.

[149] See Gran, *Islamic Roots*, 190, where he translates an extensive passage from 'Attar, "Maqama fi dukhul al-faransawiyyin lil-diyar al-misriyya," MS Adab 7574, ENL. Al-Jabarti mentions telescopes and sundials belonging to Nouet. Murphy, *Improving the Mind*, 160.

[150] Gran, *Islamic Roots*, 91.

[151] Gran, *Islamic Roots*, 103.

[152] For an overview of the Ottoman translation history of Copernicanism, see İhsanoğlu, "Introduction of Western Science to the Ottoman World." On the context for these translations, see Avner Ben-Zaken, "The Heavens of the Sky and the Heavens of the Heart: The Ottoman Cultural Context for the Introduction of Post-Copernican Astronomy," *BJHS* 37 (2004): 1–28; and see Sonja Brentjes, "Astronomy a Temptation? On Early Modern Encounters Across the Mediterranean Sea," in *Travellers from Europe in the Ottoman and Safavid Empires, 16th-17th Centuries: Seeking, Transforming, Discarding Knowledge* (Aldershot, UK: Ashgate, 2010).

Hüseyn Hüsni, whose office was under the supervision of the *hekimbaşı*,[153] completed a translation of J.J. de Lalande's textbook *Astronomie* in 1813. When 'Attar returned to Cairo in 1815 and, shortly thereafter, encountered Muhammad al-Khudari as a student, it is difficult to imagine that the only cosmography they discussed was the geocentric one that ultimately appeared in Khudari's *Commentary on the Brilliancy*. In fact, soon after 'Attar assumed the editorship of the government bulletin, it discussed the concept of the earth's rotation in print in 1830.[154]

Muhammad al-Khudari was not the only scholar of astronomy whose awareness of heliocentric astronomy never made it into writing. Recall Mustafa al-Bulaqi's habit of dropping in on the Engineering School to discuss astronomy with Mahmud al-Falaki. It is inconceivable that the latter would not have introduced Bulaqi to the heliocentric theory (if indeed he was not already familiar with it). Yet, as we have seen, Bulaqi appears to have made no mention of such issues in writing.

When E.W. Lane observed the "manners and customs" of Egypt in the 1820s, he explained the attitude of scholars toward heliocentricity in religious terms: "Those persons in Egypt who profess to have considerable knowledge of astronomy are generally blind to the true principles of the science: to say that the earth revolves round the sun, they consider absolute heresy."[155] Lane's observations can hardly be dismissed out of hand. Yet, if 'ulama' considered heliocentricity a heresy, they must not have thought it was a very important one. One searches in vain for a nineteenth-century Egyptian treatise on the theological or exegetical issues raised by heliocentrism – until 1874, when a debate among Christians in the Syrian press spilled over into Egypt, provoking the Azhar-educated viceregal official 'Abd Allah Fikri to write an Islamic *defense* of the theory. Moreover, outside of scholarly astronomical writing, 'ulama' openly discussed heliocentricity in nuanced terms. In a mid nineteenth-century commentary on a poem by the Naqshbandi shaykh Mawlana Khalid (d. 1827), the Baghdad scholar Abu al-Thana Mahmud al-Alusi (d. 1857) wrote an excursus on the term "sun," in which he related the claims of "European astronomers today" (*ahl al-hay'a al-yawm min al-ifranj*), including the heliocentric arrangement of the planets, their relative sizes and distances, and the discovery of Uranus, Ceres, and Pallas. While his stance on most of these claims is not explicit, he remarks that the placement of the sun in the fourth (geocentric) sphere is something

[153] İhsanoğlu, "Ottoman Science," 18.
[154] Crozet, *Sciences modernes,* 221.
[155] Lane, *Manners and Customs,* 217–18.

that "moderns can hardly accept" (*la yakād al-muḥdathūn an yusallimū dhālika*). Only the notion of an infinite universe was rejected by Alusi as directly contradicted by "Sharia texts" (*nuṣūṣ al-sharī'a*).[156] My point is not that all 'ulama' found heliocentricity unproblematic. However, the fact that Mahmud al-Alusi wrote extensively about heliocentricity in a commentary on a Sufi poem, while Muhammad al-Khudari omitted the topic from a commentary on an astronomical handbook, suggests that the broad question, *how did theology shape what 'ulama' thought?*, cannot be answered without attending to the more specific question, *how did genre-specific purposes shape what 'ulama' wrote?*

As I have argued above, Khudari's authorial choices were guided by the social importance of specific kinds of knowledge, namely the practices of chronology, horology, and astrology linked to the "apparent phenomena" that one could calculate with a *zīj*. As far as these practices were concerned, in order for scholars of astronomy to derive any benefit from heliocentricity, they would have needed much more than the basic concept of putting the planets in motion around the sun. They would also have required substantially different sets of tables and the technical know-how to perform calculations with them. In fact, a few late Ottoman scholars of astronomy did come into possession of such information, and they found it significant indeed. These scholars, some of whom are the subject of Chapter 4, are a telling exception to the rule. Absent the tools that linked a heliocentric model with predictive techniques, the new cosmology by itself offered little benefit to late Ottoman scholars of astronomy.

The contingent circumstances under which new astronomical models acquire significance may be further illustrated here by comparison to the case of Jesuit astronomy in Ming China. As Ben Elman has shown, although Jesuit missionaries eventually succeeded in introducing their Tychonic cosmology into the Ming astro-calendrical bureau, they were only able to do so once they had demonstrated the superiority of their methods for calculating eclipses, a crucial objective of the Ming astronomers. Basic elements of European astronomy that were of less interest to the Chinese, such as the linear conception of time, fell by the wayside.[157] Similarly, for nineteenth-century 'ulama', the utility of heliocentric astronomy had to be established; it did not arrive ready-made.

---

[156] Abu al-Thana' Shihab al-Din Mahmud al-Hasani al-Husayni al-Alusi al-Baghdadi, *al-Fayd al-Warid 'ala Rawd Marthiyyat Mawlana Khalid* (Cairo: al-Matba'a al-Kastaliyya, 1278 [1861–62]). This lithograph edition postdates the composition of the commentary; Alusi died in 1853–54.

[157] Benjamin Elman, *On Their Own Terms: Science in China, 1550–1900* (Cambridge, MA: Harvard University Press, 2005), 96.

Aside from the open question of the new model's utility, another obstacle to discussion of heliocentricity lay in the pedagogical practices that shaped Khudari's writing. The problem was not novelty per se. As I have shown, Khudari strove for and advertised novelty in his *Commentary on the Brilliancy*. His primary task as a commentator, however, was to make a specific text legible to novice as well as advanced readers. The vast majority of the work he put into the text – clarifying ambiguous language, defining terms, penning introductions on prerequisite subjects, explaining the tables, working through examples for the present date – was oriented toward this goal. A discussion of Copernicus would have been beside the point. In other words, the problem of the absence of European astronomy from *The Commentary on the Brilliancy* and similar texts is only a problem if we have a particular expectation of what scientific writing should be. If we assume that a scientific author is necessarily interested in the latest theories, then Khudari's work is mystifying; we can only think of it as hopelessly behind the times. But Khudari and many of his peers did not share this assumption. For them, the meaning of "astronomy" took shape around different priorities: the devotional value of studying the celestial phenomena, the social relevance of specific astronomical problems, and a commitment to certain pedagogical and textual practices that framed the transmission of knowledge.

This is not to say that there was no kind of scholarly astronomical writing in which a discussion of heliocentric astronomy might have figured. An entire field of scholarly astronomy, *'ilm al-hay'a*, focused on the structure of the celestial spheres. As Khudari himself seems to have noted, however, *hay'a* was not a field in which his contemporaries took much interest. Perhaps this lack of interest can be interpreted as awareness that heliocentricity had rendered the classic *hay'a* texts obsolete – but such a conclusion would place a great deal of weight on what is essentially negative evidence. A better explanation may lie in the simple truth that the intensive study of planetary models is a contingent and relatively rare occurrence in the history of astronomy. Historians are only beginning to understand its social context in Islamicate cases. Even in the more developed literature of Copernican studies, a solution to the problem, "What was the question for which heliocentrism was the answer?" remains hotly contested.[158]

---

[158] See the critical reception of Robert Westman's magnum opus, in which he argues that Copernicus' motives were astrological. Michael Shank, "Made to Order," *Isis* 105 (2014): 167–76; Noel Swerdlow, "Copernicus and Astrology," *Perspectives on Science* 20 (2012): 353–78; cf. Robert Westman, *The Copernican Question: Prognostication, Skepticism, and Celestial Order* (Berkeley, CA: University of California Press, 2011). The

The silence of Muhammad al-Khudari, Ibrahim ibn Hasan, and so many others on heliocentricity – among other developments – reveals a community of scholars remarkably secure in their methods for producing knowledge, and in the value that this knowledge contained. In this sense, their silence presents a more radical challenge to the historian than acceptance or rejection of heliocentricity would have posed. The historiography of nineteenth-century Islamic thought, as well as of late Ottoman society, is too often organized around the notion of encounter with Europe. Whether the protagonists are advocates or critics of assimilating European ideas and culture, this encounter remains at the heart of the story. Muhammad al-Khudari's silence makes sense only within a different narrative, a narrative in which Europe is present but, for important purposes, irrelevant. Ultimately, we cannot be sure what Khudari thought about heliocentricity – only that he did not deem it an important astronomical issue.

Scholars' choice of indifference toward what others, including interlocutors like Mahmud al-Falaki, considered critical knowledge suggests the need for a broader approach to the "study of not knowing" than Robert Proctor has recently called for in his programmatic essay on "agnotology."[159] For Proctor, while not knowing is a choice, it is always a choice made in opposition to knowing. This understanding follows from a specific set of histories, in which ignorance has resulted from corporate-sponsored obfuscation, national security-driven classification, or the ethical concerns of scientists. Thus, ignorance is knowledge obscured, censored, or restrained. By focusing only on such oppositional relationships between knowledge and ignorance, however, we risk forgetting a basic insight of science studies: any process of knowing is a process of knowing some things rather than others, a process of learning to see only those particular phenomena that matter to understanding a specific object of knowledge.[160] As George Saliba has remarked, the Egyptian and Syrian polemicists like 'Abd Allah Fikri who rose to the defense of heliocentricity in 1874 were ignorant of the model's historical connection to the work of Muslim scholars, including Ibn al-Shatir.[161]

---

quotation is from Bernard Goldstein, "Copernicus and the Origins of the Heliocentric System," *Journal for the History of Astronomy* 33 (2002): 219–35, quoted in Shank, "Made to Order," 167.

[159] Robert N. Proctor, "Agnotology: A Missing Term to Describe the Cultural Production of Ignorance (and Its Study)," in *Agnotology: The Making and Unmaking of Ignorance*, ed. Robert N. Proctor and Londa Schiebinger (Stanford, CA: Stanford University Press, 2008), 1–33.

[160] Collins, for example, vividly demonstrates that the skill of selective seeing is an important element of the "tacit knowledge" theorized by Michael Polanyi. H.M. Collins, *Changing Order: Replication and Induction in Scientific Practice* (London: SAGE, 1985).

[161] Saliba, "Copernican Astronomy in the Arab East," 154.

In Saliba's view, such ignorance was unfortunate, inasmuch as it led
Ottoman intellectuals to view modern astronomy as an entirely foreign
phenomenon. If their ignorance was regrettable, however, it was also
predictable. The "Copernican" system taught in the nineteenth century
bore only a nominal relationship to the planetary models of Nicolaus
Copernicus. Moreover, the very question of how Copernicus appropriated mathematical tools that Muslim astronomers had developed in geocentric models follows only from a particular historicist agenda that was
unknown to late Ottoman (or, for that matter, Victorian) science education. In other words, when it came to heliocentricity, there was much
that even those who "knew" did not know, because it did not matter to
them. As new and powerful astronomers and astronomical institutions
emerged in late Ottoman Egypt, they would choose not to know a great
deal else.

★ ★ ★

Muhammad al-Khudari al-Dimyati arrived in Cairo and joined the circles of learning at al-Azhar sometime between 1815 and 1816. The following year, he was joined by another young, talented student, named
Rifaʿa Rafiʿ al-Tahtawi. It is hard not to imagine that they knew each
other well: both of them about eighteen years old, unusually gifted,
and students of the charismatic Hasan al-ʿAttar. But only one of them,
Tahtawi, was chosen to go to Paris as imam of the first major Egyptian
"educational mission," in 1826. Only one of them, Tahtawi, became central to the historiography of nineteenth-century Egypt. His work has long
been the subject of books, chapters, and translations. As a pioneering
educational reformist and translator of French political theory and science into Arabic, he embodies the beginning of modern Egyptian intellectual history, and he holds an important place in the historiography of
modern Islam.[162]

Muhammad al-Khudari, meanwhile, went back to Damietta and is
now forgotten. Perhaps, had he not lost his hearing, he too might have
been destined for Paris, and this chapter would not be the first study
of his life and work. From the point of view of late Ottomans, however,
Muhammad al-Khudari was hardly so obscure. His *Commentary on the
Brilliancy* was studied and admired by students like Ibrahim ibn Hasan,

---

[162] Mohammed Sawaie, "Rifaʿa Rafiʿ al-Tahtawi and His Contribution to the Lexical
Development of Modern Literary Arabic," *IJMES* 32 (2000): 395–410; John
Livingston, "Western Science and Educational Reform in the Thought of Rifaʿa al-
Tahtawi," *IJMES* 28 (1996): 543–64; Troutt Powell, *A Different Shade of Colonialism*,
47–55; Hourani, *Arabic Thought in the Liberal Age*, ch. 4.

and by scholars further afield, like the anonymous author in Aleppo in the 1860s. Their perspective is perfectly intelligible. In a multiconfessional society, scholarly astronomy taught the use and meaning of the various communal calendars, and precise methods for navigating between them. By calculating the "apparent phenomena" of celestial motion, scholarly astronomy was also necessary for the purposes of astrology – a practice far more common, despite its problematic theological status, than the use of scholarly astronomy's mathematical techniques to schedule prayers. In other words, scholarly astronomy was a capacious set of practices linked particularly with the ways in which late Ottomans measured and interpreted the passage of time.

These practices were also linked with the Islamic milieu of scholarship in which they occurred. They were linked, however, less by the narrow bonds of ritual applications and theological considerations, and more by broadly shared approaches to making knowledge in specific kinds of spaces and texts. Indeed, scholarly astronomy's roots in the networks, techniques, and norms of 'ulama' go a long way toward explaining the endurance of its appeal, even as new sites of astronomical knowledge emerged in late Ottoman Egypt. As we turn our attention toward some of these new sites, we must bear in mind, of course, that they did not fill a vacuum. Neither, however, did they simply displace the older sites of astronomy, which continued – for many people – to work.

2 Astronomers and Pashas: Viceregal Imperialism and the Making of State Astronomy

Astronomy, which once flourished upon the banks of the Nile, has received the most precious encouragements from Your Highness, and your royal munificence has bestowed upon it magnificent instruments that will permit the Observatory of Cairo to retake the place in the world that the Observatory of Alexandria occupied in antiquity.
        – Isma'il Effendi Mustafa, dedicating his study of a geodetic base-bar to the viceroy Ismail Pasha in 1864[1]

The total eclipse of the sun on 18 July 1860, was visible across a long, narrow strip of the northern hemisphere. As the sky went dark, two Egyptian observers stood ready at positions along the path of totality. One, Mahmud Bey Hamdi, was in Dongola, in today's Northern Sudan, about 1,000 miles up the Nile River from Cairo. The other, Isma'il Effendi Mustafa, was in the Santuaria del Moncayo, in the Spanish Pyrenees, about 200 miles northeast of Madrid. As the crow flies, Mahmud Bey and Isma'il Effendi[2] were almost 3,000 miles apart that day. Both, however, stood where they did because they served the Ottoman Viceroy of Egypt, Said Pasha. In similar ways, they had both arrived at their distant locations thanks to the new, difficult, potentially quite rewarding path that led through the schools of Egypt's "new order" (*nizam-ı cedid*) into government employ. And in different ways, the work each did in July 1860 contributed to two important and closely related viceregal projects.

The first of these projects was to seize the opportunity "to enter Egypt into the scientific concert of Europe," as Said Pasha's scientific advisor, Edmé-François Jomard, had pitched the idea of an eclipse expedition to the viceroy.[3] Fittingly, one outcome of this proposal, Mahmud

---

[1] Ismaïl-Effendi Moustapha, *Recherche des coefficients de dilatation et étalonage de l'appareil à mesurer les bases géodesiques appartenant au gouvernement Égyptien* (Paris: Impremerie de V. Goupy et C, 1864), v–vi.

[2] I refer to Mahmud and Isma'il by the titles they bore at the moment in question. "Pasha," as they are often referred to, is a title they each acquired only late in life.

[3] E.-F. Jomard to the Viceroy of Egypt, 30 March 1860, in Mahmoud-Bey, *Rapport à son altesse Mohammed Saïd, vice-roi d'Égypte, sur l'Éclipse totale de soleil observé à Dongolah*

Figure 2.1 Mahmud al-Falaki's illustration of the total solar eclipse of 1860 as viewed from Dongola in the Egyptian Sudan. *Rapport à son altesse Mohammed Saïd, vice-roi d'Égypte, sur l'éclipse totale de soleil, observée à Dongolah (Nubie), le 18 Juillet 1860* (Paris: Mallet-Bachelier, 1861), 29. Bibliothèque nationale de France.

Bey's report on his voyage to Dongola, was eventually presented to the Académie des sciences in Paris (see Figure 2.1). Ismaʿil Effendi, meanwhile, observed the eclipse as a member of the expedition party of the Observatoire impérial de Paris itself (see Figure 2.2).[4]

The second viceregal project that occupied Mahmud Bey and Ismaʿil Effendi was cartography. The previous year, Said Pasha had tasked Mahmud Bey with producing an "astronomical map" (*kharīṭa falaki-yya*), meaning a map of the viceroy's domains based on astronomically

---

(Nubie), le 18 Juillet 1860, par Mahmoud-Bey, astronome de son altesse (Paris: Mallet-Bachelier, Imprimeur-Libraire du bureau des longitudes, de l'école polytechnique, 1861), 19.

[4] On the arrangement of Ismaʿil's participation in the expedition, see Paris Observatory MS 1060 II-A-3, especially Jomard to Le Verrier, 22 June 1860. See also Jomard to the President of the Académie des sciences, 5 November 1860, in Mahmoud-Bey, *Rapport*, 24.

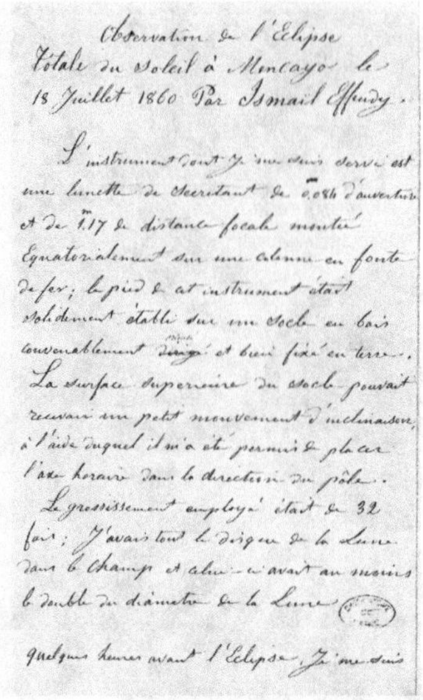
Observation de l'Eclipse
Totale du Soleil à Moncayo le
18 Juillet 1860 Par Ismail Effendy.

L'instrument dont je me suis servi est une lunette de Secrétant de 0,084 d'ouverture et de 1.17 de distance focale montée Equatorialement sur une colonne en fonte de fer; le pied de cet instrument était solidement établi sur un socle en bois convenablement dirigé et bien fixé en terre. La surface superieure du socle pouvait recevoir un petit mouvement d'inclinaison, à l'aide duquel il m'a été permis de placer l'axe horaire dans la direction du pôle.
Le grossissement employé était de 32 fois; j'avais tout le disque de la Lune dans le champ et celui-ci avait au moins le double du diamètre de la Lune.

Quelques heures avant l'Eclipse, je me suis

Figure 2.2 Isma'il Effendi's report on the total solar eclipse of 1860 as seen from Moncayo. Eclipse Papers, MS 1060 II-A-3, Paris Observatory. Courtesy of l'Observatoire de Paris.

determined positions. Dongola, the northernmost of the Sudanese territories conquered by Mehmed Ali Pasha in 1820–21, was emphatically a part of those domains. Said Pasha commissioned the map not long after deciding to maintain Ottoman-Egyptian rule in the Sudan, which he had considered relinquishing.[5] For Mahmud Bey, therefore, the grueling voyage to Dongola (much of which, south of Wadi Halfa, had to be crossed on land by camel) was about more than an eclipse. It provided a rare opportunity to determine the longitude and latitude of locations – ultimately about forty – between Cairo and *La Nubie*, as he called it.[6] For Isma'il Effendi, meanwhile, it was no coincidence that his most noteworthy contribution to the Paris Observatory's eclipse expedition was to determine the observation party's position.[7] Surveying techniques occupied a major part of his training in Paris; when he eventually returned to

[5] On Said Pasha's policies in the Sudan, see P.M. Holt and M.W. Daly, *A History of the Sudan*, 6th edition (Harlow, UK: Pearson, 2011), 50–52; and Troutt Powell, *A Different Shade of Colonialism*, 50.
[6] Mahmoud-Bey to Jomard, 21 October 1860, in Mahmoud-Bey, *Rapport*, 23.
[7] Ahmed Zéki Bey, "Notice biographique sur S.E. Ismail Pacha El-Falaki," *Bulletin de la société khédiviale de géographie*, 6th Ser. (Cairo: Impremerie nationale, 1908), 15.

Cairo, in 1864, it was to direct a newly established Viceregal Observatory that supported the making of the astronomical map.

Muhammad al-Khudari, from whose perspective we explored the practice of Muslim scholarly astronomy in mid nineteenth-century Egypt, was very interested in eclipses. Several of the claims to novelty in his *Commentary on the Brilliancy* are based on his analysis or tabular presentation of eclipse predictions. Yet it is unlikely that Khudari observed the eclipse of 1860. As an Azhari scholar living in Damietta, with no position in the viceregal bureaucracy, he could hardly have made the voyage up the Nile to the other end of the viceroy's dominion – a trip that was almost too much for Mahmud Bey, who had all the resources of the state behind him.[8] Needless to say, it is even more difficult to imagine Khudari in the Spanish Pyrenees with a team from the Observatoire de Paris. Moreover, had Khudari somehow found himself in a position to observe the eclipse, he would have seen it in a different light from Mahmud Bey or Isma'il Effendi. Khudari wanted readers of his *Commentary on the Brilliancy* to observe eclipses in order to help resolve a disagreement between Ulugh Beg (d. 1449) and Ibn al-Shatir (d. 1375) over the correct procedure for eclipse prediction. The motives that brought Egyptian astronomers to Moncayo and Dongola – joining the European "scientific concert," mapping the Sudan – did not concern him in the least.

As this contrast suggests, while Mahmud Bey and Isma'il Effendi were both known popularly by the moniker *al-falakī*, "the astronomer," their careers marked the rise of new meanings for that word in Egypt, meanings that emerged with the making of a newly close relationship between science and the Ottoman-Egyptian state in the middle of the nineteenth century. Within the growing ranks of technically trained bureaucrats who served the "Khedives" or Ottoman viceroys of Egypt in this period, few had longer and more consequential careers than Mahmud and Isma'il.[9] This chapter traces their footsteps in order to understand how astronomy entered the service of the viceroys through new sites, career paths, and networks – in sum, a new form of authoritative astronomical knowledge, situated differently from the scholarly astronomy of 'ulama' like

---

[8] Due to the stresses of the voyage, Mahmud Bey's theodolite sustained damage and could not be used, and the astronomer himself developed an eye illness (of all things) on the return voyage. Mahmoud-Bey, *Rapport*, 6; and Mahmoud-Bey to Jomard, 21 October 1860, in Mahmoud-Bey, *Rapport*, 23.

[9] Although the ruling members of Mehmed Ali Pasha's dynasty are often referred to as the "Khedives," Ismail Pasha was the first to use this title legally, having purchased the privilege from the Sultan. I favor the term "viceroy," the English translation that was used at the time (as "vice-roi" was in French), since it captures the relationship of substantial autonomy – but not sovereignty – that the dynasty enjoyed with respect to Istanbul.

Muhammad al-Khudari: in military academies and observatories, rather than in homes; in bureaucracies, rather than relationships of patronage; in Paris as well as in Cairo; in French as much as in Arabic.

Viceregal astronomy grew out of, and in turn shaped, the Ottoman-Egyptian state's efforts to position itself as an empire akin to and in concert with European counterparts, particularly the second French Empire. It is no accident that the origins of modern Egyptian state astronomy lie in the middle of the nineteenth century, a period when Ottoman Egypt grappled with fundamental questions about its territoriality, as Mehmed Ali's heirs, particularly Ismail Pasha (r. 1863–79), consolidated Egyptian rule in the Sudan and sought to expand their empire into East Africa. Eve Troutt Powell has shown that the project of viceregal imperialism in Africa gave shape to some of the earliest thinking about Egyptian nationhood, as "the Sudan helped Egyptians identify what was Egyptian about Egypt, in an idealized, burgeoning nationalist sense."[10] After the Mahdist movement seized control of the Sudan in 1881 and Britain occupied Egypt in 1882, the importance of restoring Egyptian rule in the Sudan acquired increasingly nationalist and racial overtones: by colonizing black Africans, Egyptians – "colonized colonizers," in Troutt Powell's formulation – demonstrated their fitness for national independence.[11]

The making of an Egyptian empire under these circumstances raised questions about *space* and *time* that state astronomers were particularly equipped to answer. Mahmud Hamdi and Isma'il Mustafa helped to create some of the first widely seen Arabic maps of Egypt, representing the space of the viceroy's domains with new specificity. Just as significantly, they argued for particular interpretations of the relationship between the space of Egypt, its history, and the people who now inhabited it. At times using the very instruments and techniques that they deployed in their cartography, viceregal astronomers also sowed the seeds of a new Egyptian historiography, most famously in Mahmud's studies of Alexandria and the pyramids. Through such cartographical and historiographical work, astronomers joined the viceroy's domains with the "scientific concert of Europe" by virtue of Egypt's past as well as its present.

Viewed from the specific perspective that viceregal astronomy opened onto the past, however, Egypt's history was colored by the lens of Orientalism. This was especially true when it came to the history of science. For Mahmud and Isma'il, scholarly astronomy was not

[10] Troutt Powell, *A Different Shade of Colonialism*, 51.
[11] Troutt Powell, *A Different Shade of Colonialism*, 6.

the living tradition practiced by Muhammad al-Khudari and his students, but rather an object of historical memory – and a relatively minor one. Viceregal astronomers downplayed the significance of the Islamic period for science in Egypt, focusing instead on sites like Alexandria and the pyramids, by which they linked Egypt with the Hellenic genealogy of European knowledge on the one hand, and what they argued were distinctively Egyptian achievements on the other. This Ottoman-Egyptian participation in Orientalism, as well as in the emergent field of Egyptology, contributed to the development of certain tropes that would become themes of Egyptian territorial nationalism by the early twentieth century. The privileged status of antiquity as the essential period for defining Egypt's identity, which distinguished the inhabitants of the Nile Valley from their possible Muslim or Ottoman communities, was one such notable trope.[12]

Even as viceregal astronomy shaped new thinking about Egyptian territoriality and peoplehood, it also laid a basis for people trained and employed in new institutions to answer some old questions concerning religious practice. Mahmud and Isma'il published almanacs including the times of prayer, built a public time signal that vied with the call of the muezzin, and published studies on topics such as the date of the Prophet's birth. Such intrusions onto the domain of scholarly astronomy had relatively limited consequences in the middle of the nineteenth century, however. The kinds of religious activism that would take advantage of the new relationship between science and the state had yet to emerge in earnest.

By contextualizing viceregal astronomy in relation to Egyptian imperialism, I differ from historians who have interpreted nineteenth-century Egypt as a site of "anti-imperial" science.[13] Instead, I build on the work of scholars who have highlighted and sought to explain the new forms of power that the Egyptian state began to exercise over its subjects well before the onset of European occupation in 1882. In his influential work on this period, Timothy Mitchell has shown that the introduction of a European epistemology of representation, what he called "the world-as-exhibition," rendered Egyptian spaces – the city, the village, the school – susceptible to new kinds of political and economic control between the 1820s and 1860s.[14] Focusing on the history of medicine, Khaled Fahmy

---

[12] On Egyptians in Egyptology, see Donald Malcolm Reid, *Whose Pharaohs? Archaeology, Museums, and Egyptian National Identity from Napoleon to World War I* (Berkeley, CA: University of California Press, 2002).

[13] Cf. Crozet, *Sciences modernes*, 13.

[14] Mitchell, *Colonising Egypt*, 13, 95.

and others have more recently shown that the state's centralization of political power and pursuit of military expansion depended on new techniques of disciplining the body, but that these techniques faced intense, sometimes overwhelming, resistance.[15] In other words, Egypt in the early and mid nineteenth century saw the emergence of a politics organized around the state's monopolistic claim to scientific management of its space and population, the forms of control that this claim authorized, and the many kinds of contestation that reshaped it.

Viceregal astronomers were constantly observing, measuring, and representing. The view of Egypt that they produced, however, was not *only* the panoptic view of a disciplinary state. Indeed, for the purposes of "political and economic calculation" on which Mitchell focuses,[16] the astronomical map was deemed an obvious failure. Yet, the viceregal government spent lavishly to promote it as a symbol of Egyptian modernity. Moreover, some of the ideologies that astronomers like Mahmud al-Falaki helped to shape bore consequences quite remote from the objectives of the Ottoman-Egyptian ruling elite – again, the privileging of Egypt's antiquity over its Ottoman history comes to mind. While viceregal astronomy participated in producing the "effect of order" that made Egypt colonizable, this effect was not merely coercive, but also enabling of new, unpredictable affinities and relations of authority.

While deepening our historical understanding of the relationship between science and the Egyptian state, the case of viceregal astronomy also expands the historiography of science and empire. The work that Mahmud, Isma'il, and their colleagues performed, including the instruments that they commissioned or built, participated in the making of astronomical knowledge in Europe, and even (in at least one instance) the United States. They did so, however, not within an empire that linked Egypt and the Atlantic through the center of Paris or London (for example), but rather within an empire that sought to move knowledge and objects between Paris and the Sudan through the center of Cairo. In other words, viceregal astronomy points to the importance not only of "non-Western" knowledge and objects, but also of non-Western empires, in the making of a modern science.

What follows is not a full biography of either Mahmud or Isma'il, much less a complete portrait of viceregal astronomy. Instead, drawing on the Egyptian National Archives, the archives of the Observatoire de Paris, and published work by the viceregal astronomers and their contemporaries, I analyze moments in Mahmud and Isma'il's careers that shed

---

[15] See especially Fahmy, *All the Pasha's Men,* 209–26; and Fahmy, "Medicine and Power."
[16] Mitchell, *Colonising Egypt,* 33.

light on the creation of a new relationship between state and science in Egypt. I begin with the new geography of education, following Mahmud and Isma'il in their early years through a network of schools in Cairo, the Nile Delta, and France, which were conceived within the military ambitions of Mehmed Ali Pasha and his successors between the 1820s and 1850s. The viceregal astronomers' years in Paris, especially, highlight the collaborative nature of French and Egyptian imperial geodesy. When Mahmud and Isma'il returned to Egypt in the 1860s, they deployed new embodied skills and forms of textuality, remapping Egyptian space and rewriting Egyptian history in ways that remain influential in Egypt even today. The twentieth-century reinvention of Mahmud al-Falaki as a father (or grandfather) of modern Egyptian science required certain erasures, which suggest continuities that run through the viceregal, post-Ottoman, and postcolonial Egyptian states.

From al-Hissa to Paris: A Geography of Viceregal Astronomy

Mahmud Hamdi was born in 1815; Isma'il Mustafa, in 1825.[17] In the intervening decade, the struggle for power between Mehmed Ali Pasha and Istanbul bore profound consequences not only for the relationship between Egypt and the rest of the Ottoman Empire, but also for the relationship between the Pasha and those he governed. As the Pasha began to build his own empire in the Sudan in the 1820s, and subsequently broke with the Sultan and occupied much of the Eastern Mediterranean throughout the 1830s, new kinds of public works and technical education played a crucial role in his ambitions. Whereas the management of the Nile had long been a local affair, with peasants maintaining irrigation projects near their own villages, the rebuilding of the Alexandria-Nile Canal between 1816 and 1820 entailed the forced relocation and labor of over 300,000 people, about a third of whom died of starvation,

---

[17] For Mahmud's life, see Ismail-Bey Moustapha, "Notice nécrologique de S.E. Mahmoud Pacha El Falaki," and Muhammad Mukhtar Bey, "Tarjamat hayat al-'alim al-fadil al-maghfur lahu Mahmud Basha al-Falaki," in *Notices biographiques sur S.E. Mahmoud-Pacha El Falaki* (Cairo: Impremerie Nationale, 1886). The astronomer has also been the subject of a hagiographical study by Ahmad Dimirdash, *Mahmud Hamdi al-Falaki* (Cairo: al-Dar al-Misriyya li-l-ta'lif wa-l-tarjama, 1966), as well as numerous short biographical notices, including a preface by Mahmud's grandson in Mahmud Basha al-Falaki, *al-Zawahir al-falakiyya al-murtabita bi-bina' al-ahram*, trans. Mahmud Salih al-Falaki (Cairo: Maktabat al-Anjlu al-Misriyya, n.d.). Isma'il's life has received less attention. For a list of biographical notices, see Arthur Goldschmidt, Jr., *Biographical Dictionary of Modern Egypt* (Cairo: AUC Press, 2000), 52–53, but these reproduce information contained in Zéki Bey, "Notice biographique."

disease, and other hazards.[18] The technical demands of this work seem to have provided the immediate reason for the creation of the Engineering School (*Mühendishane*) in Cairo's Citadel, the center of the Governor's personal household and power, in 1815. The school, attended by about eighty students, was equipped with "instruments of engineering, surveying, and astronomy, from England and other countries,"[19] suggesting that the new institution trained students in contemporary methods of surveying.

While the Engineering School was the first institution to grow out of the close relationship between a new scale of state-building and the need for new kinds of technical education in Egypt, this relationship became more fruitful as a result of two policies that Mehmed Ali Pasha introduced in the early 1820s. One policy, compelling peasants to cultivate long-staple cotton (rather than mostly wheat), required a new infrastructure of irrigation and transportation geared toward growing the new crop and exporting it to European markets. The other policy, mass conscription, entailed a new attention to regulating Egyptian bodies.[20] Thus, the 1820s and 1830s saw expansion in the size and curriculum of the Engineering School, as well as the founding of a medical school at Qasr al-ʿAyni, which trained doctors for the army and, more broadly, oversaw the health of a population whose physical well-being bore new relevance to the state's ambitions. However, the engineering and medical schools were only two threads in an expanding web of schools that supported Mehmed Ali's determination to wield autonomous power within the Ottoman Empire. These included a school of languages, which produced texts, largely based on French originals, in mathematics, geography, medicine, and other sciences; as well as primary and secondary schools that prepared students for training in the technical academies.[21]

It was within this new educational system that a young Mahmud Hamdi took his first steps onto the path of viceregal service. Born in al-Hissa, a village in the Gharbiyya (Western) province of the Delta, Mahmud entered the recently founded Naval School and Shipyard (al-Madrasa al-Bahriyya and Tersane) at Alexandria around the year 1831, when he was sixteen.[22] This complex of naval institutions was greatly

[18] Mikhail, *Nature and Empire*, 281–82.
[19] Mubarak, *al-Khitat*, 4:67.
[20] Fahmy has framed the introduction of these policies as the key shift in the nature of Mehmed Ali's rule. Fahmy, *All the Pasha's Men*, 11.
[21] Crozet, *Sciences modernes*, 231–70.
[22] Ismail-Bey Moustapha, "Notice nécrologique," 7; Muhammad Mukhtar Bey, "Tarjamat hayat al-ʿalim," 4. The precise date of Mahmud's entry into the naval school at Alexandria varies in biographical sources. However, the school underwent a major expansion in 1831, and he is known to have graduated in 1833.

expanded following the destruction of the Egyptian fleet at the battle of Navarino, in 1827. The devastating loss was the last that Mehmed Ali Pasha would suffer in the service of the Sultan. As the Pasha embarked upon a decade of intra-Ottoman warfare, he ordered the conscription of 8,000 youths to staff the shipyards and 12,000 more to learn seamanship.[23] Mahmud might well have been among them. Although extant accounts of Mahmud's early years agree that his older brother, a graduate of the Naval School himself, enrolled the boy because of his "nobility" (nijāba),[24] such a voluntary step would have been exceptional. According to a naval history written by the director of the Egyptian military schools in the late nineteenth century, the Naval School's officer class came from the Pasha's Mamluks and children of his other servants; everyone else was conscripted.[25] Mahmud was certainly not a Mamluk, and there is no evidence that his late father had been in the Pasha's employ.

The practice of "scholastic conscription" (tajnīd 'ilmī) was not unique to the Naval School; it was the foundation of the new educational system. Provincial administrators (mudīrs) were responsible for contributing a quota of students to the schools, just as they had to meet a quota for the army. The same officials carried out the two kinds of conscription, which led villagers to equate them. The new schools were guarded by soldiers, and students who left without permission were liable to see their families detained and held responsible until they returned.[26] It is in this light that we should view the beginning of Mahmud's technical education and government service. Even if his brother took the unusual step of enrolling Mahmud voluntarily, the context of conscription likely bore on his decision: had Mahmud not entered the Naval School, as a villager in the Nile Delta in the early 1830s, he would have borne the dreadful risk of conscription into the Pasha's army – a fate that many people maimed themselves to avoid.[27]

Upon graduating from the Naval School in 1833, Mahmud joined a small group of students transferred from Alexandria to Cairo to expand the ranks of the Engineering School.[28] He arrived in the capital – perhaps

---

[23] Isma'il Sarhank, Haqa'iq al-akhbar 'an duwal al-bihar (Bulaq: al-Matba'a al-Miriyya, 1314 [1896–97]), 1:242.

[24] Moktar-Bey, "Tarjama," 4. His father died when Mahmud was a child, making his brother his guardian.

[25] Sarhank, Haqa'iq al-akhbar, 1:231.

[26] Ahmad 'Izzat 'Abd al-Karim, Tarikh al-ta'lim fi 'asr Muhammad 'Ali (Cairo: Maktabat al-Nahda al-Misriyya, 1938), 644–46. I borrow the term "scholastic conscription" (tajnīd 'ilmī) from 'Abd al-Karim.

[27] Fahmy, All the Pasha's Men, 101–102. 'Ali Mubarak entered the Qasr al-'Ayni school out of personal ambition, but he immediately regretted the decision. Mubarak, al-Khitat, 9:40

[28] Mubarak, al-Khitat, 10:41; Crozet, Sciences modernes, 146.

for the first time in his life – at the height of Mehmed Ali Pasha's expansionist project. The 1833 Peace of Kütahya granted Mehmed Ali and his son Ibrahim Pasha the governorship of a vast swath of the Ottoman lands. Eighteen thirty-three was also the year that another group, largely composed of engineers, arrived in Cairo: the Saint-Simonian followers of Enfantin, who dreamed of connecting East and West through a canal at the Suez Isthmus.[29] Two of the Saint-Simonians became particularly important in the education of the future viceregal astronomers. One, Charles Lambert, a recent graduate of the French École polytechnique and École des mines, was to direct the Engineering School for over a decade and play a key role in developing its curriculum.[30] The other, J.-A. Yvon Villarceau, was a twenty-two-year-old, independently wealthy bassoonist, who taught music at Mehmed Ali's school of cavalry. Falling under Lambert's wing, Villarceau also began to study astronomy. The bassoonist soon tired of Saint-Simonianism and returned to France, but by the early 1840s he was working in the Observatoire de Paris, where some of Lambert's Egyptian disciples later became his students.[31]

Like other new institutions of learning, the Engineering School hardly transformed the nature of knowledge for everyone in Egypt. Even as Mahmud established himself at the Engineering School, first as a student and subsequently as a teacher, Ibrahim ibn Hasan was busy studying Muhammad al-Khudari's *Commentary on the Brilliancy* in order to compose his own, astrologically oriented commentary on the *zīj*. But the Engineering School was certainly transformative for those who experienced it, in terms both of social status and specific skills. Mahmud studied more than mathematics and astronomy; he also learned French – "presque sans maître," according to Isma'il, yet well enough to translate a mathematical textbook into Arabic.[32] Furthermore, Mahmud's appointment as an adjunct professor came with a commission at the rank of second lieutenant (*mulāzim*).[33] Over forty years later, when he retired

---

[29] On the connection between Saint-Simonian utopianism and the French technical elite in the first half of the nineteenth century, see John Tresch, *The Romantic Machine: Utopian Science and Technology after Napoleon* (Chicago: University of Chicago Press, 2012), 198. On the Saint-Simonians in Egypt, see Mitchell, *Colonising Egypt*, 16–17.

[30] Crozet, *Sciences modernes*, 187–88, and see 175–76 for Lambert's biography. On the Saint-Simonians and engineering under Mehmed Ali, see Ghislaine Alleaume, "L'école polytechnique du Caire et ses élèves: la formation d'une élite technique dans l'Égypte du XIXème siècle" (Ph.D. Diss., Université de Lyon II, 1993), 193–98.

[31] J. Bertrand, "Éloge historique de M. Yvon Villarceau, membre de l'Institut," read at the Académie des sciences, 24 December 1888, in *Éloges académiques* (Paris: Librairie Hachette, 1890).

[32] Ismail-Bey Moustapha, "Notice nécrologique," 8–9. Crozet, *Sciences modernes*, 185.

[33] Ismail-Bey Moustapha, "Notice nécrologique," 8. On the engineering career as a means of social advancement, see Alleaume, *L'école polytechnique*, 2:481–512.

and requested his pension from the government, his years of service would be reckoned from the time he assumed this position and received his commission, which marked the beginning of his ascent through the viceregal bureaucracy.[34] By 1844, he had been promoted to captain and was teaching astronomy from the works of Delambre and Francoeur.[35] When, in 1845, Lambert began taking students from the Engineering School and training them in a new observatory near Bulaq, Mahmud was naturally among them.[36]

The reasons for the Bulaq Observatory's establishment are puzzling. Mehmed Ali Pasha had ordered it built, on the site of an old French observatory, in 1839, but it was apparently not functional until 1845.[37] In later decades, the state's astronomers came to play an important role in naval training: Sulayman Halawa, for example, served as captain of one of the Khedive's steamships in the 1860s, and authored a manual on nautical mathematics.[38] At the Bulaq Observatory, however, Lambert seems to have occupied his students mostly with meteorological observations, as well as magnetic observations requested by the Royal Society in London.[39] It is difficult to explain most of these activities as answering immediate needs of the Ottoman-Egyptian state, in the way that the technical schools of this period typically functioned. Perhaps because of this ambiguity, the observatory closed after only five years in operation. ʿAli Mubarak Pasha, a close confidant and servant of the viceroys Abbas and Ismail, claimed that Abbas Pasha was appalled by the size of the educational budget that he inherited upon assuming the viceroyship in 1848: 100,000 guineas, over 11,000 of which went to the Observatory.[40] By his own account, Mubarak, charged with reducing the budget, eliminated the Observatory due to a lack of personnel who could staff it properly (*li-ʿadam wujūd man yaqūm bihā ḥaqq al-qiyām*). At the same time, he suggested rectifying this situation by sending a group of three to

[34] "Taʿrifa muqadamma min Mahmud Hamdi al-Falaki," 10 Muharram 1300 (20 November 1882), accessed through http://modernegypt.bibalex.org/collections/Documents.

[35] Ismail-Bey Moustapha, "Notice nécrologique," 9. Crozet, *Sciences modernes*, 195.

[36] Crozet, *Sciences modernes*, 195; Sami, *Taqwim*, 2:501.

[37] The only significant study of the Bulaq Observatory is Crozet's, in *Sciences modernes*, 194–99.

[38] Sulayman Halawa, *Al-Kawkab al-Zahir fi Fannn al-Bahr al-Zakhir* (Cairo, 1291 [1874–75]). For Halawa's career, see Mubarak, *al-Khitat*, 14:100–103; and Crozet, *Sciences modernes*, 115.

[39] Augustin Pellissier, *Rapport adressé à M. Le Ministre d'instruction publique et des cultes, par M. Pellissier, Professeur de philosophie, chargé d'une mission en Orient, sur l'état d'instruction publique en Égypte* (Paris: Paul Dupont, n.d.), 7–8 (from the second letter, dated 3 June 1849).

[40] Mubarak, *Al-Khitat*, 9:44; see Sami, *Taqwim*, 3:34, for the size of the Observatory budget.

France for further astronomical training. The group included Mahmud, of course, as well as a young man named Isma'il Mustafa, who had been a student of Mahmud's from the age of fifteen.[41] Although there is much that is open to question in this story,[42] the closing of the Observatory was at least partially consistent with the preferences of the new viceroy, who thought that reforms should happen more gradually than they had under Mehmed Ali Pasha, and less under the control of Europeans.[43] His more conservative policy spelled the end of Charles Lambert's 11,000-guinea-a-year observatory – but also sent Mahmud and Isma'il to Paris.

The practice of dispatching select students to schools in Europe, at the government's expense and under government supervision, began on a large scale in 1826, and was to continue into the twentieth century.[44] In France, this practice was institutionalized in the *Mission égyptienne*. According to the travel account written by the *Mission*'s first imam, Rifa'a Rafi' al-Tahtawi, the experience of Egyptian students in Paris was a highly regulated intellectual and cultural encounter: part scholarly travel, part military operation.[45] Tahtawi himself read Rousseau and Montesquieu, corresponded with Orientalists like Sylvestre de Sacy, remarked upon heliocentric astronomy and the discovery of new planets with the telescope, and commented on everything from French clothing to the French Revolution. At the same time, to ensure that they met the priorities of the viceroy, students were subject to constant supervision. Each took monthly tests and, by viceregal order, submitted a monthly account of his studies, including the number of lessons taken, with the teacher's signature beneath his name. The director of the *Mission* chastised Tahtawi when students neglected this obligation.[46] At the beginning of their studies, all students lived and studied on the same premises, which they left only on Sundays, and then only with written permission from an instructor, which they presented to the officer appointed by the viceroy to oversee them. Later, when the students attended various

[41] Ismail-Bey Moustapha, "Notice nécrologique," 8; Zéki Bey, "Notice biographique," 6. Little else is known of Ismail's personal background.

[42] According to another of Isma'il and Mahmud's prominent colleagues, it was Lambert's idea to send the Observatory's best students to France – but this account offers no explanation for the closing of the Bulaq facility. Zéki Bey, "Notice biographique," 6.

[43] Toledano, *State and Society*, 42.

[44] For the history of the missions, see 'Abd al-Hakim 'Abd al-Ghani Muhammad Qasim, *Tarikh al-Ba'that al-Misriyya ila Urubba* (Cairo: Maktabat Madbuli, 2010); and Mitchell, *Colonising Egypt*, 69–74.

[45] Rifa'a Rafi' al-Tahtawi, *Takhlis al-ibriz ila talkhis Bariz* (Bulaq: Dar al-Tiba'a al-Khidiwiyya, 1250 [1834–1835]).

[46] Tahtawi, *Takhlis al-ibriz*, 142.

specialized schools, they were permitted to go out more frequently, but again only with slips from their teachers, which they submitted to the sentry on duty (*nöbetçi*).[47]

While astronomers were presumably exempt from at least one disciplinary requirement, the nighttime curfew, the viceroy himself kept a close eye on Isma'il and Mahmud's activities.[48] As soon as Isma'il Effendi received his appointment as a member of the Observatoire impérial's staff, the director notified Jomard, "conforming to the desire that you expressed to me to be informed of the progress of Monsieur Ismail in the study of astronomy."[49] Jomard immediately communicated the news to the viceroy Said Pasha, who read it with "very lively interest."[50] Said Pasha was similarly interested in knowing how Mahmud's work held up to European standards. After the eclipse expedition to Dongola in 1860, it was the viceroy himself who requested that Mahmud's report be read and examined by members of the Académie des sciences in Paris.[51]

Isma'il Mustafa arrived at the Observatoire impérial in 1858.[52] It can hardly be coincidence that the man assigned to be his principal teacher and mentor was another former student of Lambert's – the old bassoonist, Villarceau. Now a member of the Bureau des longitudes, however, Villarceau was also an appropriate choice given the viceregal astronomers' interest in geodesy. The two became personally close: Isma'il Effendi later thanked his teacher's family for receiving him "less as a student than as a friend."[53]

The Egyptian in Paris, studying astronomy with the Frenchman who had become an astronomer in Egypt: it was one of many strange turns and reversals that characterized the increasingly complex networks of knowledge bridging France and Egypt in the middle of the nineteenth century. Until his death in 1862, the president of the *Mission égyptienne* in France, and longtime scientific advisor to the viceroys, was Edmé-François Jomard, former member of Napoleon's Institut d'Égypte

---

[47] On the disciplinary regime of the Egyptian school in Paris, see Mitchell, *Colonising Egypt*, 71–74.

[48] Tahtawi, *Takhlis al-ibriz*, 139–40.

[49] Leverrier to Jomard, 19 November 1859, in Zéki Bey, "Notice biographique," 8.

[50] Koenig Bey to Jomard, 9 December 1859, in Zéki Bey, "Notice biographique," 10.

[51] Jomard to the President of the Académie des sciences, 5 November 1860, in Mahmoud-Bey, *Rapport*, 24.

[52] Isma'il's whereabouts between 1850, when he and Mahmud were reportedly dispatched to Paris, and 1854, when he began training as an observer at the Observatoire impérial, are undocumented.

[53] Ismaïl-Effendi Moustapha, *Recherche*, VII.

and principal editor of its monumental product, the *Description de l'Égypte*.[54] When Jomard proposed the Dongola eclipse expedition to Said Pasha in 1860, he was acting on the suggestion of the astronomer Hervé Faye, who was himself the son of an engineer from Napoleon's invasion.[55] Some accounts imagined conflict between the Egyptian students and teachers they associated with the French invasion. Thus, 'Ali Mubarak's fictional narrative of an Egyptian's voyage in Europe imagines a debate, in the Royal Asiatic Society in Paris, between the Egyptian and a Frenchman who had participated in the invasion.[56] By contrast, Mahmud and Isma'il expressed only admiration for their hosts, including Jomard.[57]

The close relationship between French and Egyptian astronomers in the middle of the nineteenth century points to a larger phenomenon. Although some historians have interpreted the viceregal dynasty's investment in new scientific institutions as an example of anti-imperial science (since Egypt maintained its independence from European domination until 1882), the potential relationships between state and science can hardly be reduced to a binary of imperialism and resistance.[58] While the Ottoman-Egyptian state fought Anglo-French financial control in the 1870s and British military intervention in the 1880s, it also had its own history of seeking to increase the depth of its control over the people it governed, as well as to expand territorially. In fact, the years that Isma'il Effendi spent in Paris were a period of warm relations between an expansionist Ottoman-Egyptian state and (what was once again) the French Empire. In 1863, the viceroy Ismail Pasha even dispatched a unit of Sudanese conscripts to fight on behalf of Napoleon III in Mexico.[59] While this instance of military cooperation was exceptional, one trait that the two imperial regimes consistently shared was their interest in science – particularly, as the following section will show, at the intersection of astronomy and geodesy.

---

[54] On Jomard's relationship with Mehmed Ali Pasha, see Alain Silvera, "Edmé-François Jomard and Egyptian Reforms in 1839," *Middle Eastern Studies* 7, no. 3 (1971): 301–16.

[55] Jomard to the Viceroy of Egypt, April 1861, in Mahmoud-Bey, *Rapport*, 18.

[56] Wen-Chin Ouyang, "Fictive Mode, 'Journey to the West', and Transformation of Space: 'Ali Mubarak's Discourses of Modernization," *Comparative Critical Studies* 4, no. 3 (2007), 348.

[57] E.g., Mahmoud-Bey to Jomard, 21 October 1860, in Mahmoud-Bey, *Rapport*, 24.

[58] Cf. Crozet, *Sciences modernes*, 13.

[59] Richard Hill and Peter Hogg, *A Black Corps D'élite: An Egyptian Sudanese Conscript Battalion with the French Army in Mexico, 1863–1867, and Its Survivors in Subsequent African History* (East Lansing: Michigan State University Press, 1995).

Instruments in Motion: Viceregal Astronomy and "Global Science"

The extensive time that viceregal astronomers spent abroad bore consequences for the scientific communities in which they trained, not only for the institutions to which they eventually returned in Egypt. In 1859, Isma'il Effendi officially joined "the number of people charged with the regular service of observations, which he shares with the astronomers, adjunct-astronomers, and assistants" at the Observatoire de Paris.[60] As a member of the regular staff – no longer a student – Isma'il Effendi began to see his work appear in the Observatory's publications, particularly in the weekly *Comptes Rendus* of the Academy of Sciences. He participated not only in the eclipse expedition of 1860, but also in the observation of Comet Tempel in 1859 and the Great Comet of 1861, as well as in more routine work of the Observatory (Figure 2.3).[61] In these and other instances, Isma'il's name appeared in print next to the names of French astronomers like Villarceau and Urbain Leverrier (the observatory's director since 1854). Both Mahmud and Isma'il presented their work in European journals and academies, whose members saw them as colleagues.[62] In 1878, a "young Egyptian" at the United States Naval Observatory – probably Ibrahim 'Ismat, later a director of the Viceregal Observatory at 'Abbasiyya – assisted Simon Newcomb's research on historical eclipse observations by translating Arabic astronomical texts.[63]

---

[60] Leverrier to Jomard, 19 November 1859, in Zéki Bey, "Notice biographique," 9.

[61] For Isma'il Effendi's observations, see Y. Villarceau, "Observations de la comète de Tempel, faites à l'equatorial de la tour de l'ouest, à l'Observatoire impérial de Paris," *Comptes rendus hebdomadaires des séances de l'Académie des sciences* 49 (July–December 1859), 484; "Ascensions droites et distances polaires apparentes de la grande comète de 1861, conclues des observations equatoriales," *Comptes rendus* 53 (July–December 1861), 1036; Urbain Leverrier, "Refutation de quelques critiques et allegations portées contre les travaux de l'Observatoire impérial de Paris, et denuées de toute espèce de fondement," *Comptes rendus* 65 (January–June 1865), 111; Y. Villarceau, "Comparison des déterminations astronomiques faites par l'Observatoire impérial de Paris, avec les positions et azimuts géodésiques publiés par le Dépôt de la Guerre," *Comptes rendus* 62 (January–December 1866), 805; Y. Villarceau, "Memoire sur les observations de l'éclipse totale de soleil du 18 Juillet 1860, faites en Espagne par la Commission française (extrait)," *Comptes rendus* 67 (July–December 1868), 275–76, 278 (a full memoir of 104 pages is referenced on p. 270, but appears never to have been published, and the relevant file in the archives of the Observatoire de Paris contains only two pages of latitude observations); A. Gaillot, "Sur la direction de la verticale à l'Observatoire de Paris," *Comptes rendus* 87 (July–December 1878), 685.

[62] For Mahmud's bibliography, see Crozet, "Trajectoire," 300.

[63] Matthew Stanley, "Predicting the Past: Ancient Eclipses and Airy, Newcomb, and Huxley on the Authority of Science," *Isis* 103 (2012), 265. I suggest that the Egyptian was 'Ismat because he is the only Egyptian astronomer who worked at the USNO in this period. He also appears to have been a bibliophile, working for a time at the new Khedival Library, where he compiled the catalogue for astronomy, astronomical timekeeping,

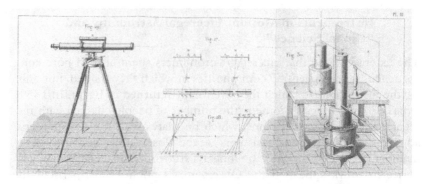

Figure 2.3 Apparatus for observing the dilation of the metals of the Brunner base-bar at different temperatures. Ismaïl-Effendi-Moustapha, *Recherche des coefficients de dilatation et étalonage de l'appareil à mesurer les bases géodesiques appartenant au gouvernement Égyptien* (Paris: Impremerie de V. Goupy et C, 1864), pl. III (detail). Bibliothèque nationale de France.

Such collaborative relationships between viceregal astronomers and the Europeans and Americans with whom they trained and worked contradict the notion of a one-way street on which "Western" science traveled to the Middle East in the nineteenth century.

Collaboration arose most intensively where the astronomical interests of viceregal Egypt aligned with those of the French Empire, as they did in the area of cartography. Isma'il Effendi's years on the staff of the Observatoire de Paris coincided with an attempt by the Bureau des longitudes to bring new precision to the map of France by re-measuring the critical Dunkirk–Barcelona meridian: the nominal basis for determining the length of the meter, the adoption of which was hotly debated in this period.[64] At the same time, Mahmud was recalled to Cairo to oversee the production of the new "astronomical map" of the Nile Valley. While Isma'il Effendi remained in Europe, his official charge from the viceregal government was to acquire the highly precise instruments of measurement that were necessary for this survey. One such instrument, which Isma'il commissioned from the instrument-maker Rigaud, was a portable

mathematics, engineering, and arithmetic. See the biographical note under the entry for 'Ismat's *Mawaqi' ghurar shuhur a'wam al-qarn al-rabi' 'ashar fi al-sinin al-ghurghuriyya* (1304 [1886–87]), in *Fihrist al-kutub al-'arabiyya al-mahfuza bi-l-Kutubkhana al-Khidiwiyya*, vol. 6 (Cairo: Matba'at 'Uthman 'Abd al-Razzaq, 1305–11 [1888–93]).
[64] See Ken Alder, *The Measure of All Things* (New York: The Free Press, 2001), especially chs. 11–12.

meridian circle, a type of transit telescope that can be used to determine latitude as well as longitude. While Villarceau drew up the specifications for the meridian circle, Isma'il was presumably involved in verifying its accuracy, and possibly in building it, as he did for the other major surveying instrument he acquired in these years, the Brunner base-bar (discussed below).[65] The result of Villarceau and Isma'il's collaboration with Rigaud was better than Villarceau had expected, leading the French astronomer to ask his Egyptian student and friend whether he could borrow the portable meridian circle for the Observatoire's own work on the Dunkirk–Barcelona line.[66] Isma'il Effendi agreed, and Villarceau subsequently used it to re-determine the position of a number of stations on the French map, including the key station of Dunkirk.[67]

Isma'il's portable meridian circle, called "Rigaud No. I" after its principal maker, led a long and itinerant life, its view of the stars linking astronomers in Egypt with those in Europe for several decades. Made at the behest and expense of the Egyptian government for a new map of Egypt, it was designed by a French astronomer, Villarceau, who had first studied astronomy in Egypt, with the same Saint-Simonian teacher, Lambert, who designed the curriculum of the Engineering School in which Mahmud and Isma'il were first educated. Isma'il Effendi himself oversaw the making of the instrument, which he lent back to his teacher to use in the mapping of France. When Isma'il eventually reclaimed it for use in Egypt, Villarceau commissioned a second meridian circle, "Rigaud No. II," for continued use by the Bureau des longitudes.

Rigaud No. I was not the only one of Isma'il's instruments to lead such a cosmopolitan existence. Another device he commissioned in Paris was a standard bar for measuring geodetic base-lines, built by the instrument-maker Brunner. To attain the precision of which the Brunner base-bar was capable, Isma'il Effendi undertook a meticulous study – almost 500 pages in length –of the dilation of the instruments' metals (platinum and brass) at various temperatures.[68] This study required Isma'il to compare the new Egyptian base-bar with the one that Brunner had used as its

[65] A.-J. Yvon Villarceau, "Longitudes, latitudes, et azimuts terrestres au moyen des observations faites au cercle meridien no. II de Rigaud," *Annales de l'Observatoire impérial de Paris: mémoires* 9 (1868), 1.

[66] Villarceau, "Longitudes, latitudes, et azimuts terrestres," 1.

[67] Rigaud No. I was also used at Strasbourg, Paris, and Talmay. "Longitudes et latitudes astronomiques," Paris Observatory Ms E6; Yvon Villarceau, "Determination astronomique de la longitude et de la latitude de Dunkerque," *Annales de l'Observatoire impérial de Paris: mémoires* 8 (1866), 210–11, 316; *Annales de l'Observatoire impérial de Paris: observations* 1863, 33.

[68] Ismaïl-Effendi-Moustapha, *Recherche*. The history of the project is outlined on pp. xi–xiii.

model, an instrument recently made for the Spanish survey.[69] Two years after the eclipse of 1860 had passed over Moncayo, Isma'il Effendi again made his way across the Pyrenees, this time to the Madrid Observatory, where he examined the two geodetic instruments in cooperation with Col. Don Carlos Ibanez, chief engineer of the Spanish map service. Even when Isma'il finally brought the base-bar back to Cairo, further transnational entanglements lay in its future. It was this same instrument, forged amid the expansionist and modernizing aspirations of the viceregal era, which was unearthed forty years later in order to measure the baseline of the cadastral survey of Egypt carried out under the British Occupation.[70]

Through their travels, uses, and reuses, the instruments of viceregal astronomy tell a story that is much more than "Egyptian." In fact, they suggest ways in which the making of science and technology in the nineteenth century was "global" not only in the well-established sense that European traders, colonists, and scholars brought non-Western knowledge and materials back to centers like Paris, London, or Kew, but also in the sense that projects and careers launched from a center like Cairo did work inside the United States Naval Observatory and the Observatoire de Paris. That being said, perhaps the expansive category of "global" does more to celebrate than to characterize the highly specific relationships that explain the design, manufacture, movement, and use of a technology like Rigaud No. I or the Brunner base-bar. Just as the scholarly astronomy of Muhammad al-Khudari was made in particular sites of learning connected by the travel of Muslim scholars and the cultivation of Islamic scholarship, viceregal astronomy forged material, political, intellectual, and personal connections between Egyptian and French imperial state-building in the mid nineteenth century – and unwittingly laid the ground for Egypt's later administration by the British Empire.

Viceregal Instruments and the Observing Body

As similar projects began to make use of similar instruments in France and in Egypt, such instruments required different embodied skills from the practice of scholarly astronomy, and hence different kinds of bodies

---

[69] The Spanish device itself was modeled on the famed "Borda No. 1," which had been used to measure the base-lines of the French meridian since Delambre. See Ismaïl-Effendi-Moustapha, *Recherche*, 150; and, on the history of the original Borda instrument, Alder, *The Measure of All Things*, 217.

[70] Timothy Mitchell, *Rule of Experts* (Berkeley, CA: University of California Press, 2002), 88.

that could produce astronomical knowledge. Scholars of astronomy like Muhammad al-Khudari, we know, understood the observational instruments that they used to be significantly and irremediably flawed. While doubting the precision of their observational instruments, however, such scholars trusted the ability of their textual instruments – mathematical tables – to preserve reputable sets of observed values and to facilitate calculation. The distinctively textual character of scholarly astronomy in a culture of learning that generally prized aural knowledge helped us to make sense of Khudari's decision to focus on astronomy specifically after becoming deaf. (See Chapter 1.)

By contrast, it is no accident that Yvon Villarceau, Isma'il Effendi's mentor at the Observatoire de Paris, was first trained as a musician. While sight was certainly a prerequisite for the viceregal astronomer, so too were particular skills of listening. Mahmud and Isma'il were trained in an era of French astronomy when practitioners learned to record the precise timing of their observations by listening to the beats of a pendulum.[71] They were trained, moreover, by a master. According to a lengthy éloge for Villarceau that was read at the Académie des sciences upon his death, his years as a bassoonist had prepared him admirably for astronomical observation, since "his ear, exercised on sixty-fourth notes, divided the second into ten equal parts."[72] As John Tresch and Emily Dolan have recently shown, Enlightenment and Romantic discourses linked science and music through their shared devotion to the skilled use of instruments. Pointing to the case of William and Caroline Herschel (both of whom were accomplished musicians, even as they achieved more fame for their work with telescopes), Tresch and Dolan argue that "looking through a telescope required the same kind of dedicated practice as the performance of fugues at a keyboard."[73] In fact, musical and observational virtuosity shared more than a general commitment to "practice"; they shared the same embodied skill of keeping time by the ear. (Chronographs, which George Airy introduced at the Greenwich Observatory beginning in the 1840s, eventually replaced this

---

[71] H. Faye, *Cours d'astronomie de l'école polytechnique* (Paris: Gauthier-Villars, 1881–83), 1:148. For Isma'il's facility with the pendulum, see Ismail Effendi Moustapha, Notebook 4: 28 June 1855–[2 November] 1856, Paris Observatory MS AF-14, at 13 August and 23 August 1855.

[72] Bertrand, "Éloge historique," 339. On the emergence of the tenth of a second as a measurable duration with diverse significance in this period, see Jimena Canales, *A Tenth of a Second* (Chicago: University of Chicago Press, 2009).

[73] John Tresch and Emily I. Dolan, "Toward a New Organology: Instruments of Music and Science," *Osiris* 28 (2013), 289.

embodied skill with a mechanical discipline that aspired to eliminate the foibles of individual judgment.)[74]

The importance of auditory skill to Villarceau and his Egyptian students, contrasted with the deafness of Muhammad al-Khudari, points to ways in which the emergence of viceregal astronomy entailed not only a new materiality of science in late Ottoman Egypt – pendulums and meridian circles, rather than quadrants and the *musātira* – but also new conceptions of the ideal astronomer and the skills he had to embody. The viceregal astronomers, especially Isma'il, were virtuoso observers, whereas Muhammad al-Khudari, so far as we know, never recorded a single observation of his own. Instead, he staked the reputation of his *Commentary on the Brilliancy* on redesigned tables, textual corrections, and clear explanations. This is not to say that scholarly astronomy gave no thought to observational skill and its embodied practices. Ibrahim ibn Hasan, for example, recommended observing a stellar transit with the left eye, rather than the right, "because the left one has more light."[75] Such a comment, however, suggests that, for scholarly astronomy, the risk of error that human bodies introduced into astronomical knowledge could be satisfactorily resolved through the application of a normative principle (the left eye sees better than the right). For viceregal astronomy, the individual body was itself an instrument to be constantly and finely tuned. While historians of Egypt have devoted much attention to the state's focus on "seeing" the bodies of its subjects in new ways (e.g., through censuses and death certificates), less is understood about the training of state functionaries' bodies quite literally to see and hear in new ways. Yet newly observant bodies were essential for certain projects of state-building, to which I now turn.

Remapping Egyptian Space and Time

Said Pasha recalled Mahmud to Egypt in 1859, granting him the rank of Bey, along with 50,000 *kuruş* "to settle his necessities and required expenses."[76] The return of the viceroy's astronomer coincided with the resumption of survey work that Mehmed Ali Pasha had initiated in 1813.[77]

---

[74] On the disciplining of observers with the chronograph, see Simon Schaffer, "Astronomers Mark Time: Discipline and the Personal Equation," *Science in Context* 2 (1988), 119.

[75] Ibrahim ibn Hasan, *Sharh Muqaddimat Mahmud Qutb al-Mahalli*, MS ṬM 225, ENL, fol. 21a.

[76] Sami, *Taqwim*, 3:338.

[77] Henry Lyons, *The History of Surveying and Land-Measurement in Egypt* (Cairo: National Printing Department, 1907), 25. Lyons was mistaken, however, in dating Mahmud Bey's assumption of cartographic duties to 1861. See Mahmoud-Bey to Jomard, 10 May 1860, reproduced in Mahmoud-Bey, *Rapport*, 21.

Mahmud Bey became director of the new Astronomical Map Authority (*maṣlaḥat al-kharīṭa al-falakiyya*), a task that was to occupy him for the subsequent decade, which marked a highpoint in the Ottoman-Egyptian government's modernizing aspirations. Flush with cash from a boom in cotton prices during the American Civil War, and increasingly with loans from European creditors, the government spent lavishly to expand the school system, along with other public works and cultural establishments, especially after the ascension of the Viceroy Ismail Pasha in 1863.

While these policies famously ended in bankruptcy and Anglo-French financial oversight in 1876, in the 1860s it was possible to envision a different future for Egypt. Ismail Pasha not only dispatched troops to fight in Mexico on behalf of Napoleon III, but also oversaw the completion of the Suez Canal and built a fleet of steamers that competed with the Peninsular and Oriental for the Red Sea hajj traffic. It was in this period that Ismail Pasha is famously said to have declared, "My country is no longer part of Africa; it is part of Europe" – a statement later quoted by the nationalist delegation (*wafd*) that sought to negotiate on Egypt's behalf at Versailles in 1919.[78] While these words might seem to stake a political claim by way of a geographic metaphor, in their original context they articulated a claim that was equally political and geographic. In the 1860s and early 1870s, the Ottoman-Egyptian state sought to remap its space and rewrite its history according to French conventions. These projects, in which viceregal astronomy played a crucial role, fashioned Egypt as a part of Europe, but – as in the case of Ismaʿil al-Falaki's meridian circle – they also entailed the participation of Egyptians in shaping the growth of transnational disciplines, rather than the mere adaptation of "Western" science to a "local" context.

In a significant coincidence, the year in which Mahmud Bey returned to Egypt to direct the astronomical map was also the year in which the term "Egyptology" first came into use.[79] The viceroy's astronomer played a noteworthy role in the emergent field, serving as the only Egyptian on the founding board of the second Institut d'Égypte, which was also established in 1859.[80] Even as he labored toward completion of the astronomical map, representing the lands that the viceroy governed in the present, he authored two historical studies that were to win him more lasting recognition than his cartographical work: the 1862 treatise *L'âge et le but des pyramides*, and the 1872 *Mémoire sur l'antique Alexandrie*. In these studies, Mahmud Bey deployed his technical skills as an astronomer and

---

[78] Gershoni and Jankowski, *Egypt, Islam, and the Arabs*, 43.
[79] Reid, *Whose Pharaohs*, 131.
[80] Reid, *Whose Pharaohs*, 131, 226, 243–44.

cartographer to rewrite the history of sites that he, like many Orientalists, believed were critical to understanding the nature of Egypt as a civilization distinct from other Ottoman, Arab, or Muslim societies. By rereading these studies in their original context – the making of the astronomical map – this section explores the relationship between cartography and Orientalism in viceregal Egypt. Bound up with the making of a map were the beginnings of a new Egyptian historiography.

Horses, Chronometers, and the Great Pyramid of Giza: Making the Astronomical Map

Mahmud Bey's Astronomical Map Authority oversaw and coordinated the work of survey parties dispatched from Cairo to measure the Nile Valley. The execution of this work both illustrated and extended the power of the viceregal government. While the surveyors were sent from Cairo, it was the responsibility of each province (*mudīriyya*) to provide local labor as well as other materials. Every triangulation party (*firqat al-muthallathāt*)[81] required horses, troops, at least one local judge or authority (*ḥākim*), and wood.[82] When the engineers were ready to begin work on a new province, the Map Authority would alert the provincial governor (*mudīr*) and order him to provide the necessary support.[83] Mapping Egypt was a test of the extent to which the growing viceregal bureaucracy could govern its subjects in terms of both breadth (from Alexandria to the Sudan) and depth (from the viceroy down to a group of fifteen conscripts and six horses).

The work of the astronomical map appears to have been the immediate cause for the reestablishment of a government observatory in Cairo in 1867 – seventeen years since the old site at Bulaq had closed. Isma'il Bey, who returned from Paris in 1864 with orders to set up the facility, served as director.[84] The new Viceregal Observatory was unambiguously

---

[81] According to Captain Henry Lyons, who assumed responsibility for survey work in Egypt under the British Occupation, Mahmud Bey's survey lacked a system of triangulation. However, the records of the Astronomical Map Authority refer repeatedly to "triangulation parties." Perhaps their work was limited to the leveling that the authority carried out relatively late in its survey, and which Lyons conceded was of a relatively high quality. Given the effort that Isma'il Effendi invested in acquiring the Brunner base-bar, however, a precise geodetic survey must have been intended at some point, even if it was never completed. H.G. Lyons, *The Cadastral Survey of Egypt, 1892–1907* (Cairo: National Printing Department, 1908); Lyons, *History of Surveying*, 26.

[82] See e.g., *Daftar sadir al-kharita al-falakiyya*, DWQ 4003-000258, #162.

[83] See e.g., *Daftar sadir al-kharita al-falakiyya*, DWQ 4003-000258, #66.

[84] Zéki Bey, "Notice biographique," 11. Construction was delayed by lack of skilled labor, and the equatorial dome was apparently not yet completed in 1869. See Ismail al-Falaki to Leverrier, 15 June 1869, MS 3711, no. 226, Leverrier Papers, Institut de France: I am grateful to Simon Schaffer for sharing his summary of this file.

a military institution. It was located in the guard tower of an old garrison (*karakol*) in 'Abbasiyya, a suburb just east of Cairo that was reserved for military schools (*al-madāris al-ḥarbiyya*), rather than civilian schools (*al-madāris al-mulkiyya*).[85] Initially, the Observatory fell entirely under the purview of the Ministry of War,[86] which maintained budgetary responsibility even when the Observatory staff later came under the administration of the Ministry of Public Instruction (*niẓārat al-maʿārif al-ʿumūmiyya*). Detailing this arrangement, the Vice Chief-of-Staff (*nāʾib al-sirdār*) of the Egyptian Army singled out the Observatory's geodetic role in order to illustrate the importance of maintaining its instruments properly.[87] Cartography, in other words, was the chief significance of the Observatory, at least to the section of the government that paid for its instruments.

Expenses were not trivial. In one year alone, the Astronomical Map Authority paid 6,000 Egyptian pounds for the "necessary materials," such as chronometers, that the Observatory and Engineering School had furnished for its work.[88] This was in addition to the cost of labor and more basic materials (such as the horses) borne by the *mudīriyya*s, to say nothing of the salaries of the surveyors. When the portion of the map covering the Delta region was completed, Ismail Pasha ordered 6,000 copies printed: 1,000 maps of each of the five Delta *mudīriyya*s, and 1,000 of a map covering the entire region. For this order, the Egyptian government agreed to pay the German printer Koffman 60,375 francs, not including the cost of packaging and shipping the maps to Egypt.[89]

Such expenses suggest that the Ottoman-Egyptian government under the viceroys Said and Ismail Pasha attached considerable importance to the astronomical map. In part, of course, the map was an administrative tool.[90] However, it is difficult to explain the printing of 1,000 copies for administrative purposes alone. As Timothy

---

[85] Zéki Bey, "Notice biographique," 11; and Survey Department (Cairo), *A Report on the Meteorological Observations made at the Abbassia Observatory, Cairo, during the year 1900* (Cairo: National Printing Office, 1902), 1; Crozet, *Sciences modernes*, 103.

[86] Taʿrifa muqaddama min Mahmud Hamdi al-Falaki, 10 Muharram 1300 (20 November 1882), accessed through http://modernegypt.bibalex.org/collections/Documents.

[87] Talab al-sirdariyya tatabbuʿ al-rasadkhana al-harbiyya, 24 Safar 1303 (1 December 1885), DWQ 0075-014381.

[88] See entries for Shawwal 1286, nos. 14, 44, and 66, *Daftar sadir al-kharita al-falakiyya* 1286–1287 (1869–70), DWQ 4003-000257.

[89] Sami, *Taqwim*, 3:1106–7.

[90] See "Laʾihat al-muhandisin" in Sami, *Taqwim*, 3:1116–22, for an example of the map's administrative use.

Mitchell has noted, the astronomical map provided the basis for a 45-square-meter relief map at the heart of the Egyptian palace at the 1867 World Exhibition in Paris.[91] Perhaps the printed edition was also used for such representational work – in schools, for example. If so, however, the map represented "Egypt" to students in a particular way: not as the vague "well-known lands" of the lower Nile Valley that the Ottoman government granted to Mehmed Ali Pasha in the "first political map of modern Egypt" in 1841, nor as the expansive territory including the Sudan and Eastern and Western deserts that only emerged with clarity in the early twentieth century.[92] Rather, the astronomical map presented Egypt as a set of provinces composed of cities, villages, branches of the Nile, canals, railway lines, and pilgrimage sites in the Delta and Cairo regions (see Figure 2.4). Far more than an Ottoman province, Egypt was to be seen as a systematically governed space characterized both by modern industry and its own, distinctively Egyptian sites of Islam. In fact, the attention that the map drew to the railway alongside pilgrimage sites (maqāmāt al-awliyā') illustrates an important relationship between the two: as Barak has shown, it was precisely in the second half of the nineteenth century that railway travel transformed the celebration of mawlids at such sites from local practices into an element of mass culture.[93] The map also represented an expanding space, since Upper Egypt and the Egyptian Sudan were meant to be included at a later date. (Although the Map Authority performed considerable work in these areas, the southern portions were never finished.)[94]

As a representation of Egypt as a systematically governed space that was more than provincial yet rooted specifically in the lower Nile Valley, and as a space that was technologically modern yet grounded in its own history, Mahmud Bey's astronomical map resembles another "remapping" of Egypt in the late nineteenth century: 'Ali Mubarak's bio-geographical dictionary, al-Khitat al-Tawfiqiyya al-Jadida.[95] Mubarak's compendium is one example to which historians of nineteenth-century Egypt have drawn attention in order to highlight the role of the viceregal

---

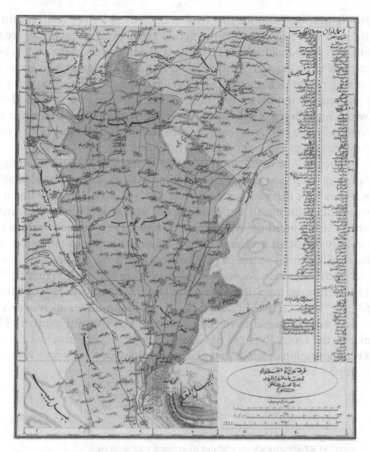

Figure 2.4 "Astronomical map" of Qalyubiyya Province. Light gray lines mark railways; crescents mark gravesites of venerated figures (*maqāmāt al-awliyā'*); dotted lines are administrative boundaries. Mahmud al-Falaki, Khartat Mudiriyyat al-Qalyubiyya (Cairo: Kaufmann, 1289 [1872]). Kroch Library, Cornell University.

state in reframing the spaces that constituted Egypt, enabling the exercise of new kinds of power over Egyptian land, streets, and bodies.[96] In

[96] Mitchell, *Colonising Egypt*; O. El Shakry, *The Great Social Laboratory*; Troutt Powell, *Another Shade of Colonialism;* Mitchell, *Rule of Experts*; Fahmy, *All the Pasha's Men*; and Fahmy, "Medicine and Power." As Ellis has pointed out, however, one limitation of this attention to the "representation" of Egypt is that it overlooks the realities that existed on the ground regardless of what bureaucrats and intellectuals in Cairo thought the nation "looked" like. Ellis, *Between Empire and Nation*.

this light, it bears notice that Mubarak enjoyed a close relationship with the man who had overseen the actual remapping of Egypt twenty years earlier.[97] Mubarak had been a student of Mahmud's in the 1840s, before surpassing his teacher in viceregal service and occupying numerous high-ranking positions under Abbas Pasha and Ismail Pasha. Mubarak played a particularly important role in the administration of schools and public works, including the railroad.[98] Toward the end of his career, when he sat down to write al-Khitat, he seems to have had Mahmud Bey's recently published map (among others) in front of him.[99]

Following Mubarak, we too can interpret the literary "mapping" of Egypt in al-Khitat in light of Mahmud Bey's astronomical mapping. Scholars have debated, for example, the extent to which Mubarak merely reproduced European notions of space and order, versus the extent to which he deployed the Arabic genre of khitat to give such notions a particularly Egyptian meaning.[100] In a sense, a similar question lies at the heart of Mahmud's map: its point of zero longitude. In the 1860s, a standard prime meridian for measuring longitude had yet to emerge.[101] The 1841 Ottoman map delineating the hereditary governorate of Mehmed Ali was based on the Paris meridian.[102] In Mahmud's map, however, longitude was measured from the meridian of that most Egyptian of landmarks, the Great Pyramid of Giza (see Figure 2.5). This choice suggests a reorganization of space that placed a certain idea of Egypt – rather than Europe – at its center.

---

[97] Ouyang's study of Mubarak's writing, for example, uses the term "mapping" or "remapping" in a figurative sense, without discussing cartography.

[98] See Mitchell, Colonising Egypt, on the relationship between Mubarak's architectural and educational projects; see also Hunter, Egypt under the Khedives, 123–38; Delanoue, Moralistes et politiques, 488f.; and Wadad al-Qadi, "East and West in 'Ali Mubarak's 'Alamuddin," in Intellectual Life in the Arab East, 1890–1939, ed. Marwan Buheiry (Beirut: AUB Press, 1983), 21–37.

[99] Al-Khitat sometimes states geographic distances with no apparent source, e.g., that the town of Bijam lies on the east side of the Sharqawiyya Canal, northeast of Basus by 2 km and southeast of Qalyub by 4.8 km. Such a statement seems to have been read from a map, and specifically a map that uses metric units. One such map that Mubarak likely used was that of the Description de l'Égypte, since he regularly cited the Description (which he called "the French Khitat") for other information. Mubarak's orthography is also much closer to that of the Description map than to Mahmud Bey's. However, in certain cases – such as the location of Bijam – Mubarak's geography more nearly matches Mahmud Bey's. Most likely, Mubarak used both maps, among others.

[100] Ouyang, "Fictive Mode," strongly adopts the latter interpretation. Cf. Mitchell, Colonising Egypt.

[101] Peter Galison, Einstein's Clocks, Poincaré's Maps: Empires of Time (New York: Norton, 2003), 119–21.

[102] Biger, "The First Political Map of Egypt," 84.

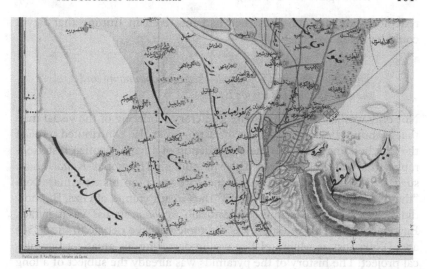

Figure 2.5 Detail from the "astronomical map" of Qalyubiyya Province. The largest X at the bottom of the map, left of center, marks the Great Pyramid of Giza and zero longitude. Mahmud al-Falaki, Khartat Mudiriyyat al-Qalyubiyya (Cairo: Kaufmann, 1289 [1872]). Kroch Library, Cornell University.

Such an interpretation, however, overlooks the fact that the *savants* of Napoleon's invasion had determined the longitude of the Great Pyramid from Paris. A map based on the latter position could easily be converted and read in France. Mahmud was well aware of this consideration, writing to Jomard that the astronomically determined points on the completed map would be connected with those determined by the French astronomer Nouet, "dans votre grande et memorable ouvrage de l'expédition."[103] So, did the adoption of the Great Pyramid as zero-longitude mark the map as particularly Egyptian, or did it merely serve to mask the incorporation of Egypt into a foreign cartography? A similar question can be asked, more figuratively, of 'Ali Mubarak's "remapping" of Egypt in *al-Khitat*. Yet, for both Mahmud Bey and 'Ali Mubarak, who lived most of their lives prior to the British Occupation (and long before the rise of anticolonial nationalism), the question might not have suggested any great tension. For these French-educated technicians, bureaucrats, and servants of the viceroys, it was precisely through the

[103] Mahmoud-Bey to Jomard, 21 October 1860, in Mahmoud-Bey, *Rapport*, 24.

redeployment of techniques they had mastered in Europe that Egypt could be made visible.

The Great Pyramid Revisited: Viceregal Astronomy and the Making of History for Egypt

Not only was *al-Khitat,* in a sense, a literary rendering of the visual map that Mahmud Bey produced, but the astronomer also contributed some of the first building blocks to the construction of a new Egyptian historiography, to which Mubarak would give monumental form in *al-Khitat.* I have suggested that Mahmud chose to measure space on his new map from "that most Egyptian of monuments," the Great Pyramid of Giza. But how did the Great Pyramid acquire its iconically *Egyptian* status? Was it, in fact, Egyptian at all? This question was the topic of fierce debate in the 1850s and 1860s, just as Mahmud Bey was undertaking his great cartographical project. The history of the pyramids was already the subject of a long tradition of European fascination and study. As early as the seventeenth century, John Greaves had sought to reveal the origins of the pyramids by measuring them.[104] Jomard himself had measured the pyramids for the *Description de l'Égypte.*[105] In the 1850s and 1860s, however, pyramidology reached a new level of intensity. Although Mahmud Bey minimized the importance of this context as a motivation for his own work,[106] it was surely relevant that the latest, sensational contribution to pyramidology, John Taylor's *The Great Pyramid: Why Was It Built?* (1859), denied the responsibility of the civilization of the Pharaohs for building the pyramids (or at least the Great Pyramid). Taylor, the Astronomer Royal for Scotland, dated the building of the Great Pyramid to around 4000 BC, and argued that it was a divine revelation, a physical corollary to the Bible.[107]

---

[104] Reid, *Whose Pharaohs,* 24–25.

[105] Jomard, "Description générale de Memphis et des pyramides," *Description de l'Égypte: Antiquités,* vol. 2 (Paris: Imprimerie Royale, 1818), 16f.

[106] Mahmoud-Bey, *L'âge et le but des Pyramides, lus dans Sirius,* Extrait des Bulletins de l'Académie Royale de Belgique, 14, 2nd Ser. (Alexandria: Imprimerie Francaise Mourès, Rey & Cᵉ, 1865), 6.

[107] John Taylor, *The Great Pyramid: Why was it Built? & Who Built it?* (London: Longman, Green, Longman, and Roberts, 1859), 205–6 and 262f. A contemporaneous review of competing theories of the pyramids' construction appears in Charles Piazzi Smyth, *Life and work at the Great Pyramid, during the months of January, February, April, and May, A.D. 1865* (Edinburgh: Edmonston & Douglas, 1867). On the relationship between Smyth and Mahmud al-Falaki, see Simon Schaffer, "Oriental Metrology and the Politics of Antiquity in Nineteenth-Century Survey Sciences," *Science in Context* 30 (2017): 173–212; see also Schaffer, second Tarner Lecture (2010), available at http://sms.cam.ac.uk/media/741069;jsessionid=2C144475E44DBF24CAF626035BBE4967 (accessed 2 July 2013).

Mahmud Bey's analysis, by contrast, emphasized the integral relationship between the design of the pyramids and ancient Egyptian civilization. The central argument of *L'âge et le but des pyramides, lus dans Sirius,* is that the pyramids were designed with an "astrological and religious purpose,"[108] such that the star Sirius – which the ancient Egyptians, according to Mahmud Bey, associated with the god of death – would shine (at its culmination) directly onto the pyramids' southern faces. Since, according to Mahmud Bey's measurements, the pyramids were built with their corners oriented toward the cardinal directions, and their faces inclined at approximately 52.5° to the horizon,[109] they must have been erected sometime around the year 3300 BC, when the relationship with Sirius would have been most exact.[110]

At stake in the measurement of the pyramids was the ownership of Egypt's past. Mahmud Bey's theory, by explaining the architecture of the pyramids in light of specifically ancient Egyptian beliefs and practices, claimed the pyramids as a part of Egyptian history at a time when these most recognizable monuments on the Nile had to be *made* Egyptian. Only thus could the Great Pyramid *become* the symbolically obvious point from which to measure the new map of Egypt. In this light, it also makes sense that Mahmud Bey sought to minimize the religious differences between ancient and contemporary Egyptians. Despite having many "gods," the ancients, according to Mahmud Bey, worshipped "a single supreme being" (*un seul être suprême*), and believed in the afterlife and divine judgment of the soul.[111]

Mahmud Bey's study of Alexandria grew out of a similar interplay of European and viceregal historiographical agendas. As Donald Reid has shown, nineteenth-century Orientalists saw Egypt as the source of multiple aspects of European identity: biblical, spiritual, and – embodied in the city of Alexandria – classical.[112] It was Napoleon III's personal interest in the history of Caesar that led him to ask the viceroy Ismail Pasha for a map of ancient Alexandria.[113] In Mahmud Bey's hands, however, the result of that request became a history of the ancient city and its environs, a map not only of the ancient location but also of the booming contemporary city, and a testament to the fitness of the current viceregal dynasty (contrasted with the Ottoman sultans) as heir to the glories of Greece and Rome. The great decline of Alexandria, Mahmud Bey

---

[108] Mahmoud-Bey, *L'âge et le but,* 28.
[109] Mahmoud-Bey, *L'âge et le but,* 12.
[110] Mahmoud-Bey, *L'âge et le but,* 26.
[111] Mahmoud-Bey, *L'âge et le but,* 13–14.
[112] Reid, *Whose Pharaohs,* 9.
[113] Crozet, "Trajectoire," 298, fn 38.

averred, had only commenced upon its "subjection" in the beginning of the sixteenth century – i.e., with the Ottoman conquest. It awaited a "new hero," Mehmed Ali Pasha, to "revive it from its ashes."[114] While a similar characterization of the Ottoman era as a dark age later became a trope of Egyptian nationalist historiography, for Mahmud, loyal servant to a series of Turcophone viceroys who acknowledged the sovereignty of Istanbul, the point seems rather to have been to celebrate the distinctive path on which the viceregal dynasty had led Egypt in contrast to previous Ottoman governors.

Even as Mahmud Bey redirected European scholarly interests into the service of the viceroys, his work retained certain premises that originated among Orientalists with whom he had conversed during his years in Europe. It is noteworthy that, among his three major historical works – the study of the pyramids, the Alexandria *Mémoire*, and an earlier study of the Arabian calendar – two concerned Egypt but not Islam, while one concerned Islam but not Egypt. This pattern conformed to the notion that "Egypt ceases to be Egypt when it ceases to be ancient," as one critic has characterized the Orientalist reduction of Egypt to its pre-Islamic history.[115] In Mahmud Bey's decision to highlight the revival of Alexandria as the symbol of the viceroys' enlightenment, there is even an echo of J. Barthélemy Saint-Hilaire, who succeeded Jomard as president of the *Mission Égyptienne* in Paris. In his 1857 *Lettres sur l'Égypte*, Saint-Hilaire had remarked:

I have confidence in the destiny of Egypt, which will become, if we succeed, the obligatory passage of European civilization in its relations with Asia. Perpetually traversed by commerce, industry, capital, and the representatives of the Christian nations, Alexandria may retake its ancient and holy mission, which has always been to reunite the most diverse religions, the most contrary beliefs: Greece and Egypt, paganism and Christianity, and today, Christian luminaries and Mahomedan coarseness (*la rudesse mahométtane*). Under this new regime, I hope that Alexandria will be able to find, if not more prosperity than it once had, at least greater peace (*repos*) as well as greater dignity.[116]

For the viceregal astronomers, however, the significance of Alexandria lay less in its image as a Mediterranean melting pot, and more specifically in its place in the history of science, which linked Egypt with the

---

[114] Mahmoud-Bey, *Mémoire sur l'antique Alexandrie* (Copenhagen: Impremerie de Bianco Luno, 1872), 4.

[115] A. Zvie, "L'Égypte ancient ou l'Orient perdu et retrouvé," in *D'un Orient l'autre* (Paris, 1991), quoted in Reid, *Whose Pharaohs*, 8.

[116] J. Barthélemy Saint-Hilaire, *Lettres sur l'Égypte* (Paris: Michel Lévy, 1857), 90. "If we succeed" refers to Saint-Hilaire's advocacy of a canal at Suez: like the Saint-Simonians of the 1830s, he conceived of the canal in moral as well as economic and strategic terms.

intellectual genealogy of the European Renaissance. Thus, in Isma'il's dedication of his work on the Brunner base-bar to the viceroy Ismail Pasha, the astronomer reached across two millennia in order to frame the instruments and institutions of viceregal astronomy not as the latest Ottoman (or Muslim) astronomical achievement, but rather as a revival of the Greek astronomy that flourished in Alexandria.

The significance of Mahmud Bey's historical studies for nineteenth-century and later Egyptian intellectuals has been well remarked. Mubarak drew on the Alexandria *Mémoire* to write his own history of Alexandria in *al-Khitat*.[117] Other viceregal historians, such as Mikha'il Sharubim, drew more heavily on Mahmud Bey's pyramidology.[118] Not that Mahmud Bey wrote these works in Arabic; they appeared originally in French and were not translated into Arabic for decades.[119] However, Egyptological studies held great interest for Egypt's Francophone class, appealing as they did to the emergence of interest in the late nineteenth century in forging a specifically Egyptian historiography out of the Pharaonic and Hellenic legacies.[120] This trend later reached its zenith in the Pharaonic nationalism of the 1920s.

What is often unremarked, however, is the connection between Mahmud Bey's Egyptology and his day job, as it were, directing the viceroy's cartography during the 1860s. Both historical projects were fundamentally cartographical. This is most obvious in the case of the *Mémoire sur l'antique Alexandrie*, which was at heart an attempt to locate and map the ruins of the ancient city. The project required an initial survey of contemporary Alexandria, in order to create a grid onto which Mahmud Bey could map whatever ruins were discovered. To this end, he supervised a triangulation of 80 km (east–west) by 30 km (north–south).[121] The Astronomical Map Authority provided the model, and to some extent even the specific people, for supporting this work. Mahmud Bey employed the same engineering corps (*hay'at al-muhandisīn*) that he usually directed on the map, and even brought along a draftsman, Amin Sabbagh, whose name appears on at least one portion of the astronomical

[117] Reid, *Whose Pharaohs*, 166.

[118] Reid, *Whose Pharaohs*, 209. See Mikha'il Sharubim, *al-Kafi fi tarikh Misr al-qadim wa-al-hadith* (Bulaq: al-Matba'a al-Kubra al-Amiriyah, 1898).

[119] See Mahmud al-Falaki, *Risala 'an al-Iskandariyya al-qadima wa-dawahiha wa-l-jihat al-qariba minha allati uktushifat bi-l-hafriyyat wa-a'mal sabr al-ghawr wa-l-mash wa-turuq al-bahth al-ukhra*, trans. Mahmud Salih al-Falaki (Alexandria: Dar Nashr al-Thaqafa, 1966); and Mahmud al-Falaki, *Al-Zawahir al-falakiyya*. Mahmud Salih al-Falaki was the first Mahmud al-Falaki's grandson.

[120] Reid, *Whose Pharaohs*.

[121] Mahmoud-Bey, *Mémoire sur l'antique Alexandrie*, 5.

map as well.[122] In addition, the government put at Mahmud Bey's disposal 200 laborers for the excavation work – not unlike the provision of troops and material that were mandated for the triangulation parties of the astronomical map.[123]

Mahmud Bey's pyramidology also drew on the techniques and instruments he possessed as a cartographer. The key measurements of the pyramids' orientation and inclination were all achieved with methods and instruments that he was accustomed to deploying, and had access to, because of his cartographic work. His value of 52.5° for the inclination, for example, was based in part on measurements of the height of the Great Pyramid, which he determined barometrically with a technique developed by the French astronomer Laplace.[124] Mahmud Bey likely imported this practice from geodesy, where the precise determination of altitudes was an essential element of triangulation, since surveyors generally need to site their triangles from elevated positions that subsequently must be adjusted to sea level.[125] The use of the theodolite to determine orientation was likewise a practice of the surveyor.

In some ways, the intersections of astronomical and historiographical practice in Mahmud Bey's work recall the breadth of learning that characterized scholars of astronomy like Muhammad al-Khudari. While both viceregal astronomy and scholarly astronomy were concerned with knowing the past, however, the pasts that they knew were different. Scholarly astronomy knew histories of particular calendars like the Arabic, Coptic, and Hebrew dates; it knew histories of significant events and their relationship to astrological phenomena; and it knew a history of its own relation to a particular set of astronomical texts, mostly Arabic (with some Persian) works dating to the fourteenth century and later. None of these histories gave a prominent role to territory – let alone to Egypt specifically – as history's subject. It was this sense of Egypt as a defined space-with-a-past that began to emerge in the simultaneously cartographical and historiographical work of viceregal astronomy. As the contrasts between viceregal and scholarly astronomy indicate, however, viceregal astronomy not only produced a particular history for Egypt, but also occluded other possible histories. While envisioning science in Egypt in terms of a revival of Egypt's Hellenic heritage, viceregal astronomers said little about more recent traditions of scientific inquiry like scholarly astronomy; from their perspective, one could pass from Ptolemy to the

---

[122] Mahmud al-Falaki, *Khartat Mudiriyyat al-Gharbiyya* (Cairo: Kaufmann, 1289 [1872]).
[123] Mahmoud-Bey, *Mémoire sur l'antique Alexandrie*, 5–6.
[124] Mahmoud-Bey, *L'âge et le but des Pyramides*, 10–11.
[125] Alder, *Measure of All Things*.

nineteenth century and miss little of significance. Perhaps it is surprising that Egypt's new astronomers claimed a more ancient pedigree than scholarly astronomy's practitioners generally did. Antiquity, however, allowed the viceregal astronomers to fashion an identity that was, in its Hellenism, emphatically European, while also, in its Pharaonicism, distinctively Egyptian.

The Textuality of Viceregal Astronomy

While astronomy provided some of the tools with which Mahmud al-Falaki made historical arguments, measurement and mathematics were complemented by particular ways of reading and appropriating texts. In this respect, too, viceregal astronomy both resembled and differed from scholarly astronomy. Well before carrying out his work on the pyramids and Alexandria, Mahmud honed particular techniques of reading and writing history during his years abroad. Even as he was conducting research on geomagnetism in Europe in the late 1850s, he published an article on the Jewish calendar, another on the pre-Islamic calendar of Arabia in 1858, and a third in 1859 comparing the French verb *avoir* with the Arabic *kāna*.[126] The textual techniques that Mahmud employed to pursue these interests emerged not from Islamic discourses of historical and lexical inquiry, as in Khudari's case, but rather from his engagement with mid nineteenth-century European scholarship on the history and philology of the Orient. This distinction helps explain why viceregal astronomers, despite certain intersections of their work with elements of Islamic practice, wielded little religious authority in the middle of the nineteenth century.

Mahmud's historical methodology was a function of his audience as well as training. He wrote his historical (and linguistic) articles in French and published them in venues like the *Journal Asiatique*. Among the works from the 1850s, only the study of the pre-Islamic calendar was ever translated into Arabic – thirty years later. Of course, as Mahmud Effendi himself illustrates, a growing elite of Egyptians read and wrote French comfortably; his work certainly had readers in Egypt, 'Ali Mubarak not least among them. Nevertheless, it is striking that Mahmud took to writing in

[126] Mahmoud Effendi, "Mémoire sur les Calendriers judaïque et musulman: du Calendrier judaïque," mémoire présenté à la classe des sciences de l'Académie royale de Belgique, 8 May 1855 (Brussels: Hayez, 1855) (see Crozet, "Trajectoire," 306); Mahmoud Effendi, "Mémoire sur le calendrier arabe avant l'islamisme, et sur la naissance et l'âge du prophète Mohammed," *Extrait du Journal Asiatique* no. 2 (1858); Mahmoud Effendi, "Identité du rôle de l'auxiliaire *avoir* et du verbe كان, lié avec un autre verbe," *Journal Asiatique*, 5th Ser., 13 (April–May 1859), 293–309.

French at a time when much Egyptian scientific work was concerned with translating French into Arabic, as he himself had done earlier in his career.[127] Isma'il published prolifically in Arabic through the 1870s, including a two-volume manual on astronomy and geodesy and regular contributions to the journal of the Egyptian civil school system, *Rawdat al-Madaris* ("The Schoolyard"). In this context, Mahmud's choice of language drew criticism even from staunch admirers. When the popular monthly journal *al-Muqtataf* published a summary and glowing review of all of Mahmud's publications (shortly after the journal relocated to Cairo, in 1885), the editors voiced one major reservation about his work: "We only wish that Arabic had a share in it, so that it might fill an Arabic mold as it has filled a French mold. Surely Arabic is more appropriate for it than other languages, and Arab schools are more fitting to acquire it than European schools."[128] Such criticism suggests that, however valuable Mahmud's work might have been to a certain Francophone elite, the use of French in his scientific and historical writing limited its readership in Egypt significantly.

Not only the language, but also the premises and style of Mahmud Effendi's historical works, addressed a European audience. The treatise on the pre-Islamic calendar of Arabia is a particularly striking example, because it could easily have been written for a different audience. Its central concern is a question with deep roots in Islamic scholarship: what kind of calendar had the Prophet Muhammad's contemporaries used? The answer has implications for the meaning of *nasī'* (Qur'an 9:37), a term denoting some form of calendrical manipulation apparently practiced by the pagans of pre-Islamic Arabia; the Qur'an's prohibition of this practice is the scriptural basis for Islam's purely lunar calendar. Furthermore, in trying to reach a definitive answer to the meaning of *nasī'*, one of the subordinate arguments that Mahmud Effendi employed was an astronomical determination of the Prophet's *mawlid* (birth-date). This argument had ritual implications, since the Prophet's *mawlid* was (and remains) an important, albeit controversial, occasion for celebration.

While the questions that Mahmud Effendi tackled in this article were present in venerable Islamic debates, they were also questions that European scholars of Islam had come to debate,[129] and it was on their

---

[127] On the use of Arabic in scientific translation and education in the middle of the nineteenth century, see Crozet, *Sciences modernes*, ch. 6; and, on the transition toward French and English, p. 369.

[128] "*Fa-inna al-'arabiyya la-aḥrā bihā min ghayrihā, wa-makātib al-'arab awlā bi-iqtinā'ihā min makātib al-ifranj.*" "Hadaya wa-taqariz," *al-Muqtataf* 9 (1885): 279.

[129] Some Muslim scholars (notably astronomers) argued that *nasī'* had been a method of intercalation akin to the Jewish luni-solar system. Scholars of exegesis (*tafsīr*), however, generally understood *nasī'* to have been the arbitrary transference of holidays, or of the

terms that Mahmud Effendi made his argument. His first move was to set aside all of the Muslim sources – historical, exegetical, philological – that address the question of the pre-Islamic calendar directly.[130] In the case of the Islamic historiographical tradition, Mahmud Effendi wrote that since it only began two or three centuries after the *hijra,* its view of the pre-Islamic era was necessarily cloudy. As for Muslim exegesis and philology, he did not think that analysis of the ambiguous word *nasī'* alone could yield a definitive answer.

The alternative strategy that Mahmud Effendi chose was to select events during the Prophet's lifetime whose Arab dates were well-attested and which were also associated with other historical occasions – especially astronomical events. The specific events that Mahmud Effendi analyzed were the death of the Prophet's son Ibrahim, the date of the Prophet's entry into Medina, and the date of the Prophet's birth.[131] Through a combination of chronological analysis and textual criticism, Mahmud determined the Julian dates corresponding to the Arab dates of these events. Using the Julian dates as a yardstick to measure the distance between the Arab dates, he was then able to argue that the Arabs must have been using a purely lunar (as opposed to luni-solar) calendar.[132] This kind of application of current astronomical knowledge to the interpretation of ancient history, and vice versa, was also a concern of European astronomers in this period (albeit more in Anglophone contexts than in France), and was precisely the kind of work that Ibrahim 'Ismat assisted at the United States Naval Observatory in the 1870s.[133]

Much of the evidence that Mahmud Effendi drew on in making this argument came from hadith, which testified, for example, that the Prophet's son had died in the month of Dhu al-Hijja on the day of a solar eclipse.[134] Mahmud Effendi's approach to the hadith, however, was not that of a *muḥaddith,* a scholar trained in the discipline of

---

sanctity of particular months, within a purely lunar system. The debate was picked up by nineteenth-century French Orientalists. See Caussin de Perceval, "Mémoire sur le calendrier arabe"; Sylvestre de Sacy, "Mémoire sur divers évènements de l'histoire des Arabes avant Mahomet," *Mémoires de littérature, tirés des registres de l'Académie royale des inscriptions et belles-lettres* 48 (1808): 484–762.

[130] Mahmoud Effendi, "Mémoire sur le calendrier arabe avant l'islamisme, et sur la naissance et l'âge du prophète Mohammed" (Brussels: Hayez, 1858), 3–5. In this discussion, I cite the Hayez edition, rather than the original extract from the *Journal Asiatique* cited above, because the former is more readily accessible. The major difference between the two texts is that the *Journal Asiatique* edition includes quotations in Arabic type, whereas the Hayez edition gives only French translations.

[131] Mahmoud Effendi, "Mémoire sur le calendrier arabe," 6f., 8f., 16f.

[132] Mahmoud Effendi, "Mémoire sur le calendrier arabe," 29–30.

[133] Stanley, "Predicting the Past."

[134] Mahmoud Effendi, "Mémoire sur le calendrier arabe," 6.

hadith transmission. Instead of recalling the texts from memory, he cited their appearance in manuscripts in the French National Library.[135] In other cases, he cited Arabic sources as they were quoted in European studies, rather than the original Arabic texts.[136] This method suggests, to begin with, how Mahmud Effendi actually did his research, but it also advertised his adherence to the stylistic conventions of European scholars. As Anthony Grafton has remarked, in practice, footnotes testify less to the veracity of an author's argument than they do to the author's membership in a particular community of scholars. By advertising the dependence of the author's work on a contingent set of questions, methodologies, and interlocutors, the footnote, Grafton argues, "separates historical modernity from tradition."[137] Perhaps for this reason, when the article was first translated into Arabic, in 1885, the manuscript citations disappeared from the text.[138]

Mahmud Effendi's methodology in this article is open to serious criticisms, some of which would be discussed publicly when the Italian Orientalist Carlo Nallino delivered a series of lectures in Cairo in 1909–10.[139] We need not be convinced by Mahmud Effendi's argument, however, in order to notice how Mahmud Effendi constructed an argument that presumably was convincing to him, and the significance that this kind of argument bore in mid nineteenth-century Egypt. In this vein, one striking characteristic of the argument is that it is more recognizable as a work of European Orientalism than as scholarship internal to Islamic discourses. In Mahmud Effendi's methodology, Islamic tradition presented a question – what calendar did pre-Islamic Arabia use? – and certain texts (e.g., hadith manuscripts) that could be evaluated as historical evidence for answering that question. The tools that one used to analyze the evidence, however, were tools that he had acquired through his travel in European scholarly communities.

Despite its concern with Islamic questions, therefore, Mahmud Effendi's article did not intervene in normative religious debates. Even when it appeared in Arabic in the 1880s, it does not appear to have sparked any discussion, in Egypt or elsewhere, on the observance of the Prophet's *mawlid*, or the meaning of *nasī'* in the Qur'an. By contrast,

---

[135] See e.g., Mahmoud Effendi, "Mémoire sur le calendrier arabe," 6.

[136] Mahmoud Effendi, "Mémoire sur le calendrier arabe," 18–19, 21.

[137] Anthony Grafton, *The Footnote: A Curious History* (Cambridge, MA: Harvard University Press, 1999), 24.

[138] Mahmud Basha al-Falaki, *Nata'ij al-afham fi taqwim al-'arab qabla al-islam wa-fi tahqiq mawlidihi 'alayhi al-salat wa-l-salam*, trans. Ahmad Zaki Afandi (Jidda: Dar al-Manara, 1992), 17f.

[139] Crozet, "Trajectoire," 302.

in the 1840s, while teaching at the Engineering School and before his nine-year sojourn in Europe, Mahmud had received public acclaim for almanacs that he published in Arabic, listing the times of prayer as well as planetary positions. These almanacs appealed to a society in which the value of timely prayer existed alongside the credibility of astrology. They were widely appreciated in mosques; indeed, it was this effort that first earned Mahmud the moniker *al-falaki*, "the astronomer," by which he was known for the rest of his life.[140] The difference between the significance of Mahmud's almanacs and the insignificance – for Muslim practice in Egypt – of his work on the Arabian calendar and the Prophet's *mawlid* was not only a function of language. In Mahmud's lifetime, the techniques of Orientalism, and indeed the position of a state astronomer, had very limited influence over the interpretation of the Qur'an or the celebration of a *mawlid* in Egypt. Half a century later, however, a popular, early twentieth-century Arabic biography of Muhammad adopted Mahmud's date for the Prophet's *mawlid*, with little discussion and apparently on no other basis than the nineteenth-century astronomer's credibility.[141] How someone like Mahmud came to wield such religious authority is a question that I will later address in the context of the transformation of science in the Arabic press in the late nineteenth and early twentieth centuries.

The Legacies and Erasures of Viceregal Astronomy

The most dramatic intervention of the viceregal astronomers in public timekeeping was not textual, like Mahmud's almanacs, but rather aural. In the summer of 1874, on the front page of the government bulletin *Vekâyi-i Mısriye* (*Egyptian Gazette*), Isma'il Bey explained a new sound, engineered by Mahmud Bey, that residents of Cairo would henceforth hear once a day. By rigging a lens to the firing mechanism of a cannon in the Cairo Citadel, Mahmud Bey had developed a time signal that the sun would trigger at noon.[142] After its inauguration made front-page news, the Citadel cannon quickly became an authoritative reference for timekeeping in Cairo. By the end of the year, a visiting British astronomer

---

[140] Crozet, "Trajectoire," 291.
[141] Muhammad al-Khudari Bey, *Nur al-yaqin fi sirat sayyid al-mursalin* (Cairo: Dar al-Basa'ir, 2008), 16. The author (d. 1927) was apparently unrelated to the Muhammad al-Khudari discussed in Chapter 1.
[142] Isma'il Mustafa Bey, "Hawadis-i dahiliye: fi sharh midfa' al-zawal," *Ruznâme-i Vekâyi-i Mısriye*, 24 Rabi-ul-Sani 1291 (9 June 1874), 1.

reported that residents of the city set their watches to noon when they heard the blast.[143]

In the echo of the Citadel cannon, we can hear the most consequential legacy of viceregal astronomy: science in Egypt had a new voice. Those who spoke with this voice of science had acquired the ability and the prerogative to do so in ways that differentiated them from Muhammad al-Khudari or his students. Their education and their languages, their politics and their finances, and the locations in which they studied and worked were all different. As their use of a cannon in the traditional center of government in Cairo suggests, the prominence of the vicere-gal astronomers came not from their place in Islamic scholarly circles, but rather from the material resources, publicity, and power that came with service to the state. Thus, in the very article in which *al-Muqtataf* lamented Mahmud's decision to write in French rather than Arabic, the journal was nevertheless able to assume that it "does not need to spread word of the excellence of Egypt's two famous astronomers and scholars, since their excellence has become known through the lands."[144] Indeed, Egyptians need not have read Mahmud or Isma'il's work, or even have heard the Citadel cannon, to have known who they were when this article appeared in 1885. The *falaki*s were at the height of their careers in the early 1880s, each serving in high-ranking government positions. Mahmud received the title of Pasha in 1881, served as Minister of Public Works in the short-lived government of the summer of 1882, and was subsequently Minister of Public Instruction from 1884 until his sudden death in 1885. Isma'il Bey was entrusted with reorganizing the Engineering School in 1883, and became Isma'il Pasha upon his retirement in 1887.[145]

It was no accident that the two astronomers were at the peak of their influence in the crucial years of crisis around the 'Urabi Revolution and onset of the British Occupation in 1882. As one study of the 'Urabi movement has observed, Mahmud and Isma'il were typical of Egypt's "European-trained experts" in their consistent support for Tawfiq Pasha,[146] whom the European powers had forced the Sultan to install as viceroy in 1879 (deposing Ismail Pasha), and whom the British Occupation restored to power after crushing the 'Urabi Revolution.

---

[143] Quoted in Jessica Ratcliff, *The Transit of Venus Enterprise in Victorian Britain* (London: Pickering & Chatto, 2008), 187 n. 26.

[144] "Hadaya wa-taqariz," 282.

[145] Zéki Bey, "Notice Biographique," 12. On Isma'il Pasha's retirement in 1887, see DWQ 75-043252.

[146] Alexander Schölch, *Egypt for the Egyptians! The Socio-Political Crisis in Egypt 1878–1882* (London: Ithaca Press, 1981), 130.

Figure 2.6 Mahmud Pasha al-Falaki at the end of his career, in 1882.
Bibliothèque nationale de France.

Mahmud Bey served in the directorate of the Land Register Office starting in 1879, overseeing cadastral work intended to prop up the government's perilous financial situation.[147] It was as an ally of the viceroy that
the recently promoted Mahmud Pasha entered the 1882 government of
Isma'il Raghib Pasha (see Figure 2.6). And, in a little-noticed but fitting
penultimate chapter in a career of viceregal service, Mahmud Pasha sat
as president of the commission that prosecuted 'Urabi's alleged followers
in the Tanta area.[148]

The prominence that Mahmud and Isma'il "the astronomers" attained
as government officials was not anomalous, but consistent with their
long careers in viceregal service. Men like Mahmud and Isma'il not only
refashioned Egypt's place in the world through their measurements,
observations, and publications; the sum of these efforts helped to establish a new relationship between state and science in Egypt. Whereas

[147] Schölch, *Egypt for the Egyptians*, 119.
[148] "*Khitab rasmi bi-sha'n akhir wazifa fi al-miri li-Mahmud al-Falaki*," 10 Muharram 1300
(20 November 1882).

scholars of astronomy like Muhammad al-Khudari enjoyed occasional and informal patronage from political elites, viceregal astronomers lived their lives, from childhood education to pensioned retirement, in service to the Ottoman-Egyptian state, which granted them extraordinary technical as well as personal resources, and shaped – and was shaped by – their work.

Yet the important role that viceregal astronomers played in the creation of a new culture of state science in Egypt has been largely erased by the ways in which their legacy has been interpreted since Mahmud and Isma'il died (in 1885 and 1901, respectively). In Egyptian historical memory, the political dimension to the *falakis'* careers has received little attention, perhaps because the reality of the two Pashas as high-ranking government officials does not sit well with their image as astronomers. Instead, Egyptian historiography has tended to celebrate the *falakis* as exemplars of Egyptian national achievement in modern science, while stripping them of their original context as viceregal servants. In a 1966 biography of Mahmud al-Falaki, for example, Ahmad Dimirdash wrote, "He did not get mixed up in political streams, but was content to raise the country's banner (*rafaʿa rāyat al-bilād*) in the field of scientific research... he was the best scientist whom Egypt produced in the nineteenth century." An even more direct erasure of the *falakis'* original political context occurs in the work of Mahmud's grandson, Mahmud Salih al-Falaki, who published the first Arabic translation of the Alexandria *Mémoire* in 1966, and also re-translated his grandfather's study of the pyramids. Where his grandfather had written in French of "our august Viceroy [Said]" and "Our well-loved prince, His Highness Ismail Pasha," Salih al-Falaki put in Arabic, "the Egyptian Government" (*al-ḥukūma al-miṣriyya*).[149]

The erasure of its original, Ottoman-Egyptian context has hardly rendered Mahmud Pasha's legacy apolitical, however. Rather, as in much nationalist historiography in Egypt, this erasure has facilitated the deployment of the astronomer's authority in the service of new agendas.[150]

---

[149] Compare Mahmoud-Bey, *L'âge et le but des Pyramides*, 6, and Mahmoud-Bey, *Mémoire sur l'antique Alexandrie*, 5, with Mahmud al-Falaki, *Al-Zawahir al-falakiyya*, 42, and Mahmud al-Falaki, *Risala ʿan al-Iskandariyya*, 51.

[150] On the erasure of Egypt's Ottoman history as part of the shift from an Ottoman-Egyptian to Egyptian-nationalist conception of the ruling elite, see Ehud Toledano, "Forgetting Egypt's Ottoman Past," in *Cultural Horizons: A Festschrift in Honor of Talat S. Halman*, ed. Jayne Warner, vol. 1 (Syracuse, NY: Syracuse University Press, 2001), 150–67. On Egyptian nationalist historiography in broader context, see Yoav Di-Capua, *Gatekeepers of the Arab Past: Historians and History-Writing in Twentieth-Century Egypt* (Berkeley, CA: University of California Press, 2009); Israel Gershoni and James Jankowski, *Commemorating the Nation* (Chicago: Middle East Documentation Center, 2004); and Anthony Gorman, *Historians, State and Politics in Twentieth Century Egypt: Contesting the Nation* (London: RoutledgeCurzon, 2003).

Mahmud Salih al-Falaki, an economist and diplomat, also went to Paris in the employ of an Egyptian state – to serve as Nasser's ambassador to France. A full-page portrait of Nasser appears at the beginning of his translation of the Alexandria *Mémoire*, for which the governor of Alexandria wrote a preface, explaining the work's significance in light of the 1952 revolution and Nasser's desire for "an all-encompassing awakening, linking our present with our past, and returning to our country its place and its glory ('*izz*)."[151] For Mahmud the grandfather, such a marriage of his scientific work to a political agenda would hardly have been scandalous. The irony, in fact, is that the original study of Alexandria was conceived in just such a marriage. Mahmud, of course, could not have foreseen the specific, anticolonial meaning that his historical work would acquire in Nasser's day. Notwithstanding this difference, however, perhaps it was the very closeness of the Nasserite historiographical project to its viceregal predecessor that drove the translator of the 1960s to strike the rhetorical vestiges of viceregal patronage from his grandfather's oeuvre. While the ideology of the Egyptian state has changed dramatically since Mahmud's lifetime, the distinctly intimate relationship between the state and its scientific practitioners is an underlying continuity that runs from its origins in the viceregal period even to the present day. Indeed, as Khaled Fahmy notes in a recent exchange among historians of science in the Middle East, the longstanding subservience of the Egyptian medical profession to the state has been on stark display since 2009, as forensic specialists have consistently provided cover for the brutality of state security services. The murder of Khaled Said, which served as a rallying cry for the January 25th revolution, is only the most notorious case.[152]

But nationalism and its historiographical erasures are not the only means by which the enduring significance of viceregal astronomy has been distorted. Other narratives have occluded the political importance of viceregal astronomy not by celebrating its scientific achievements, but by minimizing them. Such narratives of failure began to be told about the ventures of viceregal astronomy not long after the commencement of the British Occupation. In part, the notion that viceregal astronomy had failed stemmed from the absence of people to follow in Mahmud and Isma'il's footsteps and maintain the institutions, particularly the observatory, they had established. By the end of the nineteenth century, *al-Muqtataf* could refer to the 'Abbasiyya Observatory as "effectively non-existent" (*fi ḥukm al-'adam*)."[153] But a narrative of failure also developed around Mahmud

[151] Mahmud al-Falaki, *Risala 'an al-Iskandariyya*, 11.
[152] Khaled Fahmy, "Dissecting the Modern Egyptian State," *IJMES* 47 (2015), 559.
[153] "*Marsad Misr al-falaki*," *al-Muqtataf* 19 (1895), 778.

and Isma'il themselves. Captain Henry Lyons, one of the key figures to establish British cartography in Egypt under the Occupation, wrote off the work of his viceregal predecessors in a few sentences. Isma'il's measuring device, he observed, "was brought to Egypt about 1867, but was not, so far as I can ascertain, seriously taken into use."[154] Mahmud's astronomical map, meanwhile, "contained very little detail, and a smaller scale would have sufficed. ... The control by the astronomically fixed positions was not exact."[155]

Lyons' critique continues to resonate in recent scholarly accounts, in which the limitations of viceregal astronomy have exemplified the shortcomings of nineteenth-century efforts to establish modern scientific institutions in Egypt.[156] But such an account misses much of the significance of the new astronomers. Only by isolating the technical work of the viceregal astronomers from its political context can we judge their legacy in terms of the scale of a map or the use of an instrument. If Isma'il's base-bar was never used for the astronomical map; if his meridian circle did more for the map of France than it did for the map of Egypt; and if Mahmud's cartography did not satisfy the standards of a late nineteenth-century British surveyor, none of this seems to have mattered very much to Mahmud, Isma'il, or the government that they served and that promoted and rewarded them time and again. On the path of simultaneous servitude and advancement that the viceregal state placed before select subjects, few were more successful than the two astronomer-pashas, Mahmud and Isma'il. In return, they worked to shape and to achieve that state's aspirations to join the "scientific concert of Europe," and to help it survive the crises of the late 1870s and 1880s. The progeny of this symbiotic relationship, more than all the maps and instruments, was a new set of sites for authoritative knowledge-making in Egypt.

★ ★ ★

The preceding two chapters have mostly explored the worlds of scholarly and viceregal astronomy as separate terrains. Whereas the practice of scholarly astronomy took shape around places, problems, techniques, resources, and virtues particular to the community of 'ulama', viceregal astronomy was forged in the relationship between a new kind of Egyptian state and its functionaries. While "scholarly" and "viceregal" are ultimately heuristic terms – neither people nor science may be reduced to a particular social context – they nevertheless point to

---

[154] Lyons, *History of Surveying in Egypt*, 25.
[155] Lyons, *History of Surveying in Egypt*, 26.
[156] E.g., Crozet, "Trajectoire," 301.

real differences in the work of "astronomy" and its significance in these different contexts. Isma'il al-Falaki never wrote a pedagogical commentary on a *zīj*; Muhammad al-Khudari never traveled to Spain to observe the dilation of metals in a geodetic instrument. For Khudari, the study of celestial motion deepened one's sense of subservience to God and enabled the measurement and interpretation of time. Isma'il and Mahmud's astronomy did not lack for intersections with religious practice and temporality, but these intersections were part of a project that sought to define a territory, "Egypt," in new spatial and historical terms.

While the significant differences between Muhammad al-Khudari and Isma'il or Mahmud al-Falaki are an important part of the story of astronomy in late Ottoman Egypt, they hardly tell the story in its entirety. From publishing almanacs and building a public time signal, to reinterpreting a Qur'anic prohibition and the date of the Prophet's birth, viceregal astronomers did more than simply encroach on scholarly astronomy's separate terrain. They also borrowed directly from scholarly astronomical knowledge and practices. Moreover, this borrowing went in both directions, as a number of 'ulama' trained in scholarly astronomy became interested in adapting astronomical technologies and texts that were newly available in Egypt in the nineteenth century, often for reasons wholly unrelated to viceregal state-building. If we are to understand the role of astronomy not only in the forging of a new relationship between science and state, but also in the production of new kinds of religious authority, it is imperative that we understand how the practices of viceregal servants and the practices of 'ulama' increasingly overlapped with and modified each other. The following two chapters turn to this less dichotomous, more mutually constitutive relationship between Egypt's astronomers, beginning with a simple question that Mahmud al-Falaki faced when building the Citadel cannon: what time is "noon"?

Part II

Objects of Translation

3   Positioning the Watch Hand: 'Ulama' and the Making of Mechanical Timekeeping in Cairo

Writing in Cairo in 1737, the Damascene scholar 'Abd al-Latif al-Dimashqi commented on a new fashion among the city's residents:

> These watches have spread, and are tucked in most people's collars. Their forms and types have varied, and gone far in making them accurate and benefiting their craftsmen. However... their watch-hands are not free of earliness or delay, unlike the motion of the rotating [celestial] sphere, for that is the work of the Wise and Knowing.[1]

The irregularity of mechanical timepieces posed a problem that Dimashqi knew how to solve. In a handbook entitled *The Clearest Method for Correcting the Position of the Watch Hand (al-Manhaj al-Aqrab li-Tashih Mawqi' al-'Aqrab)*, Dimashqi became the first of a number of 'ulama' in eighteenth- and nineteenth-century Cairo who developed a new genre of scholarly literature: manuals and tables concerning "the position of the watch hand" or "where the watch hand falls" (*mawqi' 'aqrab al-sā'a*, pl. *mawāqi' 'aqārib al-sā'āt*). These manuals and tables, which emerged from the discipline of *mīqāt* (astronomical timekeeping), have been left almost unexamined by historians.[2] Understanding why they were written and how they were used sheds light on material, intellectual, and social continuities – as well as ruptures – that lie between the Ottoman and late Ottoman periods.

---

[1] 'Abd al-Latif al-Dimashqi, *al-Manhaj al-Aqrab li-Tashih Mawqi' al-'Aqrab*, Isl. Ms. 808, Special Collections Library, University of Michigan, Ann Arbor, p. 24.

[2] On *mīqāt*, see Sonja Brentjes and Robert G. Morrison, "The Sciences in Islamic Societies," in *The New Cambridge History of Islam*, Vol. 4, ed. Robert Irwin with William Blair (Cambridge: Cambridge University Press, 2011), 594–95; A.J. Wensinck and D.A. King, "Mīḳāt," *Encyclopaedia of Islam*, 2nd edition, ed. P. Bearman et al. (Brill Online); and François Charette, *Mathematical Instrumentation in Fourteenth-Century Egypt and Syria: The Illustrated Treatise of Najm al-Din al-Misri* (Leiden: Brill, 2003). Several of the texts examined in this article are very usefully treated in King, *In Synchrony*. However, King's analysis is focused on the mathematical content of the tables.

This chapter, along with the chapter that follows it, ventures inside an oft-overlooked world of scholars who worked with new scientific techniques, yet were not part of the new social groups or state institutions that constituted viceregal science. In fact, such scholars complicate the binary opposition of "viceregal" and "scholarly" science that the previous chapters depicted. The gulf between these milieus was admittedly substantial: astronomy, for bureaucrats like Mahmud al-Falaki or Ismail al-Falaki, comprised a different set of problems, to be addressed with different resources and in different spaces, than for ʿulama' like Muhammad al-Khudari. At the same time, there was also overlap, exchange, and circulation between these milieus. Such connections become especially evident if, in addition to following the movements of individuals and their networks, we follow the history of certain tools with which they worked.

The earliest study and use of mechanical timepieces in Cairo depended on the mathematical knowledge of ʿulama', rather than on Ottoman or colonial projects of modernization and the new classes of scientific practitioners they produced. As clocks and watches became increasingly available among elite households and ʿulama' in eighteenth-century Cairo, scholars of mīqāt played a crucial role in developing practices that linked these fashionable objects with the distinctive temporality known as ġurūbī sāʿat or "alaturka" time.[3] Far from rendering scholarly astronomy obsolete, the rise of a new, European-imported timekeeping technology allowed certain ʿulama' to claim new relevance and expand the audience to which their knowledge spoke. In this way, mechanical timekeeping was a harbinger of certain modern developments – the expansion of literacy, mass communication, and state regulation of religion among them – which historians once interpreted as having marginalized the ability of ʿulama' to "speak for Islam," but which recent scholarship has shown allowed ʿulama' to exercise authority in new ways.[4]

The specific guidance that ʿulama' provided for using clocks and watches shaped a routine of mechanical timekeeping that was particularly Ottoman and Islamic, while also increasingly concerned with precision. Exact, mathematical definition of the Islamic prayer intervals had long been a value of the highly elite, scholarly tradition of mīqāt. Manuals and tables on positioning the watch hand, however, provide some of the first evidence for the entrance of this virtue into wider social practice,

---

[3] On "alaturka" time and its meaning in late Ottoman society, see Wishnitzer, *Reading Clocks* and "'Our time': On the Durability of the Alaturka Hour System in the Late Ottoman Empire," *International Journal of Turkish Studies* 16 (2010): 47–69.

[4] Zeghal, "Religion and Politics in Egypt"; Zaman, "Commentaries, Print, and Patronage"; Zaman, *Ulama in Contemporary Islam*.

as the correction of timepieces and the performance of prayers became measured in minutes and even seconds. This popularization of *mīqāt* accelerated in the late Ottoman period, but ultimately not in the hands of 'ulama'. When the viceregal astronomers – Mahmud Hamdi, Ismail Mustafa, and their students and colleagues like Sulayman Hallawa – began to publish prayer tables and build time signals, they were adapting practices that 'ulama' had long taught, while marginalizing the 'ulama' themselves.

Recovering the role of 'ulama' in this process should change our perspective on the rich portrait of Ottoman temporality that has emerged in recent years. As Avner Wishnitzer has shown, early modern Ottomans used clocks as elements of a temporal culture that served the society's patrimonial relations. Mechanical timekeeping, by itself, did not disrupt well-established rhythms in which indefinite waiting expressed devotion to a master (or beloved), and power entailed the ability to command another's time at will.[5] Wishnitzer's work builds on a generation of scholarship in time studies that has moved away from E.P. Thompson's famous distinction between preindustrial "task-orientation" and industrial "time-discipline."[6] Rather than seeing the clock as the instrument of a single experience of modern temporality whose driving force is industrial capitalism, scholars have drawn attention to the differentiation of modern temporalities according to class, gender, and political – especially colonial – contexts.[7] Meanwhile, the old narrative of time "standardization" has come under a related set of critiques. In Egypt, for example, the prototypical technologies of standardization – not only the clock, but the telegraph and the railroad in particular – seem to have given birth instead to the notion of "Egyptian time" as a substandard approximation.[8]

If the advent of mechanical timekeeping did not disrupt the patrimonialism of Ottoman society in the eighteenth and early nineteenth centuries, however, neither was it a value-neutral development. By linking new objects with old routines, 'ulama' asserted their authority to determine the usage of such objects. In doing so, they sought to propagate a

---

[5] Wishnitzer, *Reading Clocks*. For a broader chronological perspective, see François Georgeon and Frédéric Hitzel, eds., *Les Ottomans et le temps* (Leiden: Brill, 2012).

[6] E.P. Thompson, "Time, Work-Discipline, and Industrial Capitalism," *Past and Present* 38 (1967): 60.

[7] Paul Glennie and Nigel Thrift, "Reworking E.P. Thompson's 'Time, Work-Discipline, and Industrial Capitalism,'" *Time and Society* 5 (1996): 275–99.

[8] Barak, *On Time*. For a global perspective on the "pluralization" of time in the same period, see Vanessa Ogle, *The Global Transformation of Time: 1870–1950* (Cambridge, MA: Harvard University Press, 2015), and Vanessa Ogle, "Whose Time Is It? The Pluralization of Time and the Global Condition, 1870s–1940s," *American Historical Review* 118 (2013): 1376–402.

particularly precise way of measuring not only time but also the fulfill-ment of religious duties. Far from making time "Egyptian," they believed they were making time correct.

The Mechanical Timepiece in Ottoman Cairo: Material Fashion and Scholarly Preoccupation

By the time Dimashqi wrote his handbook in 1737, mechanical time-pieces were available to both scholarly and political elites in Cairo by way of Ottoman, Mediterranean, and global pathways of trade and travel. In Istanbul, European ambassadors had been bringing clocks as gifts to the Ottoman court, often accompanied by clockmakers to main-tain them, since at least the sixteenth century.[9] Indeed, the mechan-ical devices were so popular at court that it became almost de rigueur for diplomats to bring a trove of them to distribute as gifts.[10] By the eighteenth century, European travelers were bringing clocks across the Mediterranean for scientific purposes, particularly astronomical obser-vation.[11] Ottoman scholars and craftsmen also produced mechanical timepieces. Taqi al-Din ibn Ma'ruf composed a treatise on the construc-tion of mechanical clocks, which he used in the short-lived Ottoman observatory of the 1570s.[12] However, although domestic production of clocks and watches continued, most of the devices in use throughout the Ottoman period were imported.[13] In the eighteenth century, a small number of English clockmakers dominated the trade, selling, accord-ing to a French consul in Greece at the end of the century, over 12,000 watches annually in Ottoman locales around the Mediterranean, includ-ing Salonika, Istanbul, the Syrian lands, and Egypt.[14] The most popu-lar maker of watches for the Ottoman market, the Englishman George Prior, exported more than 40,000 of them before his death in approxi-mately 1815. (This figure excludes the numerous watches produced by

---

[9] Otto Kurz, *European Clocks and Watches in the Near East* (Leiden: E.J. Brill, 1975). By the sixteenth century, mechanical clocks were also available in Persia. Stephen Blake, *Time in Early Modern Islam* (Cambridge: Cambridge University Press, 2013), 74.

[10] Kurz, *European Clocks and Watches in the Near East*, 30.

[11] Brentjes, "Astronomy a Temptation?" ch. 8, 26.

[12] Avner Ben-Zaken, *Cross-Cultural Scientific Exchanges in the Eastern Mediterranean, 1560–1660* (Baltimore: Johns Hopkins University Press, 2010), 10–21.

[13] Kemal Özdemir, *Ottoman Clocks and Watches* (Istanbul: Creative Yayıncılık, 1993), 113. On the economics of Ottoman clock importation, see Ian White, *English Clocks for the Eastern Markets* (Sussex: Antiquarian Horological Society, 2012).

[14] Louis-Auguste Félix de Beaujour, *A View of the Commerce of Greece, Formed after an Annual Average, from 1787 to 1797* (London: H.L. Galabin, 1800), 241. On English trade in the Levant in the eighteenth century, see Ralph Davis, *Aleppo and Devonshire Square* (London: Macmillan, 1967).

less reputable manufacturers who put his name on their work.)[15] Prior is said to have remarked that "the streets of the Turkish cities must be all paved with English watches."[16]

Although historians have not studied watch-use in Egypt prior to the late nineteenth century, histories of eighteenth-century Ottoman cities point to a growing market for such a fashionable object, and for manuals on how to use it.[17] It was in this period that middling social groups such as barbers, scribes, soldiers, and farmers began to participate in elite cultural practices, even composing their own histories.[18] In Cairo especially, manuscript culture thrived on the growth of a class of people who were moderately educated without being scholars.[19] Within the learned elite, meanwhile, study of the "uncommon sciences" like astronomy connected Arabophone scholars with Ottoman and Mamluk patrons.[20] A "florescence" of rigorous scholarly activity was particularly evident in logic, for which one observer noted an "inordinate enthusiasm" among 'ulama' in Egypt.[21] The year 1737 specifically, when Dimashqi authored his handbook, was a moment of economic and cultural recovery from a period of hardship earlier in the 1730s.[22] Manuals on positioning the watch hand were a part of this broader social context, not an isolated phenomenon in the intellectual life of scholars.

Mechanical timepieces, though not found literally in "most people's" collars, as Dimashqi wrote, were certainly present beyond the upper crust of eighteenth-century Cairene society. We are not dealing only with people who could afford clocks like the Swiss piece by Berthout that French savants seized from an abandoned Mamluk palace at the end of the century.[23] A 1781 copy of Dimashqi's handbook was written and

---

[15] Kurz, *European Clocks*, 78–79.

[16] Beaujour, *A view of the Commerce of Greece*, 241. The quotation is de Beaujour's, apparently paraphrasing a comment that he had heard attributed to Prior.

[17] One area of Ottoman clock history in the Arab provinces that has received attention is the building of clock towers, but this Hamidian project came only in the late nineteenth century. Zeynep Çelik, *Empire, Architecture, and the City: French-Ottoman Encounters, 1830–1914* (Seattle, WA: University of Washington Press, 2008), 146–53.

[18] Dana Sajdi, *The Barber of Damascus: Nouveau Literacy in the Eighteenth-Century Ottoman Levant* (Stanford, CA: Stanford University Press, 2013).

[19] Nelly Hanna, *In Praise of Books: A Cultural History of Cairo's Middle Class, Sixteenth to the Eighteenth Century* (Cairo: AUC Press, 2004).

[20] Murphy, "Improving the Mind."

[21] Khaled El-Rouayheb, "Opening the Gate of Verification: The Forgotten Arab-Islamic Florescence of the 17th Century," *IJMES* 38 (2006): 263–81, with quotation on p. 268.

[22] Jane Holt Murphy, "Ahmad al-Damanhuri and the Utility of Expertise in Early Modern Ottoman Egypt," *Osiris* 25 (2010): 85–104. On the economic history of eighteenth-century Cairo, see André Raymond, *Artisans et commerçants au Caire au XVIIIe siècle* (Damascus: Institut français de Damas, 1974).

[23] Murphy, "Locating the Sciences," 564.

owned by one "Mahmud, clerk in the office of the Volunteers, son of the late Sulayman Effendi, son of the late 'Abd al-Rahman Effendi, son of the late Mustafa Effendi."[24] This genealogy, along with the position of clerk (*khalīfe*), indicates membership in the Ottoman bureaucracy.[25] However, the Volunteers (*gönüllüyān*), a relatively large regiment in the middle of the eighteenth century, were several rungs down on the Ottoman military ladder; appointment to their office likely did not bring the highest level of prestige.[26] The passing of Dimashqi's work through the hands of someone like Mahmud suggests that mechanical timekeeping was only "elite" in a relatively broad sense – not the domain of one or two beys with expensive tastes, but of the significant number of people in eighteenth-century Cairo who possessed a certain degree of education and means. Given the sheer volume of clocks and watches imported, this relatively broad class of consumers makes sense.

The use of timepieces was also a part of the material culture of 'ulama' specifically. As Nelly Hanna has described, many scholars in eighteenth-century Cairo practiced trades (even giving lessons in their shops), while others married into families of craftsmen.[27] The world of books was intertwined with the world of crafts and commerce. The well-known polymath Hasan al-Jabarti (d. 1774) had at least two acquaintances known as *al-sā'atjī* (Tur. *saatçi*), "the horologist." And the famous chronicles of his son, 'Abd al-Rahman al-Jabarti, describe a mid eighteenth-century scholar named Husayn al-Mahalli who "had a shop near the gate of al-Azhar, where he would make money by selling clocks (*manākīb*) for knowing times" (*li-ma'rifat al-awqāt*).[28]

Historians have generally understood Jabarti's *manākīb* to mean "almanacs," but this is apparently a contextual guess, with no basis in the etymology of the word.[29] In fact, Mahalli was selling clocks,

---

[24] MS DM 812, ENL. The colophon identifies the copyist as Mahmud *khalīfat bāb-ı camalīyān*. Cataloguers have misinterpreted the italicized phrase as part of the copyist's name. (See *OALT* # 325 and King, *Survey of the Scientific Manuscripts*, D96.) In fact, *camalīyān* is a corruption, peculiar to Egypt, of *gönüllüyān*, the "Volunteers" of the Ottoman cavalry. The term appears with its proper Turkish spelling in the manuscript's statement of ownership. *Camalīyān* is apparently a play on the association of the *gönüllüyān* with their camels; thus, "Cameleers" instead of "Volunteers." See Thomas Philipp and Guido Schwald, *A Guide to 'Abd al-Rahmān al-Jabartī's History of Egypt* (Stuttgart: Franz Steiner Verlag, 1994), 336.

[25] Mahmud owned other scientific manuscripts, including a work on weights and measures. See King, *Survey of the Scientific Manuscripts*, D96 and Pl. CXb.

[26] Hathaway, *Politics of Households*, 33, 40.

[27] Hanna, *In Praise of Books*, 41–44.

[28] Possibly prayer times specifically. 'Abd al-Rahman al-Jabarti, *'Aja'ib al-Athar fi al-Tarajim wa-l-Akhbar*, ed. Ibrahim Shams al-Din (Cairo: Maktabat Madbuli, 1997), 1:219–20.

[29] Philipp and Perlmann, in their translation of al-Jabarti, give "almanacs," without explanation. *'Abd al-Rahmān al-Jabartī's History of Egypt*, ed. Thomas Philipp and Moshe

not books. In the 1880s, 'Ali Mubarak quoted Jabarti's passage about Mahalli with a different pluralization of the strange term: instead of *manākīb*, Mubarak used *minkābāt*.[30] The latter is a well-attested corruption of the term *bankāmāt* (sing. *bankām*),[31] which was used in Arabic to denote water clocks since at least the twelfth century (from the Persian *bingān*, or *pingān*, both of which denote either a bowl or clepsydra).[32] Closer to Jabarti and Mahalli's milieu, early nineteenth-century *mīqāt* texts from Egypt employ the term *manākib* (very close to al-Jabarti's *manākīb*) in technical contexts where they can only refer to clocks.[33] Even in the late nineteenth century, when the modern term *sā'a* was already well established, a pocket-watch could still be called a *minjāna* (at least in Morocco).[34]

To be clear, while the term *manākīb* and its variants all referred to clocks, they were not all mechanical, even in the eighteenth and nineteenth centuries. And texts "on the position of the watch hand" generally used the term *sā'a*. Nevertheless, the history of the older terms' usage highlights the dynamism of Ottoman timekeeping's material culture, and the many ways in which 'ulama' participated in that culture. While the word's etymology evokes the water clock, in the sixteenth century, Taqi al-Din used *bankāmāt* to denote all kinds of clocks, including his mechanical devices. Muddying the waters, a seventeenth-century lexicon records that "timekeepers" (*ahl al-tawqīt*) use the term "*bankāmāt*" specifically for sand clocks.[35] In the same period, the Ottoman encyclopedia *Kashf al-Zunun 'an Asami al-Kutub wa-l-Funun* (*The Removal of Doubt from the*

Perlmann (Stuttgart: Franz Steiner Verlag, 1994), 1:360. Hanna and Murphy both remark that Mahalli made his living as a copyist and bookseller. Hanna, *In Praise of Books*, 41; Murphy, "Improving the Mind," 136. Perhaps this idea originated in a footnote to a 1958 edition, which glosses *manākib* as *taqāwim* (ephemerides, or calendars). See 'Abd al-Rahman al-Jabarti, '*Aja'ib al-Athar fi al-Tarajim wa-l-Akhbar*, ed. Hasan Muhammad Jawhar et al. (Cairo: Lajnat al-Bayan al-'Arabi, 1958–1967), 2:140.

[30] Mubarak, *al-Khitat*, 15:25.

[31] Ahmad ibn Muhammad ibn 'Umar al-Khafaji, *Shifa' al-Ghalil fi ma fi Kalam al-'Arab min al-Dakhil*, ed. Muhammad Kashshash (Beirut: Dar al-Kutub al-'Ilmiyya, 1998), 94.

[32] Donald Hill, *Arabic Water-Clocks* (Aleppo: University of Aleppo Institute for the History of Arabic Science, 1981). Ben-Zaken suggests that *bankām* had an older, Latin origin. Ben-Zaken, *Cross-Cultural Exchanges*, 19.

[33] Al-Khudari, *Sharh al-Lum'a fi Hall al-Kawakib al-Sab'a*, Isl. Ms. 722, Special Collections Library, University of Michigan, p. 142; Ibrahim Shihab al-Din ibn Hasan, *Sharh 'ala Muqaddimat al-Shaykh Mahmud fi 'Ilm al-Miqat*, MS ṬM 225, ENL, fols. 11v (where *minkām* and *minkāb* are equated), 16r, and 21r. In context, the devices appear to be sand clocks.

[34] Ibrahim al-Yaziji, *As'ila ila Majallat "al-Bayan" wa-Ajwibat al-Shaykh Ibrahim al-Yaziji 'alayha*, ed. Yusuf Khuri (Beirut: Dar al-Hamra', 1993), 88.

[35] Khafaji, *Shifa' al-Ghalil*, 94. The use of *minkām* (another variant) for sand clock occurs as early as the thirteenth century; see George Saliba, "An Observational Notebook of a Thirteenth-Century Astronomer," *Isis* 74 (1983), 398n11.

Names of Books and Sciences) largely paraphrases Taqi al-Din's sixteenth-century discussion in using *bankāmāt* to denote all kinds of timepieces, and expresses a preference for the mechanical or "wheeled" type.[36] And al-Jabarti himself used another variant, *minkāmāt*, when remarking on clocks – probably mechanical pieces – brought to Egypt by the savants of the French invasion.[37]

Manuals and tables on positioning the watch hand should thus be read in part as emerging out of a long history of various modes of interaction between 'ulama' and diverse timekeeping devices, as well as in light of the spread of the new, mechanical variety across a certain range of the urban population. In eighteenth-century Cairo, 'ulama' sold clocks in front of al-Azhar, used them in the practice of astronomy, and examined them with curiosity during the French invasion. Were it not for the notion that Muslim scholars had, by this period, long ceased to possess a dynamic scientific tradition, it would hardly be surprising that certain 'ulama' also offered instruction on the correct way of using them.

The scholars who offered such instruction came to Cairo from several parts of the Ottoman Empire. While Dimashqi was from Damascus, Ramadan ibn Salih al-Khawaniki (d. 1745–46), author of a *Tables of the Position of Watch Hands at Prayer Times (Jadawil Mawqi' al-Sa'at 'ala Hasab Awqat al-Salawat)*, was from a village in the Nile Delta, as was Khalil al-'Azzazi, author of a mid nineteenth-century *Introduction on Working the Position of Watch Hands according to the Shari'a [Prayer] Intervals for any Latitude (Muqadimma fi 'Amal Mawaqi' 'Aqarib al-Sa'at 'ala Qadr al-Hisas al-Shar'iyya li-Kull 'Ard)*.[38] Muhammad ibn 'Abd

---

[36] Katip Çelebi, *Kashf al-Zunun 'an Asami al-Kutub wa-l-Funun* (Beirut: Dar Ihya' al-Turath al-'Arabi, 1995), 1:255. The passage closely follows the introduction to Taqi al-Din's treatise, *The Brightest Stars in Wheeled Clocks*, wherein "wheeled" (*dawriyya*) denotes mechanical devices, as opposed to sand or water clocks. On "wheeled" clocks, see Gerhard Dohrn-van Rossum, *History of the Hour: Clocks and Modern Temporal Orders*, trans. Thomas Dunlap (Chicago: University of Chicago Press, 1996), 54. For the original passage in Taqi al-Din, see Sevim Tekeli, *16'ıncı Asırda Osmanlılarda Saat ve Takiyüddin'in "Mekanik Saat Konstrüksüyonuna Dair En Parlak Yıldızlar" Adlı Eseri* (Ankara: Ankara Üniversitesi Basımevi, 1966), 219–20.

[37] Moreh understands *minkāmāt* to mean water clocks, in keeping with the older Arabic usage of *bankāmāt*. *Napoleon in Egypt: al-Jabarti's Chronicle of the French Occupation, 1798*, trans. Shmuel Moreh (Princeton, NJ: Markus Weiner Publishers, 2004 [1975]), 110. However, clepsydras do not appear in *La Décade égyptienne* or *La Description de l'Égypte*, except in reference to ancient Egyptian technology. Describing their own clocks, the savants only mention mechanical timepieces. M. Nouet, "Observations astronomiques faites en Égypte pendant les années VI, VII, et VIII [1798–1800]," *La Description de l'Égypte: État moderne* (Paris: L'Imprimerie impériale, 1809), 1:1, accessed 17 March 2015, descegy.bibalex.org.

[38] Ramadan b. Salih al-Khawaniki, *Jadawil Mawqi' al-Sa'at 'ala Hasab Awqat al-Salawat*, Isl. Ms. 808, Special Collections Library, University of Michigan. See also MSS DM 812 and Zakiyya 822, ENL; and Khalil al-'Azzazi, *Muqadimma fi 'Amal Mawaqi' 'Aqarib*

al-Rahman al-Nabuli, author of an 1867–68 *Almanac of the Position of Watch Hands at the Beginning of Prayer Times, [Organized] by the Coptic Months (Natijat Mawqi' 'Aqrab al-Sa'at 'ala Qadr Hisas Awa'il Awqat al-Salawat fi al-Shuhur al-Qibtiyya)*, was from a town near Tunis.[39]

All of these scholars had come to Cairo because of the appeal of the al-Azhar mosque, where they studied and eventually taught in diverse fields of knowledge. Some, of course, were known primarily for their astronomical skills. Khawaniki, a student of the prominent astronomer Ridwan Effendi, produced numerous works on different kinds of time-keeping instruments as well as the prediction of eclipses and planetary positions.[40] Others, however, had broader reputations. Dimashqi was renowned as "a believer consistent in religion, worship, and probity."[41] While regarded as an authority on astronomy (*al-falak wa-l-hay'a*), he also taught students of law (*fiqh*). He was the shaykh of the Syrian pavilion (*riwāq*) at al-Azhar before he took an ascetic turn, growing long hair, performing the hajj annually, and limiting his social contact.[42]

The textual history of Dimashqi and Khawaniki's works suggests that they remained in demand among similarly trained scholars through the eighteenth and nineteenth centuries. Dimashqi's *Clearest Method*, for example, exists in at least ten manuscript copies in addition to the 1737 autograph, several of which also include a copy of Khawaniki's tables.[43] The dates of these copies cover a range of decades into the nineteenth century. Nine of them are in Egyptian libraries; the tenth was probably acquired in Egypt.[44] Other than Mahmud of the Volunteers' Office, the known copyists of the manuscripts include Shaykh 'Uthman al-Wardani (1796–97), Ahmad ibn Ibrahim al-Sharbatli (1833–34), and Ridwan ibn Muhammad ibn Sulayman (1856).[45] Wardani and Sharbatli were also

al-Sa'at 'ala Qadr al-Hisas al-Shar'iyya li-Kull 'Ard, MS TR 204, ENL. For Khawaniki's biography, see Mubarak, *al-Khitat*, 10:90. For 'Azzazi's, see Mubarak, *al-Khitat*, 15:10.

[39] Al-Nabuli, *Natijat Mawqi' 'Aqrab al-Sa'at*, MS 317 al-Saqqa 28898, Azhar. I have not been able to examine a text on "the position of watch hands" attributed to Hasan al-Jabarti (d. 1774). See King, *In Synchrony*, 1:333.

[40] Mubarak, *al-Khitat*, 10:90.

[41] Muhammad Khalil ibn 'Ali al-Muradi, *Silk al-Durar fi A'yan al-Qarn al-Thani 'Ashar* (Beirut: Dar al-Kutub al-'Ilmiyya, 1997), 3:117.

[42] Muradi, *Silk al-Durar*, 3:117.

[43] In addition to the autograph and eight copies listed in *OALT* #281, see MS Zakiyya 822, ENL, and Isl. Ms. 808, Special Collections Library, University of Michigan. For copies including al-Khawaniki's tables, see e.g., MSS DM 812 and Zakiyya 822, ENL, and Isl. Ms. 808, Special Collections Library, University of Michigan.

[44] The Michigan copy belonged to Max Meyerhof, who lived in Egypt in the early twentieth century.

[45] MSS DM 1104 and TR 286, ENL, and Isl. Ms. 808, Special Collections Library, University of Michigan, respectively.

involved in the copying of older astronomical texts (from the fifteenth and seventeenth centuries, respectively).[46] Wardani wrote his own works on various aspects of *mīqāt*, including almanacs, and had at least one student who did the same in the early nineteenth century.[47] The most prominent of these copyists, however, was Ridwan ibn Muhammad, who was best known for his knowledge of variant Qur'an readings (*qirā'āt*).[48]

In other words, the genre of manuals and tables on "the position of the watch hand" did not emerge – whether in its material or its intellectual context – from the milieu of the new scientific institutions established in Istanbul beginning in the late eighteenth century and in Egypt beginning in the early nineteenth century.[49] Neither the presence of the timepieces nor the knowledge of how to use them was a consequence of the French invasion, the *nizam-ı cedid*, or the explicitly modernizing initiatives of Mehmed Ali Pasha and his successors. Mechanical timepieces were a longstanding feature of the material culture of Ottoman Cairo. Whether imported commercially, received as diplomatic gifts, or produced domestically, they began to appear in the elite and educated circles of the city no later than the 1730s. The growth of a literature on how to use these new objects was the work of 'ulama' for whom the correct practice of timekeeping existed within an older milieu of scholarship that included astronomical knowledge alongside the study of law, the Qur'an, and other fields. At the intersection of these material and intellectual contexts, people in Cairo learned to use their new clocks and watches in a particular way.

Correcting the Watch: Making Ottoman Time Work

Manuals and tables "on the position of the watch hand" were premised on the Ottoman *ğurūbī sā'at* (sunset hour) method of timekeeping.[50] In

---

[46] Wardani was the copyist of an 1801 copy of al-Kawm al-Rishi's *al-Lum'a* (Houghton MS Arab 249). Al-Sharbatli appears as the scribe in an 1823 copy of a treatise on timekeeping, *Dustur Usul 'Ilm al-Miqat*, Isl. Ms. 760, Special Collections Library, University of Michigan.

[47] See King, *Survey of the Scientific Manuscripts*, D107; and Shaykh Sa'd ibn Ahmad al-'Abbani, *Mu'arraba li-Sana Shamsiyya 1242*, MS 371 Jawhari 42103, Azhar.

[48] I am assuming he was Ridwan ibn Muhammad ibn Sulayman "al-Mukhallilati" (ca. 1834–93). See Abu al-Khayr 'Umar ibn al-Murabah ibn Hasan ibn 'Abd al-Qadir al-Murati, ed., *Muqaddima Sharifa Kashifa lima Ihtawat 'alayhi min Rasm al-Kalimat al-Qur'aniyya wa-Dabtiha wa-'Add al-Ayy al-Munifa* (Ismailia: Maktabat al-Imam al-Bukhari, 1427 [2006]), 41–49.

[49] Adnan Adıvar, *La science chez les Turcs Ottomans* (Paris: Librairie Orientale et Américaine, 1939), 142; Ekmeleddin İhsanoğlu, *Science, Technology and Learning in the Ottoman Empire: Western Influence, Local Institutions, and the Transfer of Knowledge* (Aldershot: Ashgate, 2004); Crozet, *Sciences modernes*.

[50] For an overview, see Wishnitzer, *Reading Clocks*, 14.

this system, sometimes called "alaturka time," the day begins at sunset, which is always called 12:00. Mechanical clocks and watches keep equal hours, but must be adjusted daily in order to keep 12:00 aligned with the natural movement of sunset over the course of the year. But sunset is the only natural phenomenon to which the clock is pegged in this way. Unlike the sundial, which keeps apparent solar time throughout the day, the mechanical clock on the *ġurūbī sāʿat* system would show one time for apparent noon in February, and quite another in July. Both times, however, would be a function of the number of equal hours since sunset. For modern readers, this synthesis of mechanical and natural rhythms may seem almost like a reversal of the clock's "intended" purpose of measuring fixed, arbitrary units of time. For Ottomans, however, it was precisely through this adaptation that mechanical timekeeping came to make sense.

E.W. Lane, the British Orientalist who lived in Egypt during the 1820s and 1830s, describes a timekeeping practice that closely resembles *ġurūbī sāʿat* convention:

The Egyptians wind up and (if necessary) set their watches at sunset; or rather, a few minutes after; generally when they hear the call to evening-prayer. Their watches, according to this system of reckoning from sunset, to be always quite correct, should be set every evening, as the days vary in length.[51]

While Lane's description of people adjusting their watches by the call of the muezzin confirms the prevalence of *ġurūbī sāʿat* time in Egypt, it masks a great deal of history that made the synchronization of mechanical, ritual, and natural rhythms possible. By the time that Lane was in Cairo, some of the city's social groups had been using mechanical timepieces for almost a century. During this period, *ġurūbī sāʿat* had become an experience of temporality that was linked to the performance of specific practices by particular kinds of people.

In Istanbul, as Avner Wishnitzer has remarked, the "timekeeper house" (*muvakkithane*) was a crucial site for the regulation of *ġurūbī sāʿat* timekeeping.[52] In this space, often attached to a mosque, the *muvakkit* (Ar. *muwaqqit*), traditionally responsible for determining the times of prayer, also assumed responsibility for maintaining accurate clock time.[53] Ottoman watch-users, passing by the *muvakkithane*, would correct their timepieces according to the *ġurūbī sāʿat* system. There is little evidence, however, that *muvakkithanes* served the same, popular role in Cairo that

---

[51] Lane, *Manners and Customs*, 220.
[52] Wishnitzer, *Reading Clocks*.
[53] For an overview of the history of the *muwaqqit*, see King, *In Synchrony*, 1:631–75.

they did in Istanbul in this period. Manuals and tables "on the position of the watch hand" show that 'ulama' facilitated the *ğurūbī sā'at* use of mechanical timepieces by deploying their knowledge of astronomical timekeeping to establish a new practice – positioning the watch hand – in relation to two much older practices that were performed in Cairo not only at sunset, but throughout the day: measuring the altitude of the sun above the horizon, and listening to all five of the daily calls to prayer (*adhān*).

'Abd al-Latif al-Dimashqi's *Clearest Method* links the positioning of the watch hand with the first of these older practices. It enables its user to set a mechanical timepiece correctly, according to *ğurūbī sā'at* time, at any moment when the sun is visible. The handbook consists of two parts: a short introduction with brief instructions, followed by a lengthy set of tables. The tables give the time, in equal hours since sunset, as a function of the degree of the sun in the ecliptic and the angular altitude of the sun above the horizon. (See Figure 3.1 for an example of the tables.) The information incorporated into these tables was knowledge long available in Cairo, which had been a center for the elaboration of astronomical timekeeping (*mīqāt*) at least since the tenth-century scholar Ibn Yunus laid the foundation of the massive "Cairo Corpus" of tables.[54] Dimashqi appears to have relied on the astronomical parameters contained in this corpus to produce his own tables.[55]

Users of Dimashqi's manual need not have possessed the deep learning in astronomical timekeeping that its author did, but they must have had certain skills. Dimashqi himself reported that he wrote the manual at the request of "one of the inexperienced people delving into this craft (*ṣinā'a*),"[56] and he tried to make it as easy as possible for such a novice to use. For determining the position of the sun in the zodiac, for example, Dimashqi provided an arithmetical technique based on the correspondence between the Coptic calendar and the solar year. He also provided precise instructions for navigating the tables – everything down to the color in which he had labeled each column. Nevertheless, users of these tables must still have been able to measure the solar altitude, for which Dimashqi stipulated that one would need an altitude arc (*qaws al-irtifā'*), the graduated quadrant of which he suggested drawing on the back of the manual itself.[57] In other words, Dimashqi assumed not only that his

---

[54] For an overview, see King, *In Synchrony*, 1:206–8.

[55] King, *In Synchrony*, 1:333.

[56] Dimashqi, *al-Manhaj al-Aqrab*, Isl. Ms. 808, Special Collections Library, University of Michigan, p. 24.

[57] Dimashqi, *al-Manhaj al-Aqrab*, Isl. Ms. 808, Special Collections Library, University of Michigan, p. 24.

Figure 3.1 A table from a nineteenth-century copy of an eighteenth-century manual for the correction of timepieces in Cairo. The user would turn to this table after observing the sky with a quadrant and finding that the sun was 18° above the horizon. The top row of zodiacal houses (Virgo–Capricorn), the names of which are written upside down, is to be used with the leftmost column of numbers; the second row of houses (Aries–Sagittarius) corresponds with the rightmost column of numbers. 'Abd al-Latif al-Dimashqi, *al-Manhaj al-Aqrab li-Tashih Mawqi' al-Aqrab*, Table 18, p. 43, Isl. Ms. 808, University of Michigan Library (Special Collections Library), Ann Arbor.

reader would already know how to take a solar altitude using such an instrument, but also, and more remarkably, that his reader would be able to fashion the instrument himself.[58] Such tactile skill could hardly have been assumed of a complete novice. In a similar vein, it is also noteworthy that Dimashqi arranged the tables according to a convention that saved space but added a level of complexity. Instead of drawing a column for each zodiacal house, he combined complementary houses into one column, which should be read top-to-bottom for one house, and bottom-to-top for the other. This common shortcut relies on the fact that the day is of equal length, for example, when the sun is in the 25th of Aries (in the spring) and the 5th of Virgo (in the fall). Such a conventional arrangement of the tables made them considerably shorter and more elegant, but assumed that users were trained to understand them.

The overall impression is that "one of the inexperienced people delving into this craft" may not have been a master like Dimashqi, but he was still a student of astronomical timekeeping. It is even possible that this student was quite experienced in astronomy, and that the new "craft" into which he was delving was specifically the practice of horology. Perhaps the student who requested this manual was someone like Husayn al-Mahalli, a scholar trying to make a living in part by selling timekeeping devices. The manual was not open to all who could read (already a small audience). It required knowledge and skills that would likely have been acquired through study with a master teacher such as Dimashqi himself. The handbook thus represents Dimashqi's deployment of an Islamic scholarly tradition to enable the regulation of new timekeeping devices by a larger, but still closely defined, group of people who had basic training in this tradition.

But observation of the sun was only one, relatively technical practice through which 'ulama' could deploy their knowledge of astronomical timekeeping to facilitate the use of mechanical timepieces according to ǧurūbī sāʿat convention. In fact, most manuals and tables did not provide the correspondence between the position of the watch hand and the altitude of the sun, but rather between the position of the watch hand and the times of prayer. Such tables, though still assuming an audience that was literate and kept track of the precise calendar date, did not require observational skill. Thus, Muhammad ibn ʿAbd al-Rahman al-Nabuli wrote his *Almanac of the Position of Watch Hands at the Beginning of Prayer*

---

[58] For a description of one type of instrument that was commonly used for taking altitudes in this period, the *rubʿ al-muqanṭarāt*, see William H. Morley, "Description of an Arabic Quadrant," *Journal of the Royal Asiatic Society of Great Britain and Ireland* 17 (1860): 322–30.

Times for 1867–68 at the request of a vaguely defined "group" *(jamāʿa)*, to whom he did not attribute any particular qualifications.[59]

Copies of Dimashqi's handbook commonly include an undated *Tables of the Position of Watch-Hands at Prayer Times (Jadawil Mawqiʿ al-Saʿat ʿala Hasab Awqat al-Salawat)* by Dimashqi's contemporary, Ramadan al-Khawaniki.[60] From a certain perspective, this pairing of texts seems unnecessary. Dimashqi's handbook has its user find the solar altitude and degree of the sun in the ecliptic. The times of prayer can be determined by these data alone (for a given latitude), as was commonly done with the astrolabe or astrolabic quadrant, for example.[61] Instead, users of Dimashqi's tables apparently translated their solar observations into a conventional time based on equal hours since sunset, and then used Khawaniki's tables to find the times of prayer in these conventional terms. Perhaps this procedure made sense for a *muwaqqit* who had to tell a muezzin when to issue the call to prayer. If the timekeeper was concerned that his watch had a rate of error, for example, he could have used Dimashqi's handbook to synchronize it with *ğurūbī sāʿat* time at any point during the day, and then determined the times of prayer simply by examining the timepiece alongside the accompanying table by Khawaniki.

The terminology of this new subgenre of astronomical timekeeping, however, suggests an additional use for tables such as Khawaniki's. Throughout these manuals and tables, the Arabic word for "time" is typically linked with that for "prayer" (e.g., *awqāt al-ṣalawāt*), whereas what the watch or clock displays is termed "the position of the watch hand" *(mawqiʿ ʿaqrab al-sāʿa)*. All of the "prayer tables" in this genre, from Khawaniki's in the early eighteenth century to Nabuli's in the 1860s, literally concern not the time at which one should pray, but in what position the watch hand should fall when the time of prayer comes. Thus, the times of prayer are always the same: *fajr* (dawn), *ḍuhr* (noon), *ʿaṣr* (afternoon), *maghrib* (sunset), and *ʿishāʾ* (evening). What changes is the position of the watch hand corresponding to these times on a given day.

As this distinctive terminology suggests, tables like Khawaniki's could have been used by *muwaqqit*s and muezzins for knowing when to pray by

---

[59] Nabuli, *Natijat Mawqiʿ ʿAqrab al-Saʿat*, 51.
[60] MS DM 812, ENL; MS Zakiyya 822, ENL; and Isl. Ms. 808, Special Collections Library, University of Michigan.
[61] On determining the prayer intervals in an Egyptian manual on the astrolabic quadrant written in 1901, see Ibn Abi al-Fadl, *al-Riyad al-Zahirat*, 14–15. On the diverse ways in which Muslim astronomers historically expressed the times of prayer, see King, *In Synchrony*, 1:204–8.

the watch, but they could also have served a larger number of users who wanted to set their watches by the call to prayer. Lane described people in Egypt setting their watches once a day by the call to prayer at sunset. These tables enabled the same procedure for every time of prayer, not only *maghrib*. Thus, the "prayer table" was not necessarily, or at least not exclusively, about knowing when to pray. It was also about taking a cue from the call to prayer of the muezzin in order to set one's watch. We can read the common pairing of Dimashqi's and Khawaniki's tables not as a redundancy, but rather as a way of offering two distinct methods of knowing how to set the timepiece: one, by visual observation and measurement of the sun's position; the other, by listening for the muezzin.

From Object into Instrument

As scholars of Ottoman temporality have remarked, *ğurūbī sā'at* timekeeping challenges the notion that clock time is necessarily removed from the natural rhythm of the solar day.[62] However, the specific practices illustrated by manuals on positioning the watch hand also call into question a dichotomy, common in time studies, that opposes premodern time consciousness – which is religious or "mythical," and therefore "vague" or "impressionistic" – with modern time, which is based on more regular, "abstract measurement."[63] The teaching of 'ulama' on positioning the watch hand in Ottoman Cairo shows us a type of technical religious practice – the intersection of Islamic astronomical timekeeping (*mīqāt*) with Islamic prayer (salat) – through which the watch itself was made increasingly regular in its measurement, without the emergence of a "time" abstracted from the rhythms of ritual and celestial motion.

Indeed, the history of the "position of the watch hand" genre suggests that the very understanding of mechanical clocks and watches as instruments of precise timekeeping was only established (in Ottoman Cairo) at the nexus of *mīqāt* and salat. After all, although I have used the word "timepiece" as a catchall for various kinds of clocks and watches, the measurement of time is not the only way to use such an object. Clocks and watches bore multiple kinds of aesthetic and social value in Ottoman culture (as they do in twenty-first-century America, among other places). As Ian White and others have vividly documented, European clockmakers made considerable efforts to appeal to their conception of

---

[62] Wishnitzer, *Reading Clocks.*
[63] Cf. Pierre Bourdieu, "Time Perspectives of the Kabyle," in *The Sociology of Time*, ed. John Hassard (London: Macmillan, 1990), 219–37.

Ottoman aesthetic standards.[64] The roman numerals of the chapter ring were replaced with Eastern Arabic numerals. On ornate pieces (usually clocks), floral designs replaced figural art.[65] For watches, particular attention was devoted to the casing. A three-case design was standard, with the entirety, as Dimashqi remarked, being tucked into the "collar" where Ottomans often carried personal effects. As late as the 1870s, a British engineer who worked on one of the Ottoman railways commented on the care and deliberateness with which a "Turk" would remove a watch from the many layers of clothing, bag, and case within which it rested, only to find that it was not working, or working only approximately: "so to learn the time of day within an hour or so a quarter of an hour will be wasted."[66] Although this account is evidently part of an essentializing discourse through which Victorians imagined the "time-mindless Oriental,"[67] we can also read it as implying that watches had value apart from their utility as instruments of precise timekeeping – which is surely correct.[68] The vogue of clocks and watches in the Ottoman court, in particular, was in large part a function of their value as objects of curiosity, art, and what White has called "power projection."[69] According to Dimashqi, watches in early eighteenth-century Cairo came in a variety of forms and types (*ikhtalafa ashkāluhā wa-anwāʿuhā*), including some that only had hour hands.[70]

In the introduction to his manual, Dimashqi explained his motives by remarking that "watch hands are not free of earliness or delay."[71] But concern for earliness or delay may not have been shared by everyone with a watch in Cairo in 1737. Dimashqi, along with the rest of the scholars who produced these manuals and tables, was linking the object of the watch to a specific way of thinking about the measurement of time. "Approximation is not like precision, for it contains a kind of carelessness," wrote Muhammad al-Nabuli in the preface to his tables, which took the extraordinary measure of giving the times of prayer to the half-minute (*al-taqrīb laysa*

---

[64] White, *English Clocks*; Özdemir, *Ottoman Clocks and Watches*.

[65] Kurz, *European Clocks and Watches*, 46. For the broader context of Ottoman attitudes toward figural art, see Finbarr Barry Flood, "Lost Histories of a Licit Figural Art," *IJMES* 45 (2013): 568.

[66] H.C. Barkley, *Bulgaria before the War: During Seven Years' Experience of European Turkey and Its Inhabitants* (London: John Murray, 1877), 180–81. A longer quotation is also reproduced in White, *English Clocks for the Eastern Markets*, 15.

[67] Barak, *On Time*.

[68] See also de Beaujour, *A View of the Commerce of Greece*, 242.

[69] White, *English Clocks for the Eastern Markets*, 28.

[70] Dimashqi, *al-Manhaj al-Aqrab*, Isl. Ms. 808, Special Collections Library, University of Michigan, p. 24.

[71] Dimashqi, *al-Manhaj al-Aqrab*, Isl. Ms. 808, Special Collections Library, University of Michigan, p. 24.

ka-l-tadqīq idh fīhi naw' tasāhul).[72] 'Ulama' who devoted themselves to the study of astronomical timekeeping were particularly committed to the value of mathematical precision and accuracy in the performance of certain practices, from the observation of the sun to the fulfillment of the duty of prayer.

Of course, the many kinds of value that a clock or watch might carry – precision timepiece, art, social symbol, novelty – are not mutually exclusive. In a context where such objects were only just coming into broad use, however, it should not be taken for granted that they would be used to measure time to the minute, let alone the half-minute. The manuals and tables on positioning the watch hand helped to establish that usage. And they embodied a claim that it was 'ulama' who could provide, through their tradition of *mīqāt*, the knowledge of how to perform such measurement correctly – thereby bridging scholarly discourse, material fashion, and religious and technical practice. This phenomenon undermines a teleological understanding of the history of temporality, in which religious time is supplanted by modern time. In fact, a "modern" emphasis on precision emerged in part from a very old conception of piety, which was long cultivated among a small number of scholars but now became increasingly normative.[73]

A Cannon Fired by the Sun

When 'Abd al-Latif al-Dimashqi composed the first manual in Cairo "on the position of the watch hand," 'ulama' were the preeminent social group in the city who practiced a tradition of knowledge that dealt with precision timekeeping. By 1874–75, when an anonymous scribe wrote the last copy of Dimashqi's *Clearest Method* that remains extant today, 'ulama' held a considerably less privileged position.[74] Due to the emergence of new scientific institutions in the context of the viceregal dynasty's political ambitions, as discussed in the previous chapter, 'ulama' now competed with new kinds of technically trained actors – in particular, when it came to matters of timekeeping, with the viceregal astronomers Mahmud al-Falaki and Isma'il al-Falaki. In many ways, Mahmud and Isma'il helped to perpetuate and even popularize the timekeeping

---

[72] Nabuli, *Natijat Mawqi' 'Aqrab al-Sa'at,* MS 317 al-Saqqa 28898, Azhar, 50v.
[73] Historians of precision have generally emphasized the new political and commercial contexts of the nineteenth century. More remains to be done to understand the religious history of precision. Cf. M. Norton Wise, ed., *The Values of Precision* (Princeton, NJ: Princeton University Press, 1995).
[74] See MS TM 88, ENL.

practices pioneered by scholars of *mīqāt*. In doing so, however, they transformed both the social and material experience of timekeeping in Cairo, displacing 'ulama' as timekeeping authorities, and erasing the role of 'ulama' in developing the very practices that the viceregal astronomers co-opted.

As early as the 1840s, Mahmud Hamdi published almanacs, including prayer tables, said to be widely appreciated in mosques.[75] Notwithstanding the encroachment of such state functionaries, however, 'ulama' in Cairo continued to address the timing of prayer in their own way, and specifically in relation to mechanical timekeeping, by producing manuals and tables "on the position of watch hands." One mid nineteenth-century scholar, Khalil ibn Ibrahim al-'Azzazi, authored a manual that instructed other scholars in the technique of producing such tables. Muhammad al-Nabuli was 'Azzazi's contemporary, and may well have used the latter's manual in producing his *Almanac of the Position of Watch Hands at the Beginning of Prayer Times* for 1867–68.[76] 'Azzazi and Nabuli were Azhari 'ulama' who had spent no time in the new, viceregal institutions of science, yet their work was hardly isolated from developments in the period. Nabuli's emphasis on "precision" as opposed to "approximation," and his decision to provide the times of prayer to the half-minute, can be understood as part of a context in which 'ulama' competed with new actors (as well as among themselves) for credibility. Moreover, his use of half-minute intervals assumed, without comment, an audience possessing mechanical timepieces with minute-hands. On a more technical level, 'Azzazi's manual invokes a value for the obliquity of the ecliptic that he purports to have taken from "the new, contemporary observations" (*al-arṣād al-jadīda al-ḥāliyya*), and which appears (based on its size and precision) to be based on observatory data from the 1830s.[77] The manuals and tables produced by these 'ulama' illustrate the ongoing adaptation and persistent relevance of a scholarly scientific tradition and its

---

[75] Crozet, "Trajectoire," 291.

[76] Nabuli, *Natijat Mawqi' 'Aqrab al-Sa'at*. Nabuli's known works date to the 1860s, during the last years of 'Azzazi's life. See Nabuli, *Fath al-Mannan*, which the author completed in 1862 (as stated on p. 46 of the 1907 print edition); and Nabuli, *Kashf al-Hijab 'an Murshid al-Tullab*, MS 386 'Arusi 42765, Azar, an 1863–64 commentary on another work that he wrote in 1862–63.

[77] 'Azzazi expresses the value in sexagesimal notation, but it works out to $23° \, 27' \, 42''.52$. The closest recorded observations that I have found appear in John Pond, *Astronomical Observations Made at the Royal Observatory at Greenwich* (London: T. Bensley, 1830), 87. Government observatories like Greenwich published such observations in nautical almanacs, to which it is plausible that 'Azzazi might have had access, although he would have needed a translator's assistance.

practitioners even as new scientific institutions and technologies became established in Egyptian society.

However, Nabuli's 1867–68 almanac is the latest composition that deals specifically with "the position of the watch hand." Even the long manuscript trail of Dimashqi's *Clearest Method* disappears around the same time; as noted, the latest extant copy is from 1874–75. This date coincides intriguingly with Mahmud al-Falaki's inauguration of the Citadel cannon in the summer of 1874.[78] As discussed in the previous chapter, the Citadel cannon quickly became an authoritative source of timekeeping in Cairo. By the end of the year, the British astronomer Charles Orde Browne, in Cairo to observe the transit of Venus, reported that "watches are ill used by constant attempts to force them to agree with a gun in Cairo which the Sun itself fires at noon."[79] At least among the elite households and bureaucrats with whom Brown consorted, watches were now set at noon, rather than at sunset; and they were set by a cannon, rather than by the call to prayer.

Despite the coincidence of the emergence of this new time signal with the writing of the last extant copy of Dimashqi's *Clearest Method*, the Citadel cannon surely did not spell the immediate death of the "position of watch hands" genre or of the relevance of 'ulama' to the practice of mechanical timekeeping. In Egypt, as in other parts of the Ottoman Empire, the practice of *ǧurūbī sā'at* endured among some classes well into the twentieth century.[80] Now branded as "Arab" or – in Turcophone areas – "Turkish" (alaturka) time, it became part of an increasingly contentious politics of temporality, in which one's way of telling time could serve as a marker of cultural difference and even political allegiance.[81]

The enduring importance of the Ottoman-Islamic temporal routines promulgated in the 'ulama''s manuals on positioning the watch hand is also evident in the way that Mahmud Hamdi inscribed these routines into the technology of the Citadel cannon itself. The rise of astronomically regulated, public time signals was a global phenomenon in the late nineteenth century, but one in which "time" typically denoted *mean time* – an approximation of local time that neglects fluctuation in the length of the

---

[78] Isma'il Mustafa Bey, "Hawadis-i Dahiliye: Fi Sharh Midfa' al-Zawal," p. 1.

[79] Quoted Ratcliff, *Transit of Venus Enterprise*, 187n26.

[80] In Egypt, almanacs continued to print "Arab" time alongside "European" time at least as late as the 1930s. See Mahmud Naji, *Natijat al-Dawla al-Misriyya li-Sanat 1354 Hijriyya* (Cairo: al-Matba'a al-Amiriyya, 1935).

[81] Barak, *On Time*; Wishnitzer, "'Our Time.'"

day over the course of the solar year. By the 1870s, many regions and even nations were moving toward the adoption of standard mean time, so that "noon" in much of New England, for example, was signaled at the time of mean noon at the location of the Harvard College Observatory.[82] This flattening of the temporal landscape through the elimination of local times facilitated new forms of transportation, commerce, and communication – most famously the railroad. The Citadel cannon deviated from this trend. Instead of signaling mean noon (let alone a standard mean noon) based on an observatory clock, Mahmud went to some trouble to design a time signal that would fire at the moment of apparent local noon. In rigging the Citadel cannon, Mahmud was not merely imitating a European and American trend. He was also perpetuating the Ottoman practice of pegging the motion of watch hands to daily changes in the sun's motion through the sky, even as the moment of synchronization became noon rather than sunset. The Citadel cannon points to the way in which 'ulama' faced new competition as knowledge practitioners in nineteenth-century Egypt, but it also shows how their Ottoman and Islamic ways of keeping time were, in some ways, amplified by this development.

While the timekeeping practices that state astronomers helped to popularize were borrowed, in part, from 'ulama', we should not underestimate the transformation that these practices underwent in the process. More than mere convention was at stake in the shift from setting the watch at sunset to correcting it at noon. The viceregal astronomers' time signal originated in a different physical and social location from the call to prayer. Tellingly, Mahmud and Ismail, who trained and made their careers within the bureaucracy of the militarizing viceregal state, chose for their time signal a piece of artillery in the traditional center of government in Cairo. In doing so, they adopted a longstanding method of announcing moments of public significance. In the nineteenth century, cannon fire from the Citadel already marked the departure of the pilgrimage caravan from Cairo, celebrated major political appointments, announced the beginning and end of the fast during Ramadan, and (at least by some accounts) served as the signal for the "massacre of the Mamluks" in 1811.[83] As a daily time signal, the Citadel cannon continued to serve as an announcement from one of the city's political centers, but

[82] Ian Bartky, *Selling the True Time: Nineteenth-Century Timekeeping in America* (Stanford, CA: Stanford University Press, 2000).

[83] See Eugène Gellion-Danglar, *Lettres sur l'Égypte contemporaine (1865–1875)* (Paris: Sandoz et Fischbacher, n.d.), 45, 18; Trowbridge Hall, *Egypt in Silhouette* (New York: Macmillan, 1928), 29, 53 (although the latter work's account of nineteenth-century events is unattributed and must be taken with several grains of salt).

now as a marker of the routine rather than the extraordinary. The rapidity with which this signal appears to have displaced the call to prayer as the reference point for setting one's timepiece in Cairo illustrates one way in which viceregal astronomers were able to draw on their political resources to claim a role previously occupied by 'ulama' and muezzins.

An essential element of this shift in social authority was the material transformation that gave voice to it. It is worth asking how exactly "the sun itself" could fire a cannon. Browne was speaking literally, although his remark omitted the human work that enabled such an unusual phenomenon. According to Ismail al-Falaki's account, Mahmud affixed a convex lens to the cannon, such that the focal point of the lens fell on the gun's fuse. The lens was attached to the cannon with rotating legs, such that the lens could be adjusted within the plane of the meridian so as always to face the sun's zenith on a given day. As the sun crossed the meridian, the convex lens would focus its rays on the cannon's fuse, and the gunpowder would ignite. The key to this setup was a graduated semicircle bearing the dates of the Coptic calendar at the angles with which the lens should be aligned, on a given date, in order to focus the rays of the sun at noon. The person operating the cannon would merely position the legs of the lens in line with the day's date on the arc, and the gunpowder would ignite at the proper moment.

In other words, the shift from setting watches by the call of a muezzin to setting them by the blast of a gun was not merely a shift in social relations of authority – 'ulama' out, technocrats in – but also a more radical attempt to objectify social authority in terms of material agency: "the sun itself," not Mahmud or Ismail, would tell when to set one's watch. The astronomical knowledge on which this technology depended – knowing the angular altitude of the sun at particular times on particular Coptic dates – was mathematically equivalent to the knowledge contained in scholarly tables on positioning the watch hand. In the transformation of paper grids into a metal arc, however, lay a significant shift not only in who was responsible for this knowledge, but in the kinds of work that their "time" could perform. In the first decade of the twentieth century, when British surveyors began trying to synchronize time signals in a number of Egyptian cities with the Greenwich mean time system, the Cairo signal was none other than the Citadel cannon – no longer fired by the sun itself, but rather triggered within a telegraphic network centered on the Anglo-Egyptian Survey Department's observatory in Helwan. No one involved in building the Citadel cannon in 1874, just thirty years earlier, had been interested in global time reform. But the object they created and embedded in the aural landscape of the city could be co-opted and repurposed within a different material network and political project.[84] The objectification of timekeeping was a prerequisite for the standardization of time.

Conclusion

The Citadel cannon serves as an example of the way in which new state actors appropriated, but also transformed, practices long cultivated by Muslim scholars.

Mechanical timepieces did not enter Cairene society thanks to vice-regal modernization, let alone the later contexts of British colonialism, anticolonial nationalism, or the ascendance of the middle-class "*efendi-yya*." Rather, watches and their casings were first tucked into Egyptian collars under the circumstances of the eighteenth century: diplomatic travel and etiquette, elite taste for novelties, and a culture of learning still dominated by 'ulama'. In this context, the vogue of mechanical time-keeping became an opportunity for scholars trained in *mīqāt* to reinforce their claims to knowledge and to begin to popularize their highly precise conventions of time measurement. In contrast to the later dynamic of technology and temporality that On Barak has chronicled in the colonial period, such scholars could hardly have thought that they were rendering timepieces particularly "Egyptian," or even necessarily Ottoman. Rather, they were making timepieces "correct": incorporating new objects into their knowledge traditions, and placing the use of these objects in proper relation to existing routines (the solar day and the schedule of prayer) and actors (scholars and muezzins).

Amid the circulation of new objects with rotating hands, 'ulama' occupied a kind of center, helping to position not only the physical hand of the watch, but the use and meaning of the watch itself. The genre of "the position of the watch hand" belongs, therefore, not only to the history of science, technology, and temporality, but also to the history of authority in Islam. Long before technologies of mass communication gave 'ulama' new forms of influence and authority in the twentieth century, the spread of clocks and watches in the eighteenth and nineteenth centuries, far from rendering the tradition of *mīqāt* obsolete, opened an opportunity for its 'ulama' practitioners to deploy their knowledge for a new audience.

The institutionalization of science in the viceregal state bore two consequences for this process that appear to stand in tension with each other. On the one hand, viceregal technocrats like Mahmud al-Falaki adopted and even promoted norms and practices that originated in the scholarly

---

[84] Sound has only recently begun to attract systematic attention in Middle East studies. See the roundtable, "Bringing Sound into Middle East Studies," *IJMES* 48 (2016): 113–55.

discipline of *mīqāt*: they published prayer tables and built a time signal that facilitated a regimen of timekeeping linked precisely to the real solar day. By the end of the 1870s, more people in Cairo were regulating their watches according to the minutes of the real solar day than ever before. On the other hand, this practice came to rely on a new network of actors and objects. Where watch-users, muezzins, and scholars had once coordinated the positioning of the watch hand informally, using manuscripts, quadrants, and the call to prayer, such networks came to be organized around state astronomers, printed tables, and a cannon. The popularization of *mīqāt* depended upon the transformation of its materiality and the displacement of its old practitioners. As pious precision became the norm, its origins faded from view.

4 Positioning the Planets: Translating French Planetary Tables as Ottoman-Islamic Knowledge

Joseph Jérôme de Lalande was well known. By the time he died in 1807, he had been a leading light of French science for half a century.[1] Professor of astronomy at the Collège de France, director of the Paris Observatory and of the Bureau of Longitudes, he was famous for his ambition to create the largest and most accurate star catalogue in history. He was also famous for his atheism and for his affairs with women. He was famous for eating caterpillars, and he was famous for being ugly. This was the Lalande of Paris and of the other European cities where he traveled, where he was a member of academies, and where his readers were part of the Continental and Anglophone circles within which the work and reputation of a *savant* and *philosophe* traveled by course.[2]

There was, however, another Jérôme de Lalande. This Lalande did not belong to the Republic of Letters, but he moved in equally exalted circles, conversing with the likes of Ibn al-Shatir, the great fourteenth-century astronomer of Damascus; Ulugh Bey, of the fifteenth-century Samarqand Observatory; and Jai Singh, the eighteenth-century Mughal scholar. As the author of a *zīj* (astronomical handbook), this Lalande was particularly good at predicting the movement of the planets within a geocentric cosmology. This was the Jérôme de Lalande who emerged from a series of late Ottoman translations of the French Lalande, a series that reached across a hundred years and spanned the Eastern Mediterranean. In the 1780s, parts of Lalande's magnum opus, *L'astronomie*, were translated into Turkish for use in naval instruction in Istanbul. In 1808, his *Abrégé d'Astronomie* was translated into Arabic by prominent Greek Christians in the port of Damietta.[3] In 1813, a court astronomer in

---

[1] On Lalande, see Simone Dumont, *Un Astronome des lumières: Jérôme Lalande* (Paris: Vuibert/Observatoire de Paris, 2007); Guy Boistel et al., *Jérôme Lalande* (Rennes: Presses universitaires de Rennes, 2010); Alder, *Measure of All Things*, 76–81.
[2] On the social context of knowledge circulation in the Enlightenment, see Dourinda Outram, *The Enlightenment*, 2nd edition (New York: Cambridge University Press, 2005), 13–25.
[3] J.J. de Lalande, *al-Mukhtasar fī 'ilm al-falak*, trans. Basili Fakhr et al., Ms Arabe 2554, BnF.

Istanbul, acting on the Sultan's express orders, again translated parts of Lalande's *Astronomie* into Turkish.[4] When the latter translation became available in the provincial capital of Aleppo in the 1860s, an anonymous Muslim scholar translated it into Arabic.[5] Finally, Husayn Zayid, a scholar teaching at the al-Azhar mosque in Cairo, composed what he called a "new astronomical handbook" (*zīj jadīd*) based on the "widely known" work of Lalande in 1887.[6]

But *how* was Lalande's work "widely known" by Ottomans? By illuminating the movement of French astronomy into an Ottoman-Islamic context, the Lalande translations grant insight into broad questions concerning the mobility of knowledge. Chapter 2 showed that a non-Western empire like the Ottoman-Egyptian state of the mid nineteenth century could be a center of scientific circulation, appropriating and even propagating new technologies. Here, I extend this argument to consider how practitioners of scholarly astronomy saw themselves as participants in a kind of global science, which was rooted in use of the *zīj*. The scholars who translated Lalande into Turkish and Arabic were not, for the most part, members of the new technically trained elite, like Mahmud or Ismaʻil al-Falaki. Rather, they were people who worked in the established spaces of Ottoman-Islamic scholarship, like the court astronomer Hüseyin Hüsnü and the Azhari shaykh Husayn Zayid. Despite the novelty of using Lalande in the context of scholarly astronomy – notably because of his grounding in a heliocentric, Newtonian cosmology – his late Ottoman translators recognized much that was familiar in his work, which they translated according to the genre-specific conventions, purposes, and computational practices of the *zīj*. In at least one case, they even translated it into a geocentric cosmology. In other words, the ability of Lalande's astronomy to work in late Ottoman contexts was not a matter of reproducing the Observatoire de Paris in Aleppo or Cairo, or what Shapin and Ophir termed "the success of certain cultures in creating and spreading the very means and contexts of application."[7] Rather, it was a matter of the extent to which Ottomans could render Lalande

---

[4] Hüseyin Hüsnü, *Tercume-i zij-i Lalande*, MS T 6553, Istanbul University.

[5] Anonymous, *Tahdhib al-anam fi taʻrib lalandnama*, MS TR 182, ENL. The translator's introduction belies İhsanoğlu's claim that this title is also by Hüsnü. Cf. İhsanoğlu, *OALT*, #419, p. 582. Other manuscript copies, which I have not been able to examine, are said to be held in Damascus (Zahiriyya Library) and Aleppo, which accords with the evidence in the Cairo manuscript that the text originated in Aleppo. I am grateful to Avner Ben-Zaken for bringing the Cairo manuscript to my attention.

[6] Husayn Zayid, *al-Matlaʻ al-saʻid fi hisabat al-kawakib al-sabʻa ʻala al-rasd al-jadid* (Cairo: al-Matbaʻa al-Baruniyya, 1304 [1887]).

[7] Ophir and Shapin, "The Place of Knowledge," 16.

into languages, genres, and practices that were already, and in their own ways, transregional.

The efforts of Lalande's Turkish and Arabic translators point to the active role that 'ulama' played in the teaching of new sciences, amid hard-fought battles over the "reform" of education in late Ottoman society.[8] Zayid's Lalande-based "new $z\bar{\imath}j$" of 1887, in particular, sheds new light on the controversy over introducing natural sciences at al-Azhar during the 1880s and 1890s.[9] While some 'ulama' opposed the teaching of these sciences within the courtyards of the mosque, their resistance was not overcome by bureaucratic fiat or the imposition of new expertise from the civil or military schools. Informal networks of scientifically inclined 'ulama' played a crucial role.

Thus, like the previous chapter, the current discussion emphasizes that new technologies did not break the hold of established actors on astronomical knowledge. If anything, the availability of new planetary tables, far from spelling the end of the $z\bar{\imath}j$, brought scholars like Husayn Zayid into a more prominent position than they had previously occupied at al-Azhar – much as practitioners of $m\bar{\imath}q\bar{a}t$ gained new relevance with the spread of mechanical timepieces. By demonstrating the role of 'ulama' in integrating French astronomy into Ottoman and Ottoman-Egyptian institutions of learning, the Lalande translations also continue to complicate the social geography of "scholarly" and "viceregal" science that emerged in the first two chapters of this book. Finally, planetary tables resembled mechanical timepieces in their instrumental nature. Despite being made of paper and ink, the tables of *Astronomie* were tools, which is to say they were objects that were used as much as they were read. Indeed, it was precisely as a tool that Lalande's work struck late Ottomans as significant.

Why Lalande?

The choice of Lalande, among all the luminaries of European astronomy in the eighteenth century, underlines the highly specific and pragmatic purposes that the translations were intended to serve. Had the Ottomans been interested in celestial mechanics, they would surely have focused their efforts on interpreting Laplace. Had their curiosity centered on spectacular discoveries, they might have chosen Herschel (who first

---

[8] On late Ottoman educational reform, see Fortna, *Imperial Classroom*. On Egypt specifically, see Gregory Starrett, *Putting Islam to Work: Education, Politics, and Religious Transformation in Egypt* (Berkeley, CA: University of California Press, 1998).

[9] On the reform of al-Azhar, see Gesink, *Islamic Reform and Conservatism*.

observed Uranus) or perhaps Halley. In terms of technical achievement, Lalande's reputation rested on his ambitious star catalogue. He was best known, however, for his public roles, both as an employee of the French state and as a longtime lecturer. In an extraordinary feat of political dexterity, Lalande served prominently in the royal, republican, directory, and imperial governments of France, overseeing the French nautical almanac (*Connaissance des temps*) and, eventually, the Observatoire de Paris. Meanwhile, he filled his classroom at the Collège de France with hundreds of students from across the Continent, and his three-volume *Astronomie* was the most popular textbook on the topic in late eighteenth-century Europe.[10] It was likely Lalande's reputation as both an effective pedagogue and practitioner of specifically nautical astronomy that brought his work to the attention of the Ottoman state.[11]

Ottomans had a long history of translating European astronomy. In the most elite circles, explanations of the Copernican and Tychonic systems became available in the seventeenth century, through the same channels of diplomatic gift exchange by which mechanical clocks first came to the empire. In the eighteenth century, discussion of the Copernican system also entered scholarly literature through encyclopedic writing like the work of İbrahim Hakkı.[12] Meanwhile, Ottoman ambassadors in Europe started to take notice of the value of new astronomical tables, particularly the work of Giovanni Cassini at the Paris Observatory.[13]

Translation of European astronomical texts began to acquire an institutional form in the late eighteenth century, when a string of battlefield defeats spurred the Ottoman state to reconsider the training and organization of its military. The most radical outcome of Ottoman military reform was the replacement of the Janissary corps with a regimental infantry: first, the abortive "new order" army (*Nizam-ı Cedid*); later, and more permanently, the "trained victorious Muhammadan troops" (*Muallem Asakir-i Mansure-i Muhammediye*).[14] New forms of training began not in the army, however, but in the navy – with good reason.

---

[10] Alder, *Measure of All Things*, 85; Curtis Wilson, "The problem of perturbation analytically treated: Euler, Clairaut, d'Alembert," in *Planetary Astronomy*, ed. Taton and Wilson, 103.

[11] Specifically in Egypt, Crozet has found that Mehmed Ali's translators generally focused on pedagogical works in the sciences, rather than "first order" works. Crozet, *Sciences modernes*, 253.

[12] On early Ottoman translations of Copernicanism, see Ben-Zaken, "The Heavens of the Sky and the Heavens of the Heart," 5; and İhsanoğlu, "The Introduction of Western Science."

[13] Adıvar, *La science chez les Turcs Ottomans*, 150–51.

[14] Virginia Aksan, *Ottoman Wars, 1700–1870: An Empire Besieged* (Harlow, UK: Longman/Pearson, 2007).

In the disastrous 1768–74 war with Russia, among the most dramatic Ottoman defeats was the loss of an entire fleet in the Aegean to a Russian force that had sailed all the way from the Baltic.[15] It was likely in the aftermath of this debacle that Mustafa III (r. 1757–74) requested "the best books of European astronomy" from the Académie royale in Paris.[16] The Académie fulfilled this request with Lalande's *Astronomie*, among other works. As early as the 1780s, a European visitor saw a Turkish translation of Lalande used in the lessons of the *mühendishane*, a marine academy (literally, "geometer-house") soon to be renamed the "Imperial Naval *mühendishane*" (*mühendishane-i bahri-i hümayun*).[17]

Tellingly, the next translation of Lalande originated in the court of Mahmud II (r. 1808–39), who eventually oversaw the destruction of the Janissary corps and centralization of the army in 1826. In 1813, Mahmud, concerned with the state of Ottoman navigational knowledge, directed the court's vice-astronomer (*müneccim-i sāni*), Hüseyin Hüsnü, to render the work of Lalande into Turkish.[18] Perhaps the eighteenth-century translation had already been lost amid the political upheaval of the intervening years. Or perhaps it had relied on an early edition of Lalande's tables (e.g., 1760), whereas Hüsnü could use the 1806 edition.[19] In either case, Lalande drew the attention of Ottoman astronomers specifically as astronomy became part of the state's efforts to improve the technical capacities of the military.

The predictive power of Lalande's astronomical tables made them appealing for purposes beyond navigation, however, and hence to translators and astronomers with less direct relationships to the Ottoman state. In Aleppo, the anonymous author who translated Hüsnü's Turkish Lalande into Arabic in the 1860s appears to have been a traditionally learned Muslim scholar (*'ālim*). He wrote with rhyming prose, formulas such as "God's is the achievement of the one who said...", and references

---

[15] Donald Quataert, *The Ottoman Empire: 1700–1922*, 2nd edition (Cambridge: Cambridge University Press, 2005), 40.

[16] J.F. [*sic*] Montucla, *Histoire des mathématiques: nouvelle édition*, vol. 1 (Paris: Henri Agasse, Year 7 FRC [1798–99]), 400.

[17] L'Abbé Toderini, *De la Littérature des turcs, traduit de l'Italien en François*, trans. l'Abbé de Cournand (Paris: Poinçot Libraire, 1789), 165. For a history of these institutions, see İhsanoğlu, "Ottoman Science."

[18] *OALT* #419, p. 581. While İhsanoğlu dates the translation only to before Receb 1241/ 1826, in MS Istanbul Univ. T 6553, it is stated to have been done in 1228/1813, and examples are calculated for 1229/1814. The Aleppine translation specifies that Hüsnü finished his translation in Rajab 1229/June 1814, a discrepancy for which I cannot account, but which is not analytically significant.

[19] When Toderini visited the *mühendishane* in the 1780s, he noted that the edition of Lalande was already not the most recent one, which he recommended to the instructor. Toderini, *De la Littérature des turcs*, 165.

to Arabic poetry and classifications of knowledge that circulated in prominent eighteenth-century scholarly works, specifically commentaries on logic and Hanafi law (the Islamic law tradition favored by the Ottoman state).[20] Moreover, both the Aleppo and Cairo works, unlike Hüsnü's 1813 translation, omitted the tables for the motion of Uranus. Since Uranus is (generally) not visible to the naked eye, this choice points first to the likelihood that the Aleppo and Cairo translators did not expect their tables' users to possess telescopes. But it also suggests that the tables may have been used largely for astrological prognostication, since Uranus was not a feature of the classical cosmos within which Ottoman astrologers interpreted the "judgments of the stars." Like the slightly later work of Husayn Zayid in Cairo, the Aleppo translation appears to have been meant not for institutions of military education, but for practitioners of scholarly astronomy.

Such practitioners would have used these new, Turkish and Arabic versions of Lalande for much the same purposes for which they had long employed astronomical tables: to compute calendars and predict planetary motion. Zayid's "new $z\bar{\imath}j$" includes procedures and worked examples for these problems, as well as for a broader set of problems that adhere much more closely to the structure of a $z\bar{\imath}j$ such as Muhammad al-Khudari's than to the structure of Lalande's *Astronomie*. Thus, Zayid's first section treats chronology and the conversion of dates between the various calendars used in Egypt.[21] The sections that follow address the solar position, including problems related to general timekeeping; lunar motion, eclipses, and planetary motion; and significant times such as conjunctions, horoscopes, and prayer intervals.[22] Although he was working from tables calculated for the longitude of Paris and arranged according to the Julian calendar, Zayid rendered them into tables for the longitude of Cairo and the *hijrī* calendar, "to facilitate benefit and avoid additional labors" for his readers.[23] In these ways, both the Aleppo and Cairo works fit the conventions of the $z\bar{\imath}j$ genre, and would have been legible to scholars trained in *mīqāt*.

This is not to say that the scholarly translators of Lalande gave no voice to the novelty of their project. As in the Ottoman state translations,

[20] Compare the Aleppo text's poem on the ten principles of any science, *Tahdhib al-anam*, 6, with Muhammad al-Sabban, *Hashiya 'ala Mullawi lil-sullam* (Cairo: Dar al-Kutub al-'Arabiyya al-Kubra, 1332 [1913–14]), 34; and compare the poem in *Tahdhib al-anam*, 2, with Muhammad Amin al-shahir bi-Ibn 'Abidin, *Hashiyat radd al-muhtar 'ala al-durr al-mukhtar*, vol. 1 (Beirut: Dar al-Fikr, 1399 [1979]), 28.

[21] Zayid, *Matla'*, 8.

[22] Zayid, *Matla'*, 12, 15, 16, 17, 19, 21, 22, 23.

[23] Zayid, *Matla'*, 3.

however, what was important about Lalande's novelty was its degree of precision and accuracy. Thus the Aleppo text celebrated ten ways in which Lalande's tables surpassed previous work – all ten of them pertaining to the technical arrangement and predictive power of the text.[24] The Aleppine scholar drew special attention to the advent of logarithmic functions in Lalande's procedures, and their novelty is also highlighted in Zayid's text, where they are some of the only values written in numeral form, rather than in the old *abjad* notation, which used the numeric values of the Arabic alphabet.[25]

The repeated choice of Lalande as an object of translation points to the fact that Ottoman as well as Ottoman-Egyptian astronomers appropriated European astronomical books as tools of calculation and prediction. Both for new institutions of state military education beginning in the late eighteenth century and for Muslim scholars in the nineteenth century, certain aspects of European science that attracted systematic attention were novel for their accuracy and precision, but familiar in form. From Aleppo to Cairo, the tables and procedures of Lalande's *Astronomie* were recognized as the latest *zīj*.

Scientific Translation in Process

In order to understand how late Ottomans put Lalande to use as a *zīj*, it is necessary to understand something of the process by which they rendered him into Arabic and Turkish. Some insight into this process may be gleaned from a translation of Lalande's *Abrégé d'Astronomie*, carried out in Damietta in 1808 by a circle of Greek Christians who were involved in translating a variety of works of French and English literature into Arabic in the early nineteenth century.[26] The leading figure in this group was Basili Fakhr, a merchant of Syrian origin, who bore the ecclesiastic title of "logothetes," an honorific granted to those who defended the Church in polemical writings.[27] His lack of astronomical training may explain why he chose to translate Lalande's introductory *Abrégé*, which was popular in Europe among amateur practitioners in the eighteenth century, including women.[28] The edition of the *Abrégé* that Fakhr

---

[24] *Tahdhib al-anam*, 3.
[25] *Tahdhib al-anam*, 3.
[26] J.J. de Lalande, *al-Mukhtasar fi 'ilm al-falak*, trans. Basili Fakhr et al., Ms Arabe 2554, BnF. On the Damietta translation circle, see Hill, "First Arabic Translations."
[27] See Basili Fakhr, *al-Jawahir al-Fakhriyya 'an al-'illa al-ibthaqiyya* (Jerusalem: Matba'at al-Qabr al-Muqaddas al-Batriyarkiyya, 1861).
[28] Dumont, *Astronome des lumières*, 93.

worked from was itself an Italian translation, likely the one printed in Padua in 1796.[29]

As the text's "translator," Fakhr suffered, by his own admission, from what would seem to be key deficiencies: "ignorance of the fundamentals of astronomical terminology" (al-jahl fī qawāʿid al-musammaʾāt al-mīqātiyya), and a limited understanding of Italian (al-ʿajz fī maʿrifat kunh al-lugha al-īṭāliyāniyya).[30] In order to overcome these linguistic and technical barriers, Fakhr worked with one of the many consular agents in Damietta, a Venetian named Giovanni Lavagetti. (Fakhr was himself a consul for the French.) Lavagetti explained the Italian text to Fakhr, perhaps in a mixture of spoken Italian and Arabic. Fakhr recorded this explanation in Arabic as best he could, and handed it over to a third and final member of the translation project, ʿIsa Petro (Ysa Petros). A priest from Jerusalem and an experienced translator, Petro revised Fakhr's vernacular rendering of the Venetian's explanation to produce a final text that met Arabic literary standards.[31]

This collaborative model of translation was not unique to Damietta or to the Greek Church networks of the Eastern Mediterranean. In the organized movement of scientific translation pioneered at the Ottoman-Egyptian medical school and school of languages in the 1830s, a translation "triangle" of technical expert, polyglot translator, and Arabic editor was the norm.[32] This arrangement points to the several kinds of knowledge that late Ottomans used to translate science and exposes the reality that these different kinds of knowledge were often contributed by different kinds of people. In addition to the technical ability to explain the original text and the linguistic ability to carry out that explanation in another language (two skills possessed by the Venetian, in this case), also critical was the ability to write in good literary prose in the target language (Petro's role). What function, then, was left for Basili Fakhr? He could not read the text fully in its original language, and he could not write fluently in the target language. He could, however, understand the vernacular Italian and Arabic of the Venetian and write it down in Arabic, even if his writing was unpolished. It was this creole capacity, as well as his material resources, that placed a man with no French, no astronomy, dubious Arabic, and limited Italian at the center of the translation (see Figure 4.1).

---

[29] J.J. de Lalande, Compendio di astronomia, ed. Vincenzo Chiminello (Padua: Tommaso Bettinelli, 1796). A first Italian edition appeared in 1777, but Fakhr states that the copy from which he worked was written in 1795, which accords with the 1796 publication of the second edition.

[30] Lalande, Mukhtasar, Ms Arabe 2554, BnF, p. 2a.

[31] On Petro, see Hill, "First Arabic Translations," and Hill, "Early Translations," 181–83.

[32] Crozet, Sciences modernes, 306.

Figure 4.1 Basili Fakhr, introduction to J.J. de Lalande, *al-Mukhtasar fi ʿilm al-falak*, trans. Basili Fakhr et al., Ms Arabe 2554, BnF, vol. 1, pp. 1–2. Bibliothèque nationale de France.

If the various kinds of knowledge involved in scientific translation could be performed by different individuals, it follows that scientific translation was not only a vehicle by which science traveled, but also a record of personal encounters *through* which knowledge was exchanged. When the Venetian, the Dimyati, and the Jerusalemite came together around a text printed twelve years earlier in Padua, the Arabic Lalande that they produced did not rest on any single person's coherent interpretation of the original. Translation was not an act of solitary reading and writing: a lone scholar with an open book on one side, and a blank page on the other. Rather, translation was a compound of technical, oral, and literary skills contributed by different kinds of authors. Husayn Zayid, in Cairo, collaborated with someone who read French as well as Arabic.[33] It is likely that the anonymous scholar in Aleppo also worked with one or more other scholars.

In her influential work on reading Darwin in Arabic, Marwa Elshakry understands scientific translations as produced by "intertwined sources

---

[33] Zayid, *Matlaʿ*, 3.

of authority." From the coining of terms and the repurposing of extant vocabulary, to the framing of new ideas in terms of historical precedent and contemporary significance, the acts that compose translation allow – and sometimes require – the translator to draw upon a range of intellectual resources, not only an original text. Building on this understanding, I suggest that some instances of late Ottoman scientific translation were irreducibly polyvocal. Behind the mask of a single authorial voice, the authority of knowledge emerged from multiple perspectives. And, at least in the case of astronomical manuals, these perspectives were embedded less in the *reading* of the text than in its *use*. This dynamic becomes dramatically evident when considering the question of whether translating Lalande entailed specific cosmological commitments.

Geocentric Asteroids: The Cosmology of Lalande in Translation

Lalande's work was among the first instances of heliocentric astronomy to be incorporated in the regular pedagogy of major sites of Ottoman learning, notably including al-Azhar in Cairo. However, this aspect of its significance would be easy to overlook in the translations themselves. Lalande's Ottoman translators explained the importance of the Frenchman's work to their audiences in considerable detail, but these explanations focused on highly specific issues of accuracy, precision, and ease of use. Heliocentricity appears, at most, as an implication of certain mathematical procedures, while in one case – the Aleppo translation – the geocentric cosmos is explicitly affirmed alongside these procedures. Translated as a tool, Lalande did not require substantive reconsideration of cosmological commitments.

The Aleppo translation provides the most surprising example of this phenomenon. In a section entitled, "explanation of the spheres and their epicycles" (*al-aflāk wa-tadāwīrihā*), the translator explains that the cosmos is composed of nine spheres.[34] The earth is at their center. The sun is in the fourth sphere. The eighth sphere contains all the fixed stars and rotates daily. The ninth sphere, which controls the rotation of the others, is understood to be the divine "throne" (*al-'arsh*) referenced in the Qur'an. Models for the motion of the sun, moon, and planets contain various combinations of the classic Ptolemaic devices: eccentric orbits, the deferent, and of course, the epicycle – the orbit which is itself in motion on an orbit, for which the Ptolemaic system was famous (or

---

[34] *Tahdhib al-anam*, 8–9.

infamous). This cosmology does not differ substantially from the geocentric worldview described in Muhammad al-Khudari's *zīj* commentary in Chapter 1.

And yet, the anonymous Aleppine scholars also remarks, "the scholars of Paris have discovered other planets," including Juno, Ceres, Pallas, and Vesta (asteroids or dwarf planets first observed in the early nineteenth century), as well as Uranus (*awrānūs*).[35] However, awareness of such newly discovered celestial bodies does not seem to have challenged the author's cosmology of geocentric spheres. Thus, the anonymous scholar used geocentric devices to describe the motion even of Uranus:

As for the anomaly of the superior planets, which are Uranus, Saturn, Jupiter, and Mars, subtract the mean longitude of the planet from the mean longitude of the sun, and the remainder is the anomaly of the planet. ... And the anomaly is a motion on the circumference of an epicycle (*ḥaraka ʿalā muḥīṭ falak tadwīr*).[36]

In keeping with the common motives of Ottoman interest in Lalande's tables, the discovery of new planets was celebrated as an example of French astronomy's precision and accuracy, without comment on their cosmological implications. Far from shattering the closed, geocentric spheres of scholarly astronomy, Uranus moved quietly on an epicycle.[37]

While no such explicit cosmological translation occurs in Husayn Zayid's "new *zīj*" of 1887, neither did the Cairene scholar discuss the heliocentric model behind his computational procedures. Instead, in the book's introduction, Zayid explicitly justified the use of new astronomical tools as a matter of ensuring accuracy. The values in older tables were no longer reliable, he explained:

With the passage of time, their observation became obsolete, as a result of which some discrepancies, which should not be neglected by the meticulous, in addition to change and substitution that entered them because of their passing from hand to hand over this long period, were introduced to the one who is calculating. It is therefore necessary for one who strives for the truth to discard these ancient observations and to rely on what is newly done in recent times, even if its method (*maʾkhadh*) is not lacking in difficulty.[38]

---

[35] *Tahdhib al-anam*, 11–13.
[36] *Tahdhib al-anam*, 14.
[37] The translator's statement, "the center of the sun is the basis of the [celestial] motions (*aṣl al-ḥarakāt*)," does not reference a heliocentric model, but rather a mathematical relationship between the sun's motion and the motion of other celestial bodies. *Tahdhib al-anam*, 13. For a typical definition of the sun's "center" (*markaz al-shams*), see Dallal, *An Islamic Response to Greek Astronomy*, 66–67.
[38] Zayid, *Matlaʿ*, 3.

This remark constitutes the entirety of the book's reflection on method; it is the closest Zayid came to commenting on the way his cosmological premises diverged from those of geocentric scholarly astronomy. Perhaps he was merely being cautious. The comment itself, however, suggests that the method was mere baggage to him, a cumbersome thing one had to endure, because it came with the best tables.

To some degree, cosmological premises were implicit in the mathematical procedures that students of Zayid's book would have used. Zayid's procedure for predicting the position of a planet was to calculate the planet's heliocentric position (what Zayid calls its "true longitude as seen from the sun") before translating this position into apparent, geocentric terms (its "true longitude as seen from the earth").[39] By contrast, scholars using the corresponding procedure in Muhammad al-Khudari's *zīj* commentary made the necessary adjustments to a planet's circular orbit around the earth, in order to predict its observed position in the celestial sphere. However, while cosmological difference is implicit between these two mathematical procedures, Zayid himself drew no attention to this implication. Even the phrases, "as seen from the sun" and "as seen from the earth," are noncommittal. The classical discussion of cosmography that appeared in Khudari's text, for example, and again, surprisingly, in the Aleppo text, is absent – but with no replacement.

In both Aleppo and Cairo, scholars were able to translate and use the latest in French planetary tables without disrupting the geocentric cosmos in which scholarly astronomy had always functioned. One way of understanding this unusual celestial mechanics is in terms of the mechanics of late Ottoman scientific translation. If we think of the process of rendering French into Turkish into Arabic (or French into Italian into Arabic) as a polyvocal record of knowledge exchange, rather than as a unitary vehicle of science moving from one place to another, it becomes possible to read contradiction as conversation. Maybe the person who revised the Arabic of the Aleppo manuscript thought that the technical translator had simply forgotten the explanation of the spheres and their epicycles – a stock piece of the *zīj* – and wrote it into the introduction on his own initiative. What appears to us as a single text was not necessarily a single text inside any one author's head.

In practice, the plausibility of such a text rested on substantive similarities between Lalande's *Astronomie* and the scholarly *zīj*, beginning with their objects. Husayn Zayid, the anonymous Aleppine translator, Muhammad al-Khudari, and Lalande all shared much the same purpose

[39] Zayid, *Matlaʿ*, 19–20.

in their respective works: to enable users of their tables to calculate the position of the celestial bodies *as they appeared to them*. Thus, Lalande's French tables included a "table of movements of the sun and moon."[40] Of course, Lalande himself did not think that the sun was in motion around the earth, but, for the sake of convenience, that was how he arranged his tables. In other words, the abstraction of astronomical practice from a physical conception of the cosmos was not some kind of Ottoman-Islamic strategy intended to extract the benefits of modern science while concealing its "real" implications. To the contrary, the specific European sources selected for translation, and the particular genre into which they were translated, shared basic similarities in purpose.

They also shared remarkable similarities in technique. Bearing in mind that Lalande's *Astronomie* and its late Ottoman translations were texts less read than *used*, it is illuminating to compare late eighteenth-century French methods for calculating planetary motion with the methods of the *zīj* in terms of pen-and-paper practices.[41] Table 4.1 translates a worked example from Muhammad al-Khudari's *zīj* commentary, which he who wrote in the 1820s but which relied, ultimately, on the tables and procedures of Ibn al-Shatir's fourteenth-century *zīj*. (See Figure 4.2 for an image of the calculations in the manuscript.) In the example, Khudari calculated the position of the sun at the vernal equinox. (By definition, it should be entering the first of Aries.) Table 4.2 translates a comparable worked example from the Aleppo Lalande. In each example, the astronomer calculated an initial value for the sun's position according to its mean motion, and proceeded to add or subtract a number of corrections, found in tables, in order to reach a more precise value for the sun's actual position as seen from the earth and measured within the zodiacal belt.

When we set aside all the parallel steps in these columns, as well as steps that, for contingent reasons, are only in the Aleppo/Lalande example but that would have been completely familiar to Muhammad al-Khudari (like a correction for the equation of time), only one line remains in the Aleppo/Lalande procedure that Khudari would have found novel: the "equation of arguments." This correction of −32 seconds to the sun's position reflects the sum of the calculations in six additional columns, each of which accounts for the gravitational effect of another object in the solar system on the earth's orbit. These calculations are absent

[40] Jerome de Lalande, *Astronomie*, 3rd edition (Paris: Desaint, 1792), vol. 1, "Tables des mouvements du soleil et de la lune pour le méridien de Paris."

[41] On mathematical physics as pen-and-paper practice, see Andrew Warwick, *Masters of Theory: Cambridge and the Rise of Mathematical Physics* (Chicago: University of Chicago Press, 2003).

Table 4.1 *Muhammad al-Khudari's calculation of the position of the sun on 20 Rajab 1239 (1824), 12:00 noon, using techniques derived from Ibn al-Shatir's zīj (Damascus, fourteenth century)*

	House	Degrees	Minutes	Seconds
Step 1: calculate the sun's mean motion (*wasat*) by adding values from tables of the sun's motion for the epoch, year, month, and date.				
+	8	10	17	39
+	9	4	18	32
+	5	24	27	34
+		18	43	38
=	11	27	47	23
Step 2: calculate the sun's apogee (*awj*) by adding values from tables of the sun's apogee for the epoch, year, month, and date.				
+	3	7	56	10
+	0	0	7	46
+	0	0	0	29
+	0	0	0	3
=	3	8	4	28
Step 3: Subtract result in (2) from (1) to find the "anomaly" and the corresponding value of the sun's "equation."				
(Mean motion)	11	27	47	23
− (Apogee)	3	8	4	28
=	8	19	42	55
Corresponding value in table of the sun's equation	0	2	1	52
Step 4: Add result in (3) to (1) to find the sun's position				
(Mean motion)	11	27	47	23
+ (Equation)	0	2	1	52
=	11	29	49	15

from Muhammad al-Khudari's work, because they follow from a post-Copernican, indeed Newtonian, understanding of celestial motion.

While Muhammad al-Khudari would have found the conceptual basis of these additional calculations to be novel, however, in pen-and-paper terms he would have found them familiar. In order to produce these six columns, the Aleppo scholar read another line of terms within Lalande's tables for the sun's mean motion; found the sum of these terms; used another set of

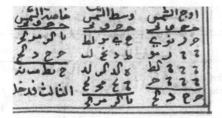

Figure 4.2 Initial steps of Muhammad al-Khudari's calculation of the vernal equinox for 1239 [1824]. Muhammad al-Khudari, *Sharh al-Lum'a fi Hall al-Kawakib al-Sab'a*, p. 90 (detail), Isl. Ms. 722, University of Michigan Library (Special Collections Library), Ann Arbor.

Table 4.2 *Aleppo/Lalande calculation of the position of the sun on 20 March 1870, 9:44pm*

	House	Degrees	Minutes	Seconds

Step 1: calculate the sun's mean motion (*wasat*) by adding values from tables of the sun's motion for the epoch, year, month, date, hour, and minute.

	House	Degrees	Minutes	Seconds
	9	9	38	14
+	0	0	18	24
=	9	9	56	38
+	1	29	8	20
=	11	9	4	58
+	0	18	43	38
=	11	27	48	36
+	0	0	22	11
=	11	28	10	47
+	0	0	1	48
=	**11**	**28**	**12**	**35**

Step 2: calculate the sun's apogee (*awj*) by adding values from tables of the sun's apogee for the epoch, year, month, and date.

	House	Degrees	Minutes	Seconds
	3	10	0	7
+	0	0	41	26
=	3	10	41	33
+	0	0	0	10
=	3	10	41	43
+	0	0	0	3
=	**3**	**10**	**41**	**46**

(continued)

Table 4.2 (*cont.*)

	House	Degrees	Minutes	Seconds
Step 3: Subtract result in (2) from (1) and find the value of the sun's "equation" corresponding to the difference.				
(Mean position)	11	28	12	35
– (Apogee)	3	10	41	46
=	8	17	30	49
Corresponding value interpolated from table of the sun's equation	0	1	53	24
Step 4: Add result in (3) to (1). Add equation of arguments, equation of time, and longitudinal correction to find the position of the sun.				
Mean motion	11	28	12	35
+ (Sun's equation)	0	1	53	24
	0	0	5	59
Step 5: Additional corrections				
– (Equation of arguments)	0	0	0	32
=	0	0	5	27
+ (Equation of time)	0	0	0	8
=	0	0	5	35
– (Correction for the longitudinal difference of Aleppo from Paris)			5	35
=	0	0	0	0

tables to find the values corresponding to these sums; and added these corresponding values to reach the bottom term in each column. The "equation of arguments," −32 seconds, is the sum of these sums. The Copernican and Newtonian revolutions transpire in six columns of arithmetic that amount, in the final calculation, to a single line of subtraction.

The work of *doing* Lalande, in other words, was both familiar – simply more tabular calculation – and devoid of physical reference. The six "Newtonian" columns are labeled with numbers, not with physical explanations (e.g., "nutational effect of the moon," "perturbation of Jupiter," etc.). They might "represent" these physical concepts in some sense, but the practitioner, busy keeping the units straight in sexagesimal addition, need not have borne such significance in mind. Whether using Ibn al-Shatir or Lalande, calculating planetary positions was a practice of addition, subtraction, and the manipulation of tables. Of course, none of this is to minimize the importance of heliocentricity. A correction of −32 seconds was significant. From a practical perspective, however, it was possible to take advantage of these techniques while placing newly discovered planets on epicycles in a geocentric cosmos.

Merit and Truth: Lalande in Late Ottoman Debates on Progress

When translators of Lalande placed his techniques within the cosmology of scholarly astronomy, they also placed his accomplishments within a particular history of astronomy, which bore implications for late Ottoman debates about the extent to which modern science represented – or indeed demanded – a break with the past. Thus, Hüseyin Hüsnü's 1813 Istanbul translation of Lalande placed the need for a new *zīj* within a story about the constant progress of observational precision. This narrative begins with praise for Ulugh Beg, but notes that his fifteenth-century observations were made obsolete by new observational instruments in the early eighteenth century: "In sum, it is apparent and evident among the masters of precision that [Ulugh Beg's *zīj*] is corrupt and flawed, because of its great difference from the *zīj* of Cassini, which is closest to the truth among the works of moderns."[42] As noted above, an Ottoman ambassador had acquired the tables of Giovanni Cassini, director of the Paris Observatory, in likely the first instance of Ottoman acquisition of European astronomical data in the eighteenth century. These tables made Ulugh Beg's outdated.

Yet, it was not long before even Cassini's observations came to seem obsolete. Hüsnü continued, "at the time of the making of Cassini's *zīj*, which is in every way preferable and superior to the aforementioned *zīj* [of Ulugh Beg], there were still some hidden celestial secrets yet cloaked and hidden under the veil of invisibility…" The best *zīj*, wrote Hüsnü, was the "new *zīj*" of the "the observer of the city of Paris, the *philosophe* Lalande."[43] Once again, Ottoman recognition of Lalande's tables as belonging to the genre of the *zīj* points to the familiarity of the Frenchman's work both in terms of its goals (predicting planetary positions) and practices (tabular calculation). But translating Lalande as a *zīj* also placed the Ottoman need for French science within a specific narrative of progress. In this narrative, Lalande was necessary not because he represented a kind of astronomy that was essentially different from what had previously been available to the Ottomans, but because his work was

---

[42] Hüsnü, *Tercume-i zij-i Lalande*, Istanbul Univ. Ms. T 6553, p. 2.
[43] Hüsnü, *Tercume-i zij-i Lalande*, Istanbul Univ. Ms. T 6553, p. 2.

the most recent example of a kind of astronomy that in fact was very old. One translated Lalande from French in the nineteenth century for the same reason that one translated Ulugh Beg from Persian in the fifteenth century. "Mathematical works are constantly reaching the degree of... perfection."

The implications of placing Lalande in the history of the *zīj* were further developed by the anonymous translator in Aleppo in the 1860s, whose introduction is only very loosely based on the Turkish text by Hüsnü. For the Aleppine translator, the history of the *zīj* comprised some twenty-five major works, including the work of Ptolemy, al-Biruni, al-Tusi, al-'Urdi, al-Battani, Ibn Yunus, Ibn al-Shatir, Ulugh Beg, Cassini, Jai Singh, Muhammad al-Khudari al-Dimyati's *zīj* commentary (with which we are familiar), the British Nautical Almanac, and Lalande.[44] On the one hand, this history of the *zīj* resembles the kind of progressive narrative later popularized by the great Victorian and twentieth-century histories of astronomy. "Knowledge and the sciences increase daily," wrote the Aleppo translator, in one of the few lines that he borrowed directly from the Turkish copy's introduction.[45] He celebrated the thousands of new stars that the "scholars of Paris" had observed – possibly a reference to Lalande's own work. On the other hand, the specific narrative of progress differed from the familiar narrative of Ptolemy, Copernicus, Kepler, Newton – a history organized around the development of planetary models. Instead, it was a narrative of progress organized around the genre of the *zīj*, which is to say around practices of tabular computation and prediction.

Telling the history of astronomy through the genre of the *zīj*, the Aleppo translator conveyed a story of progress without rupture. According to the anonymous scholar, contemporary science, particularly in Europe, is superior to earlier knowledge in the sense that it was closer to the truth: "If merit belongs to the one who begins (*al-bādi'*), the latter one (*al-thānī*) is more correct, especially when the scholars (*ḥukamā'*) of Paris have achieved renown like the sun of the equinoxes at midday."[46] In this

---

[44] *Tahdhib al-anam*, 4. The reference to the British Nautical Almanac is my interpretation of the author's citation of "*Qūnsānsnāma*," which he attributes to a scholar in London. I have not been able to find a British astronomer of the eighteenth or nineteenth century whose name might correspond to *Qūnsāns*. However, the French Nautical Almanac was (and is) called *Connaissance des temps*. Assuming that the Aleppo author was correct in describing the text as English, my best guess is that he simply gave the Nautical Almanac the name of its French counterpart. Of course, it is also possible that he was referring to the *Connaissance des temps* itself, though it is harder to imagine why he would have thought it was from London.

[45] *Tahdhib al-anam*, 1.

[46] *Tahdhib al-anam*, 1.

scheme, however, knowledge moves along a single axis from earlier scholars (*al-mutaqaddimūn*), to later scholars (*al-muta'akhkharūn*). The latest *zīj* makes earlier ones obsolete, but there is no moment when astronomy changes in a radical way. For this reason, while the most recent knowledge is "more correct" (*ahaqq*) than its predecessor, the latter retains a certain "merit" or *fadl*, a term that also connotes "excellence" or even something like "virtue" (e.g., *fadīla*).

The anonymous scholar elaborated on the nature of this "merit" with a quotation from Umayyad poetry:

> Had I, before her tears, for love of Su'da wept,
> I would have healed my soul before regret.
> But she wept before me, her tears moved me
> to tears, and I said: "credit (*fadl*) belongs to the one who was first"[47]

With these lines, which are attributed to one of the Umayyad caliphs, the translator made an analogy between the relationship of new and old scientific knowledge and a certain type of lovers' quarrel. The poet's beloved, Su'da, cries first, and by virtue of the precedence of her tears, wins out. Thus, the "merit" of the older scientific tradition is likened to a specifically feminine kind of power, as imagined by a man: the power to subvert, to undermine, or to demand recognition from the normatively more powerful – whether normative power resides in the most recent science, or in a male lover who is also the Caliph.

The Aleppo translation also used poetry to emphasize the genealogical "merit" belonging to earlier astronomy. The historical *zīj*s to which the translator traced Lalande not only came before, but in some sense produced, current achievements:

> Like the sea: the cloud waters it and has no
> superiority to it, for it is from the sea's water.[48]

---

[47] *Tahdhib al-anam*, 2. These lines are typically attributed to Yazid ibn Mu'awiyya. See Paul Schwarz, *Escorial-Studien zur Arabischen Literatur- und Sprachkunde* (Stuttgart: Kohlhammer, 1922), 58, with translation on p. 63. Schwarz renders "merit belongs to the one who was first" (*al-fadl lil-mutaqaddim*) interpretively: "To the one who first enters the path, glory (Ruhm) is due as leader."

[48] *Tahdhib al-anam*, 2. I have not found an attribution for this line, but it is quoted in the late eighteenth-century commentary of Ibn 'Abidin on an oft-studied work of Hanafi fiqh. Since the anonymous author also quoted a passage from a work of the scholar Muhammad al-Sabban from the same time period (*Hashiya 'ala Mullawi lil-sullam*), Ibn 'Abidin is a likely source. See Ibn 'Abidin, *Hashiyat radd al-muhtar 'ala al-durr al-mukhtar*, I:28.

In a battle for credit, the present cannot claim victory over the past from which it flows. Lalande differs from the work of Ulugh Beg or Jai Singh no more than the water of the clouds differs from the water of the sea.

With this poetically argued discussion of "merit" versus "truth," the anonymous scholar deployed the translation of Lalande in a high-stakes debate in late Ottoman society. Aleppo in the 1860s was a city in which the *Tanzimat* were in full swing. The city's first official gazette, appearing in both Turkish and Arabic, began to publish in 1864.[49] Among both ruling and provincial elites, one of the key controversies of the *Tanzimat* and Hamidian eras concerned the nature of progress. While some demanded a thoroughgoing adoption of "Western" (*alafranga*) style in everything from dress to governance, others advocated a more selective process of appropriation with continuity.[50] Within this debate, new kinds of science and institutions of science education were often associated with the most radical elements. Thus, the ideology of the Committee of Union and Progress (CUP) military officers who seized power in 1908, and who forged the basis for the secularism of the later Turkish Republic, has been traced to the embrace of materialism and Comtean positivism in certain military circles of the mid nineteenth century, especially at the Imperial Medical College.[51] In these circles, new scientific knowledge was understood to have inaugurated a fundamentally new era. In a widely read poem entitled "The Nineteenth Century," the late Ottoman statesman Sadullah Pasha declared, "The foundations of old knowledge have collapsed."[52] The Aleppo translator's invocation of Umayyad poetry can be read as a refutation of such radical mantras. While acknowledging the progress of science, the translator set himself against those in the late Ottoman milieu who saw such progress as both authorizing and demanding a radical departure from older modes of inquiry and social organization. For the anonymous scholar, translating Lalande was a way of acquiring the most recent knowledge, while retaining the merit of what came first.

Yet, as decades passed, and the project of translating Lalande moved from Istanbul in 1813 and Aleppo in the 1860s to Cairo in the 1880s, the accuracy and precision of Lalande's tables entered a significantly different narrative about scientific progress. In Husayn Zayid's translation – alone

---

49 Watenpaugh, *Being Modern*, 41–42.
50 Watenpaugh, *Being Modern*; and see Wishnitzer's discussion of the role of time consciousness in the performance of "modernity": Wishnitzer, *Reading Clocks*, 169–70.
51 Hanioğlu, "Blueprints for a future society"; Yalçınkaya, *Learned Patriots*, 180–81.
52 Sadullah Paşa, "Ondokuzuncu Asır," *Mecmua-i Ebuzziya* 46 (Cemazi-yelevvel 1302/1 April 1885): 1453–55, quoted in Hanioğlu, "Blueprints for a future society," 34; see also Yalçınkaya, *Learned Patriots*, 181.

among all the late Ottoman usages of Lalande – science became a criterion of stark geographic and temporal difference:

The traces of [astronomy] have almost disappeared among us. The early scholars
(*al-mutaqaddimūn*) wrote many great books on astronomy, but the later ones (*al-
muta'akhkhirūn*) neither followed their example nor walked in their ways in this
matter, until astronomical knowledge passed to the Western lands, and nothing
was left in the East but the remains of books past their time.[53]

This zero-sum conception of scientific cultures was new. The Aleppo
author understood contemporary European astronomy to be the best in
the world, but he saw it as part of an uninterrupted tradition beginning
in antiquity and largely elaborated by Muslim or Muslim-patronized
scholars, including scholars of the eighteenth and nineteenth centuries
(e.g., Jai Singh and Khudari). In this narrative, the only distinction lay
between the earlier and the later, and even this classification was less
dichotomy than genealogy, within which the earlier scholars bore a kind
of moral superiority (*faḍl*). Zayid's language added a new dichotomy,
one that emphasized a geographic – implicitly cultural – divide, with one
side ascendant and the other in decline. "The East" has long abandoned
astronomy, which now resides in "the Western lands."

The history of translating Lalande in the Ottoman Empire suggests
that Zayid's narrative requires some explanation. Previous scholars had
read European astronomy, indeed greatly valued it, without seeing it
through the lens of such a geographic and cultural dichotomy. It can
hardly be coincidence that Zayid, alone among Lalande's late Ottoman
interpreters, wrote in the context of European occupation: by 1887,
British troops had been in Cairo for five years, and Sir Evelyn Baring was
firmly ensconced as Consul General. Yet, British administrative influence had not yet begun the pervasive expansion that was to characterize
the following two decades. Zayid's stark critique of his fellow 'ulama'
emerged from the more specific context of debates over pedagogy at
al-Azhar. These debates pre-dated, but received new impetus from, the
onset of colonial rule.

Who Brought Lalande to al-Azhar?

The second half of the nineteenth century was a time of intensive
debate over the institutions of learning in Egypt, particularly al-Azhar.[54]

---

[53] Zayid, *Matla'*, 3.
[54] On the evolution of al-Azhar in this period, see Chris Eccel, *Egypt, Islam, and Social
Change: al-Azhar in Conflict and Accommodation* (Berlin: K. Schwarz, 1984).

Beginning even under Mehmed Ali Pasha, members of the new bureaucratic elite had begun to voice criticism of the Azhari 'ulama'. Such critics saw both the content and technique of Azhari pedagogy as outdated. They advocated for change in subject matter, including the teaching of natural sciences, as well as greater use of print, a centralized administration, and standardization of teaching qualifications. These proposed changes responded to real crises at al-Azhar, including overcrowding, lack of funds, and ethnic tension among the students. As recent scholarship has shown, the notion that the "reform" of al-Azhar pitted the state against conservative 'ulama', who blindly resisted any and all change, overlooks the fact that many "conservative" 'ulama' advocated and indeed shaped the course of pedagogical change in this period. Controversy arose, however, regarding the nature and degree of change necessary, especially as 'ulama' sought to preserve their autonomy from the state.[55] Given the preeminence of al-Azhar in the world of Sunni Muslim learning, these debates bore consequences not only for Cairo or Egypt, but for students and scholars who came from as far away as Indonesia and West Africa.

Published in 1887, Husayn Zayid's new $z\bar{\imath}j$ appeared at a critical juncture in these debates. Over a decade of growth in the Egyptian periodical press had spread the reformist vision of education and progress. New educational institutions, particularly the Dar al-'Ulum (est. 1872), were graduating students who competed with Azhari scholars for jobs.[56] With the restoration of the Khedive Tawfiq under the British Occupation, Shaykh al-'Abbasi al-Mahdi, who had issued the first Azhar "code" in 1872, resumed his position as Shaykh al-Azhar and issued a revised code in 1885. Under al-Mahdi's successor, the teaching of natural sciences at al-Azhar was expressly authorized in 1888.[57]

Azhari scholars played an indispensable role in making possible the development of natural science instruction inside the mosque. As Chapter 1 showed, while astronomy was not generally studied within the spaces of al-Azhar, 'ulama' associated with al-Azhar had long pursued the study of astronomy elsewhere, in the informal scholarly networks that were centered on but hardly limited to the mosque. These networks included a loosely connected group of scholars who worked to

---

[55] Gesink, *Islamic Reform and Conservatism*, critiques the assimilation of modernist polemics into Orientalist scholarship. Cf. Afaf Lutfi al-Sayyid Marsot, "The Beginnings of Modernization among the Rectors of al-Azhar, 1798–1879," in *Beginnings of Modernization in the Middle East: The Nineteenth Century*, ed. William R. Polk and Richard L. Chambers (Chicago: University of Chicago Press, 1966), 267–80.

[56] Lois Armine Aroian, *The Nationalization of Arabic and Islamic Education in Egypt: Dar al-'Ulum and al-Azhar*, Cairo Papers in Social Science 6, no. 4 (Cairo: AUC, 1983).

[57] On the relevant fatwas, see Gesink, *Islamic Reform and Conservatism*, 117–18.

incorporate the tools of recent, especially French, scientific work into the genres of Islamic scholarship. They stretch from Mustafa al-Bulaqi, who would walk between the mosques of old Cairo and the new Polytechnic in Bulaq in the 1830s, to 'ulama' who computed new almanacs in the early twentieth century. Zayid himself was a student of Khalil al-'Azzazi, who taught arithmetic at al-Azhar in the middle of the nineteenth century and was one of the authors of the manuals on positioning the watch hand. (For the details of 'Azzazi's network, see Chapter 1.) Perhaps bearing in mind the potentially divisive implications of authoring a new scientific text amid the debates over education at al-Azhar, Zayid made a point of noting 'Azzazi's explicit approval of his work.[58]

This loose network of 'ulama' points to the fact that the natural sciences were not "introduced" to al-Azhar by outside bureaucrats, or even by reformist 'ulama', like Muhammad 'Abduh, who had no training in scholarly scientific traditions. Rather, the teaching of natural sciences at al-Azhar depended on the translation work of 'ulama' who participated in a longstanding practice of appropriating new technologies and texts as part of Islamic scholarship. Indeed, the notion that natural science could have been taught at al-Azhar absent some initiative among scientifically trained 'ulama' overlooks the reality that such sciences – like any kind of knowledge – required authors and teachers who could make them legible to Azhari students. This was the role played by scholars like Zayid, whose new *zīj* of 1887 was used as part of the implementation of science instruction at al-Azhar in the following year.[59]

Zayid's text was not alone. With the passage of another Azhar code in 1896, natural sciences became more than permissible; they were now mandatory subjects at al-Azhar and its affiliated institutions.[60] Although historians have highlighted the mosque's reliance on "outside" instructors – imported from the civil schools – to teach the new subjects, the books that students used tell a more complex story.[61] Only geography appears to have been taught mainly from "modern books" chosen by instructors from the civil schools.[62] Arithmetic, algebra, and timekeeping and astronomy (*al-mīqāt wa-l-hay'a*) were taught using twenty-six books by premodern as well as eighteenth- and nineteenth-century scholars like Ridwan Effendi, Zayid, and 'Azzazi. The prominence of such venerable texts was not some revival of long-neglected learning; the books in

---

[58] Zayid, *Matla'*, 3.
[59] Mustafa Bayram, *Tarikh al-Azhar* (Cairo, 1903), 35.
[60] Gesink, *Islamic Reform and Conservatism*, 155.
[61] Cf. Gesink, *Islamic Reform and Conservatism*, 160.
[62] Bayram, *Tarikh al-Azhar*, 35.

question were precisely those that had been most commonly copied and studied by 'ulama' throughout the nineteenth century, as evidenced by their prevalence among the scientific manuscripts of the period.[63] Put another way, the novelty of the natural sciences at al-Azhar in the late nineteenth century lay not in the kinds of books, techniques, or even teachers involved, but rather in the *relocation* of these books and techniques. Texts that 'ulama' once chose to learn outside the mosque now became required reading inside the mosque.

Husayn Zayid's work underscores some of the tensions in the discourse around this "reform" of al-Azhar in the late nineteenth century. The narrative of progress in which Zayid placed Lalande's achievement was centered on a dichotomy of Eastern decline and Western progress: precisely the narrative popularized by critics of al-Azhar's traditionalism in order to legitimize reform. Evidently, this critique was powerful enough to motivate practitioners of scholarly astronomy to undertake substantial technical work, like Zayid's composition of a "new *zīj*" based on French astronomical tables. Ironically, such efforts call into question the very narrative that Zayid and the reformists peddled. 'Ulama' did not face, by necessity, a choice between East and West; it was possible for them to place European science within the history and genres of their own scientific practice. While the most accurate tables were to be found originally in French, 'ulama' still possessed the resources of an active scientific tradition with which to render such tables into viable knowledge in their own context.

Conclusion

In the long history of translating Lalande in the Ottoman Empire, the very late emergence of an association between Lalande's specific achievement and "the West" is evidence of the fact that the equation of modern science with "Western science" only developed over the course of this very period.[64] But it also points to the way knowledge moved in this context. 'Ulama' rendered French planetary tables mobile by placing them in the genre, languages, cosmology, and history of their own global science. From the perspective of such 'ulama', the latest *zīj* came from Paris in much the same way that Ibn al-Shatir had been a Damascene, Ulugh Beg

[63] The new Azhar curriculum relied heavily on mathematical works by both al-Sibt al-Mardini and Ibn al-Ha'im, which were among the most widely copied manuscript works on their subject throughout the nineteenth century. See Crozet, *Sciences modernes*, 211.
[64] Marwa Elshakry, "When Science Became Western: Historiographical Reflections," *Isis* 101 (2010): 98–109.

had worked in Samarqand, Muhammad al-Khudari in Egypt, and Jai Singh in India. There was much talk of accuracy, precision, logarithms, and the arrangement of tables. There was very little talk of heliocentricity or scientific revolution.

The use of French planetary tables in this manner belies a misperception that has troubled the historiography of the late Ottoman Empire. It has sometimes been remarked that late Ottoman interest in science was limited by its focus on the applications, rather than "theory" or "worldview," of European science.[65] This characterization has even been deployed as a kind of deep explanation for the failure of the post-Ottoman Middle East to develop a culture of research-productive science. But such a narrative overlooks the way in which scientific translation often works, not only when one society appropriates texts from another, but even when practitioners in one scientific field learn to work with knowledge developed in another. In such contexts, the forging of makeshift tools and "creole" languages is not a limitation, but an essential way in which local knowledge becomes mobile.[66] Late Ottoman 'ulama' worked with French planetary tables in much the way that Cambridge mathematicians first worked with Einstein's theory of relativity: by finding a shared set of problems, they created a science that was both different and familiar.[67]

It is worth emphasizing that this kind of translation was performed specifically by 'ulama', and not by the new technical elite. Lacking the kind of scholarly background possessed by Husayn Zayid or the anonymous Aleppine translator, new elites had neither the motive nor the means to render Lalande legible' as scholarly astronomy. But such elites were not the only translators active in late Ottoman society. Like mechanical clocks, French planetary tables were assimilated into Ottoman-Egyptian society in a way that depended upon and promoted knowledge particular to 'ulama'.

And yet, the very moment that might have been the climax of this movement – the use of Husayn Zayid's "new *zīj*" to introduce planetary tables into the formal instruction at al-Azhar – became, instead, the moment when scholarly astronomy was said to have died. The first

[65] Crozet, *Sciences modernes*; Ekmeleddin İhsanoğlu, "Some Critical Notes on the Introduction of Modern Sciences to the Ottoman State and the Relation Between Science and Religion up to the End of the Nineteenth Century," in *Science, Technology and Learning in the Ottoman Empire* (Aldershot: Ashgate Variorum, 2004), 239.

[66] Peter Galison, *Image and Logic: A Material History of Microphysics* (Chicago: University of Chicago Press, 1997).

[67] Warwick, *Masters of Theory*. I am grateful to Will Deringer for suggesting this comparison.

generation of Azhari students for whom natural sciences were a requirement learned from Zayid himself that "astronomical knowledge passed to the Western lands, and nothing was left in the East but the remains of books past their time." Here, too, lay a certain resemblance to the case of clocks and watches, in which the role of 'ulama' faded from view even as their practices of precision timekeeping began to achieve widespread currency. In both cases, a peculiar gap seems to have opened between the growing role of 'ulama' in adapting new technologies for Ottoman-Egyptian use, and the declining credibility of 'ulama' in Ottoman-Egyptian conversations about science. The origins of this fissure lay in broader debates about modern knowledge and the "reform" of Islam.

Part III

Islam, Science, and Authority

5 Orbits of Print: Astronomy and the Ordering of Science and Religion in the Arabic Press

In 1876, a prominent Egyptian writer and educational official, 'Abd Allah Fikri Bey, published a booklet entitled *A Treatise on Comparing Some Texts of Astronomy with What Appears in the Sharia Texts*.[1] Fikri's *Treatise* argued passionately and methodically for the agreement of European astronomy with canonical Islamic teachings. Written to intervene in a debate in a biweekly journal in Cairo, the *Treatise* saw publication twice more that year, in journals in Cairo and Beirut, as public debate over astronomy ricocheted across the Levant. The highest levels of the Egyptian government got involved, as did students at the Syrian Protestant College in Beirut, a printer in Alexandria, and at least one official of the Greek Patriarchate – in addition to their readers.

The debate in which Fikri participated took place amid a decade of political crisis in Egypt. The 1870s were years of intense struggle for power between the Khedive Ismail, his family, his ministers and bureaucrats, the imperial government in Istanbul, and the governments of Europe. The year 1876 was one of escalation in that struggle, bringing it toward the boiling point it would reach at the end of the decade. In March, the Cave Report popped the bubble of the country's finances, leading to the collapse of Egyptian credit and the imposition of European financial control through the Caisse de la dette publique.[2] Between 1879 and 1882, a coalition of Egyptian army officers, peasant landowners, and urban guild leaders and 'ulama' led by Colonel Ahmed 'Urabi sought to wrest power from the Ottoman-Egyptian and European elites, leading to the British Occupation of the country beginning in 1882.[3]

The turbulent political climate that Fikri navigated in the 1870s helps to explain the most striking aspect of his *Treatise*'s intervention in the

---

[1] 'Abd Allah Fikri, "Risala fi muqaranat ba'd nusus al-hay'a bi-l-warid fi al-nusus al-shar'iyya," *Rawdat al-Madaris* 7, no. 5 (15 Rabi' al-Awwal 1293 [1876]): 1–23. (The *Treatise* has its own pagination, separate from that of the regular issue.)

[2] On the Egyptian bankruptcy, see Owen, *The Middle East in the World Economy*, 122f.

[3] Cole, *Colonialism and Revolution*, 22.

debate over astronomy, which was his insistence that it should not have been a debate at all. According to Fikri, criticism of astronomical knowledge was typical of the "poor behavior" (sū' al-'āda) current among certain people who, when they encounter a beneficial piece of writing, focus on the one aspect with which they disagree, or, not finding one, invent it.[4] Even worse than these "scum of the earth" (huthālat al-huthāla), Fikri went on, are those who simply follow their lead. It is due to such behavior that people have become reluctant to publish, and "knowledge is fading" (yadmahill al-'ilm).[5] As a solution, Fikri proposed "a great scientific society" (jam'iyya 'azīma 'ilmiyya), a committee of "eminent scholars, people of expertise, insight, and knowledge in accordance with the homeland (watan) and the love and true service thereof, to whom everyone would submit their writing."[6] If the society found their work acceptable, they would issue it a note of approval (taqrīz) and permit its publication. If not, they would prohibit publication and explain to the author his error.

If Fikri's "great scientific society" seems like a disproportionate response to an argument about astronomy, the fact that such a response made sense to him suggests that more than astronomy was at stake. Fikri's "great scientific society" would have done more than prevent the ignorant from criticizing the knowledgeable. Amid sociopolitical upheaval that played out – for the first time in Egyptian history – in a rapidly expanding medium of Arabic print, Fikri saw the publishing of science in the periodical press as a crucial area in which to define who was knowledgeable about what, and where the boundaries of their competence lay.

The final chapters of this book will seek to explain the role that astronomy played in the changing practice of Islam in the early twentieth century, as the Ottoman Empire came to an end, the British Occupation tightened and then nearly lost its hold on Egypt, and a new middle class debated the future of its political and religious communities. Chapter 6 accounts for the relatively uncontested standardization of prayer timing in this context, while Chapter 7 reinterprets a hotly contested debate over the standardization of the Islamic calendar. Both debates, however, played out largely in print, and shared certain premises that had come to govern discussions of science and religion in much of the Arabic press during the late nineteenth and early twentieth centuries. The most basic of these premises was that scholars of Islam and scientific experts belonged to categorically distinct social groups. On top of this premise, however, arguments such as Fikri's helped to construct a particular

---

[4] Fikri, Risala, 20–21.
[5] Fikri, Risala, 22.
[6] Fikri, Risala, 22.

relation of authority, in which religious scholars were to interpret their texts to agree with the claims of scientific experts. This was no "differentiation" of science from religion, to borrow the language of secularization theory.[7] To the contrary, the increasing distinction of science from religion in terms of social groups was coupled with an unprecedented imbrication of science with the interpretation of religious texts.

Defining science and religion in the press entailed new representations of scientific authority, as well as new histories of science. In the print journals of the late nineteenth and early twentieth centuries, a generation of Arabic readers learned to see what science looks like, and where it comes from – as well what science does not look like, and where it does not come from. Notably, astronomers and religious scholars could be clearly distinguished in part because the Arabic press scarcely acknowledged a living tradition of astronomy among 'ulama'. This omission allowed for Islam to become the "Sharia texts" of Fikri's title, while astronomy lay outside the borders of traditional scholarly competence.

The growing appeal of such representations (and their omissions) was not, of course, the work of a single man, even as influential a bureaucrat and author as 'Abd Allah Fikri Pasha. The 1876 debate serves as a lens through which to view the way in which the writing of astronomy in the press became a site for ordering the relationship between scientific and religious knowledge, allowing for "reconciliation" (tawfiq) between the two according to specific terms. As the diversity of the participants in this debate suggests, arguments about science in the press were trans-confessional. Questions, texts, and interpretive strategies circulated among Muslim, Orthodox, Coptic, and Protestant writers and readers. Tracking the development of astronomy in the press beyond 1876, I compare the representation of astronomy in two apparently very different journals: al-Muqtataf, the preeminent Arabic journal of science in the late nineteenth and early twentieth centuries, and al-Manar, the influential journal of the Muslim reformist Muhammad Rashid Rida. Despite substantial differences between the respective ideological agendas of their publishers, al-Muqtataf and al-Manar were remarkably similar in promoting the cultural authority of new scientific actors and institutions, while confining the practice of science among 'ulama' to a place in historical memory. This understanding of science was a common denominator between al-Muqtataf's publishers' agenda of social progress, and Rida's agenda of Islamic educational and political reform.

[7] Casanova, *Public Religions*, 6.

Science in the Press: Global and Ottoman-Arab Contexts of Print and Reading

Science in the press experienced a global boom in the nineteenth century. From Russia and China to North America, interest in science was a major engine for the larger growth of print publishing in this period.[8] Rapid growth stemmed from the availability of increasingly fast and affordable print technologies, alongside the expansion of literate classes and their institutions: schools, libraries, and intellectual societies. Science publishing in this context was an arena in which different social groups competed for cultural authority. In Victorian England, for example, the men of science (most famously Thomas Huxley) fought hard to arrogate to themselves the exclusive privilege of speaking for science. Their victory, however, was hardly assured. Older kinds of scientific authorship, for example among clergy and women, maintained an important place even as their authority gradually diminished.[9]

Such contests over authority underscore the analytical pitfalls of conflating science in the press with "popular science" at a time when the distinction between "popular" and "professional" science was only just emerging.[10] In fact, the growth of science writing in a press that sought to reach relatively broad audiences was crucial to the construction of a professional–popular boundary, which may also be understood as the product of parallel processes of professionalization in science *and* writing in the nineteenth century.[11] In the case of Ottoman publishing, it is particularly difficult to specify the meaning of "popular" as opposed to professional science, since the category had no Turkish or Arabic equivalent in this period. *Al-Muqtataf,* for example, declared itself a "scientific journal" (*majalla 'ilmiyya*) on its masthead.[12] Debates about science in

---

[8] On science in the nineteenth-century European press, see Bernard Lightman, *Victorian Popularizers of Science* (Chicago: University of Chicago Press, 2007); James Secord, *Victorian Sensation* (Chicago: University of Chicago Press, 2000); Ruth Barton, "Just before 'Nature': The Purpose of Science and the Purpose of Popularization in Some English Popular Science Journals of the 1860s," *Annals of Science* 55 (1998): 1–33; Susan Sheets-Pyeson, "Popular Science Periodicals in Paris and London: the Emergence of a Low Scientific Culture, 1820–1875," *Annals of Science* 42 (1985): 549–72. For non-Western contexts, see the discussion of Chinese scientific magazines in Elman, *On Their Own Terms,* 310f.; and James Andrews, *Science for the Masses: The Bolshevik State, Public Science, and the Popular Imagination in Soviet Russia, 1917–1934* (College Station, TX: Texas A & M University Press, 2003), chs. 1 and 2.

[9] Lightman, *Victorian Popularizers.*

[10] For an overview of this problem, see Lightman, *Victorian Popularizers,* 9–13.

[11] Lightman, *Victorian Popularizers.*

[12] A fundamental problem with the popular–professional dichotomy is that, even in a highly professionalized context, scientists draw on the analytical and imaginative resources of their broader culture. See Stephen Hilgartner, "The Dominant View of

the press were a crucial way in which the meaning of science and the borders of its authority were stabilized in late Ottoman society.

Understanding the role of the press in forging the cultural authority of science as a profession requires attention to the material circumstances of print production alongside a close reading of printed texts. Late Ottoman printing differed greatly from the Victorian press, for example, in which science publishing was an increasingly commercial pursuit, shaped by the pressures of business.[13] Although private Ottoman groups had been using print on a very small scale since the sixteenth century, widespread printing only began in the first half of the nineteenth century, in presses that were financed and run by the state. In Cairo and its environs, the principal state press at Bulaq was complemented by presses at the Citadel and at several of the state technical academies, where print was used largely to produce instructional books, as well as the official gazette.[14]

The first flowering of private printing in the Ottoman Empire began during the *Tanzimat* period. In Cairo, a small number of private consortiums began to use lithographic presses in the 1850s, producing classics of Arabic literature as well as popular Islamic tracts on a commissioning model. Thanks to the Egyptian government's sale of much of its printing equipment in the 1860s, such presses were able to expand into typography and produce some of the first private periodicals in Egypt in the 1870s.[15] During the authoritarian rule of Sultan Abdülhamit II (r. 1876–1909), whose name became almost a synonym for censorship, the degree of autonomy from Istanbul that Egypt enjoyed (especially after 1882) offered a particularly inviting climate to publishers.[16] Both Cairo and Alexandria were home to a flourishing periodical press in the last quarter of the nineteenth century, in which Egyptians worked alongside a growing community of journalists from other Ottoman lands, particularly Syria. While many of the leading figures in this burgeoning print

Popularization: Conceptual Problems, Political Uses," *Social Studies of Science* 20 (1990): 519–39. Consider also the importance of "popular science" in Fleck's model of the modern scientific "Thought Collective": Ludwik Fleck, *Genesis and Development of a Scientific Fact*, trans. Fred Bradley and Thaddeus J. Trenn (Chicago: University of Chicago Press, 1979), 112.

[13] Secord, *Victorian Sensation*, 437f.

[14] On the emergence of print production in Cairo, see Kathryn A. Schwartz, "Meaningful Mediums: A Material and Intellectual History of Manuscript and Print Production in Nineteenth-Century Ottoman Cairo" (Ph.D. Diss., Harvard University, 2015).

[15] Kathryn A. Schwartz, "The Political Economy of Private Printing in Cairo as Told from a Commissioning Deal Turned Sour, 1871," *IJMES* 49 (2017): 25–45.

[16] On the exaggerated reputation of Hamidian censorship, see Donald Cioeta, "Ottoman Censorship in Lebanon and Syria," *IJMES* 10 (1979): 167–86.

culture were private individuals, however, the press remained deeply entangled with the state, as well as with other institutions and actors for whom commerce was a secondary consideration. In addition to independent ventures, the periodicals of the 1870s and 1880s included government organs, missionary publications, and "semi-private" journals receiving subsidies from the British or Egyptian governments.[17] The leading "scientific journal" al-Muqtataf, for example, had been founded under the auspices of the Syrian Protestant College in Beirut. After its publishers emigrated to Cairo in 1884, however, their work was supported by a subsidy they received from the British to publish their daily newspaper, al-Muqattam.[18]

Late Ottoman science in the press, like much science writing in the nineteenth century, evinced an optimistic faith in the progress of science and technology as a vanguard for the progress of society. One of the first private Ottoman periodicals was Mecmua-i Fünûn, the journal of the Ottoman Scientific Society, which introduced readers not only to developments in European science, but also to the ideology of materialism and progress.[19] By the end of the century, a whole coterie of young Ottoman intellectuals, many of them educated in the radical milieu of the Imperial Medical College, had adopted scientific materialism as a creed.[20] Founding members of the CUP such as Abdullah Cevdet were avid followers of European science in the press, in which they believed they had found the essence of the European civilization on which they sought to model a reformed Ottoman Empire.

Istanbul and Ottoman exiles in Europe were the main nodes of this radical, Turcophone science in the press, with which 'Abd Allah Fikri, who read and wrote in Turkish as well as Arabic, was certainly familiar. However, in Fikri's immediate milieu, the mainly Arabophone press of Egypt and the Syrian lands, science writing differed in ideology as well as language. If the Turkish writers of science in the press tended to be imperial elites and rising military officers who enlisted science as the basis for a concrete political agenda, the early writers of science in the Arabic press tended to be missionary students, provincial elites, and litterateurs who conceived of their work as part of the self-styled "nahḍa"

---

[17] For this taxonomy of the early Egyptian periodical press, see Ami Ayalon, The Press in the Arab Middle East: A History (New York: Oxford University Press, 1995), 41–49.

[18] On the relationship between the British and the "Muqattamites," see E. Roger Owen, Lord Cromer: Victorian Imperialist, Edwardian Proconsul (Oxford: Oxford University Press, 2004), 252.

[19] Hanioğlu, Brief History, 94.

[20] M. Şükrü Hanioğlu, The Young Turks in Opposition (New York: Oxford University Press, 1995), 20.

or renaissance of Arabic letters in the late nineteenth century. This was a political project in its own way, but in the short term, although Arabic writers shared the faith of their Turcophone compatriots in the progress of humanity through science, most of them were focused on achieving progress through education rather than direct political action.

Such substantive differences between the Arabic and Turkish treatment of science in the press did not mean that they were disconnected worlds of publishing. To the contrary, Ottoman-Arab elites looked to Istanbul as a major source of patronage and a center of circulation for Arabic print. For example, 'Abd Allah Fikri's interpretation of Qur'anic verses in light of modern astronomy drew directly on a book called *Efkâr ül-Ceberût fi tercümet-i Esrar il-melekût*, which was published in Istanbul in 1848, not long before Fikri's first visit to the Ottoman capital.[21] The book is a Turkish translation of, and commentary on, an Arabic text by an Azeri prince, Abbaskulu Ağa Bakülü Kudsi (also called Bakikhanov), who had traveled extensively through Eastern Europe and the Caucuses while in Russian service, and was a mainstay of the literary and philosophical societies of Tiflis in the 1840s. (He is best known for his poetry, and for his history of Shirvan and Daghestan.)[22] Stopping in Istanbul on his way to perform the hajj in 1846, Bakülü presented Sultan Abdelmecid I with a brief Arabic text, *Asrar al-Malakut*, which offered an introduction to "new astronomy" (*al-hay'a al-jadīda/heyet-i cedide*), specifically an overview of the solar system, followed by discussion of Qur'anic verses and hadith that can be interpreted to agree with it.[23] The Turkish translation, *Efkâr ül-Ceberût*, done by Şerif Halil Elbistani at the request of Grand Vizier Reşid Paşa, substantially expanded on Bakülü's discussion both of astronomy and of the Qur'an and hadith.[24] Fikri explicitly cited the book's analysis of two Qur'anic verses,[25] but several

---

[21] Bakülü Kudsi, *Efkâr ül-Ceberût fi tercümet-i Esrar il-melekût* (Istanbul: Dar üt-Tıbaat ül-Amire, 1265 [1848]).

[22] For the life of Bakülü, see Abbas Qoli Aqa Bakikhanov, *The Heavenly Rose-Garden: A History of Shirvan & Daghestan*, trans. Willem Floor & Hasan Javadi (Washington, DC: Mage, 2009), vii–xvii, especially at p. xii. See also Audrey L. Alstadt, "Nasihatlar of Abbas Kulu Agha Bakikhanli," in *Central Asian Monuments*, ed. H.B. Paksoy (Istanbul: ISIS Press, 1992), accessed online at http://eurasia-research.com/erc/007cam.htm. I am grateful to Rebecca Gould for her help contextualizing this fascinating figure: see her article, "Cosmopolitical Genres and Geographies: Poetry and History in the Nineteenth Century Caucasus," *Comparative Literature* 70 (2018, forthcoming).

[23] İhsanoğlu, "Introduction of Western Science."

[24] For example, see Kudsi, *Efkâr ül-Ceberût*, 213, where the translator's commentary elaborates on the notion that the sun might revolve around another star. The commentary cites a recent (mistaken) report in the Ottoman Press that an Irishman named Hamilton had determined the location of this star. Cf. Robert Perceval Graves, *The Life of Sir William Rowan Hamilton*, vol. 2 (Dublin: Hodges, Figgis & Co., 1885), 546.

[25] Fikri, *Risala*, 7–8.

more of the exegetical references in Fikri's *Treatise* can be found either in the Arabic or Turkish sections of *Efkâr ül-Ceberût*.[26] Thus, by the time "new astronomy" reached Fikri's readers, it had moved from the Russian imperial schools in which Kudsi was educated, through the Russian– Qajar frontier, to Istanbul, and finally to Cairo and Beirut, while passing linguistically from Russian to Arabic (possibly via Persian), from Arabic to Turkish, and finally back into Arabic. The translation of science into Arabic in the late Ottoman period took place along multiple axes, not only encounter between Western Europe and Arabs. Istanbul and the Turkish language were crucial links between Ottoman-Arab intellectuals and a larger geography of print.

The multiple printings of Fikri's treatise point to the fact that the Arabic periodical press, like late Ottoman publishing in general, was stimulated by the beginnings of a middle-class readership, particularly in the cities of the Eastern Mediterranean like Cairo, Alexandria, Beirut, and Aleppo. This readership was the product of the expansion of both civil and military schools during the *Tanzimat* period, and in Egypt particularly during the reign of Mehmed Ali's descendant the Khedive Ismail Pasha. Missionary schools, expatriate merchant communities, and the increasing prosperity of the region's religious minorities also fed the growth of a class of people who came to define themselves through participation in "modern" practices of schooling, reading, writing, and sociability.[27] As the existence of a flourishing "women's press" in this period suggests, this class was by no means limited to men.[28] Mere quantitative assessments of "literacy" and print runs understate the significance of this emerging culture. Late Ottomans who could not read commonly consumed newspapers by hearing them read in communal gatherings, while those who could not write sent letters through the services of a scribe.[29] Especially as print publishing began to enter the domain of private or semi-private enterprise in the late nineteenth century, this broadly construed "readership" debated the norms that should govern a new form of public discourse. In this context, an argument about astronomy became

---

[26] Compare, for example, Fikri's citation of Baydawi's *tafsir* on Q2:255 (*Risala*, 12), with Bakülü's citation of the same (*Efkâr ül-Ceberût*, 221). Fikri's discussion of Razi's interpretation of the word "falak" (sphere/orbit) owes a similar debt to Elbistani's commentary on Bakülü's text. (Compare *Risala*, 11–12, with *Efkâr ül-Ceberût*, 186 and 215.)

[27] Watenpaugh, *Being Modern*.

[28] Beth Baron, *The Women's Awakening in Egypt: Culture, Society, and the Press* (New Haven, CT: Yale University Press, 1994).

[29] Hoda Yousef, *Composing Egypt: Reading, Writing, and the Emergence of a Modern Nation, 1870–1930* (Stanford, CA: Stanford University Press, 2016). On the "pattern of reading papers as a collective experience," see also Ayalon, *The Press*, 156.

an argument about who had the right to speak about different kinds of knowledge.

'Abd Allah Fikri: Regulating Science and Religion in the Press as Ottoman-Egyptian Political Reform

What exactly was the "poor behavior" that provoked 'Abd Allah Fikri to advocate a new form of censorship in the Egyptian press? The controversy had its roots in 1872, when a four-part series on astronomy appeared in *al-Jinan*, a journal founded by the Syrian Christian litterateur, pedagogue, and encyclopedist Butrus al-Bustani, whose objective was to spread "universal knowledge – scientific, cultural, historical, industrial, commercial... as in the foreign countries where their benefits have become evident."[30] Framed by articles on the latest news from European countries, as well as puzzles, stories, and historical narratives, the series on astronomy offered an overview of current astronomical understanding of the solar system, stars, and comets, written by Salim al-Bustani (the founder's son).[31]

Some readers were displeased. In Alexandria, a Syrian emigrant, Salim Ilyas al-Hamawi, wrote a lengthy critique of Bustani's series. Entitled *The Decisive Proofs for the Lack of Rotation of the Terrestrial Sphere*, the piece drew mostly on biblical as well as physical arguments. When *al-Jinan* declined to publish it, the author printed it in 1873 in his own journal, *al-Kawkab al-Sharqi*, Alexandria's first Arabic periodical.[32] The journal was soon closed by Khedival order; Ismail Pasha was not ready to permit the publishing of independent newspapers in Arabic.[33] But Hamawi kept his press, and he reprinted *The Decisive Proofs* as its own booklet sometime in 1875 or 1876.[34]

Meanwhile, a parallel debate was brewing in Islamic terms in Cairo. *Wadi al-Nil* was a leading "semi-private" journal of this era, not exactly a government organ but sponsored directly by Ismail Pasha. From 1867 to 1874, it appeared semiweekly under the editorship of 'Abd Allah Abu al-Su'ud, a student of Rifa'a al-Tahtawi and his successor as director of

---

[30] Quoted in Ayalon, *The Press*, 35.

[31] *Al-Jinan* 1872, no. 14, 476–83; no. 15, 518–23; no. 16, 553–59; no. 17, 588–96.

[32] See Salim al-Hamawi, *Al-Barahin al-qat'iyya 'ala 'adam dawaran al-kura al-ardiyya* (Alexandria: Matba'at al-Kawkab al-Sharqi, 1293 [1876]), 2; and Ayalon, *The Press*, 42.

[33] Filib Di Tarrazi, *Tarikh al-Sihafa al-'Arabiyya*, vol. 3 (Beirut: Al-Matba'a al-Adabiyya, 1914), 48–9.

[34] Hamawi, *Al-Barahin al-qat'iyya*, 2. The book's cover says 1875, while the colophon gives 1876 as the date of publication.

the School of Languages (*madrasat al-alsun*).[35] Abu al-Suʿud shared his
mentor's interest in geography and related sciences, and published a text
providing a scientific explanation for lunar and solar eclipses, including
a description of the earth's spherical shape and characterizing the earth's
"dome" (*qubbat al-ard*, i.e., the sky) as an illusion. Again, certain readers
took exception to these claims, arguing in this case that they contradicted Islamic (rather than Christian) tenets. It was in response to these
critiques that ʿAbd Allah Fikri composed his defense of "astronomical
texts," which appeared first in the pages of *Wadi al-Nil* itself. Fikri published an expanded version in *Rawdat al-Madaris*, an organ for the various
schools under the Department of Education, for which Fikri worked.[36]
Founded under the editorship of Tahtawi himself, the journal had passed
into the hands of his son, ʿAli Fahmi, director of the Educational Presses
(Matbuʿat al-Maʿarif). By publishing Fikri's work first in *Wadi al-Nil* and
then, revised and expanded, in *Rawdat al-Madaris*, the viceregal state was
using the power of its presses to double down on its defense of "new"
science against criticism.

But the Arabic press of the 1870s was becoming a more open space for
debate, one which neither the government in Cairo nor the government
in Istanbul fully controlled. Several months after Fikri's *Treatise* appeared
in *Rawdat al-Madaris*, *al-Jinan* revisited the topic of astronomy, this time
to publish a piece by one Nasir Effendi Khuri, who critiqued the notions
of the earth's rotation and revolution around the sun.[37] A fierce argument ensued, ricocheting across Syria and Egypt. *Al-Muqtataf* came to
the defense of the earth's motion, critiquing the piece in *al-Jinan*.[38] The
latter's publishers soon disavowed Khuri's piece, claiming that they had
published it only as an example of "the old astronomical views," which
now "were falsified by instruments, testing, and close examination" (*al-
ālāt wa-l-fahs wa-l-tadqīq*).[39] Archimandrite Gabriel Jibara, the representative of the Antiochian Greek Patriarch in Beirut, came to Khuri's
defense in *al-Muqtataf*, arguing against the earth's motion using Old
Testament verses, previous articles in *al-Jinan*, and Hamawi's *Decisive*

[35] Ayalon, *The Press*, 41. *Rawdat al-Akhbar* (1874–78) was a continuation of the same
newspaper. For a discussion of Abu al-Suʿud and *Wadi al-Nil* in the context of the introduction of telegraphy in Egypt, see Barak, *On Time*, 117–19.
[36] On *Rawdat al-Madaris* as a window onto continuities between "traditional" and new
education in Egypt, see Hoda Yousef, "Reassessing Egypt's System of Dual Education
under Ismaʿil: Growing ʿIlm and Shifting Ground in Egypt's First Educational Journal,
Rawdat al-Madaris," *IJMES* 40 (2008): 109–30.
[37] Nasir Effendi Khuri, *al-Jinan* 20 (1876), 71.
[38] "Dawaran al-ard," *al-Muqtataf* 1 (1876), 141.
[39] "Jumla falakiyya," *al-Jinan*, no. 23 (1876), 808.

Proofs.[40] Among *al-Muqtataf's* Egyptian readers was Riaz Pasha, the Director of Education (*nāzir al-maʿārif*) and Fikri's bureaucratic superior. Perhaps annoyed that a controversy he had thought was just resolved in the Egyptian press should now bubble up in the journals that came in from Syria, Riaz Pasha sent Fikri's *Treatise* to the *Muqtataf* editors, who happily published it (now in its third printing) as a response to Jibara. It was seemingly no great leap for Fikri's work on new astronomy and Islamic sources to pass from its origins within Egyptian educational circles into a debate among Syrian Christians. By late 1876, that debate included Muslim, Greek Orthodox, and Protestant Christian participants based in Beirut, Alexandria, and Cairo, using biblical and Qur'anic texts to debate a question that some believed had long been settled by "instruments, testing, and close examination." This was the context in which ʿAbd Allah Fikri saw the exchange as an opportunity to argue for a more orderly press regime.

In a sense, Fikri's proposal was an effort to translate longstanding norms of scholarly manuscript culture into the new arena of print. The concept of *ijāza* (permission) evoked the traditional means by which 'ulama' authorized their students to teach specific books that they had mastered. The note of approval or "blurb" (*taqrīz*) was also a common means by which scholars had supported each other's work.[41] In manuscript culture, however, the *ijāza* and the *taqrīz* were ways in which one individual lent informal credibility to another. Fikri's proposal would have transformed these practices into bureaucratic instruments. It is particularly revealing that Fikri, drawing on vocabulary that Tahtawi had developed to render the French concept of *patrie* into Arabic, cited love and service of the homeland (*watan*) as the principal motives of the scientific society's potential members.[42]

Fikri's proposal reflected the general ambivalence of late Ottoman elites toward the power and dangers of the press, but also the viewpoint of a more specific Ottoman-Egyptian elite that was becoming increasingly assertive in the 1870s. The story of ʿAbd Allah Fikri's life is the story of the rise of new bureaucrats, their ambiguous relationship with the viceregal family, and the changes that they brought about in the nature of Ottoman-Egyptian politics.[43] The grandson of a prominent Maliki scholar

[40] Ghabri'il Jibara, "Thubut al-ard," *al-Muqtataf* 1 (1876), 171–74. See al-Hamawi, *Al-Barahin al-qatʿiyya*, 2.

[41] Franz Rosenthal, "'Blurbs' (*taqrīz*) from fourteenth-century Egypt," *Oriens* 27/28 (1981): 177–96.

[42] Hourani, *Arabic Thought*, 78–81. Fikri used "*mahabbat al-watan*" rather than *hubb al-watan*.

[43] For Fikri's life, see Muhammad ʿAbd al-Ghani Hasan, *ʿAbd Allah Fikri* (Cairo: al-Mu'assasa al-Misriyya al-ʿAmma lil-ta'lif wa-l-anba' wa-l-nashr, n.d.); Fikri's obituary

and the son of an engineer in Mehmed Ali Pasha's army, Fikri was born in Mecca in 1834, while his father was serving in the Hijaz. Returning to Egypt as a boy, Fikri pursued his studies at al-Azhar, rather than the government schools through which his father had achieved prominence.[44] The learning that Fikri acquired at al-Azhar, including the usual areas of Arabic language, Islamic jurisprudence, *hadith*, and Qur'an commentary, but also philosophy, is evident in the facility with Islamic scholarly sources that Fikri later displayed in his treatise on astronomy.[45]

But Fikri's path to political power opened with his decision to study Turkish. Possibly it came easily to him because it was his mother's language. (Fikri's father had met his mother during the Morean campaign.)[46] By 1851, at the age of seventeen, Fikri was able to take a position in the Turkish Bureau (*kalem*) of the office of the Khedive's "First Lieutenant," where he served mainly as an Arabic–Turkish translator.[47] Soon, Fikri joined the "viceregal entourage" (*al-maʿiyya al-saniyya*) under Said Pasha, and became a trusted aide to his successor Ismail. The latter chose Fikri to accompany him to Istanbul on official visits and to oversee the education of his children in Arabic, Persian, and Turkish. Among Fikri's students was the future Khedive Tawfiq, with whom he also spent time in Istanbul. His close association with the viceregal household facilitated Fikri's rise through the Ottoman-Egyptian government. In 1871 he entered the administration of the native primary schools (Diwan al-Makatib al-Ahliyya) as the deputy to ʿAli Mubarak, the leading figure in educational reform of his generation. When Tawfiq Pasha assumed the governorship of Egypt after Ismail's deposition in 1879, among the new Khedive's first bureaucratic appointments was to make his old teacher deputy director of the Department of Education. Fikri assumed the top position itself in January 1882.

Despite Fikri's close relationship to the viceregal household, particularly Tawfiq, his loyalties were complex. The turbulence of Egyptian

---

in *al-Muqtataf* 15 (1890): 9–16, 81–89; and Muhammad ʿAbduh, biographical introduction (*tarjama*) to *al-Athar al-Fikriyya*, by ʿAbd Allah Fikri Basha, ed. Amin Fikri (Bulaq: al-Matbaʿa al-Kubra al-Amiriyya, 1897), 4–12.

[44] Hasan, *ʿAbd Allah Fikri*, 8.

[45] Fikri studied with Shaykh ʿAli Khalil al-Asyuti, who was noted for knowledge of philosophy. ʿAbduh, introduction to *al-Athar*, 5.

[46] On Fikri's parents having met in the Morea, see ʿAbduh, introduction to *al-Athar*, 4. On the disastrous Morean Campaign, see Fahmy, *All the Pasha's Men*, 55–60.

[47] Fikri was seventeen years old by the lunar reckoning of the *hijri* calendar, which was how Egyptian Muslims of Fikri's generation kept track of birthdays (to the extent that they did so at all). By the solar reckoning of the Gregorian calendar, which Egyptians began to use in the 1870s, Fikri would have been several months younger, and just shy of seventeen.

politics in the 1870s was partly a result of the financial crisis and European encroachment, but it can also be seen as a consequence of sixty years of state expansion that had brought more Egyptians – like Fikri – into the government and destabilized the position of the old Ottoman elite.[48] It was precisely in this decade that men like Fikri's patron, Riaz Pasha, began to wield power independently of the viceroy, paving the way for the radical experiment in governance that the 'Urabi movement briefly introduced in 1881 – but also facilitating the British Occupation of the country beginning in the summer of 1882. Fikri himself was sufficiently implicated in the 'Urabi movement to be briefly imprisoned after its defeat.[49] According to 'Urabi himself, Fikri had taken up the cause of the rebels so enthusiastically as to suggest the killing of his former student the Khedive.[50] Although this account should be taken with a grain of salt ('Urabi was embittered by Fikri's reconciliation with Tawfiq), what is clear is that Fikri was part of the class of bureaucrats who, during the 1860s and 1870s, came to constitute a base of power that was at least partially independent of the viceroy and his household.

In this light, it bears notice that Fikri's proposal for a "great scientific society" came at a time when censorship took the form of direct vice-regal funding, licensing, and banning of periodicals.[51] Fikri's proposal might have made the press a more disciplined space, but it would also have transferred the exercise of power from the hands of the viceroy to the hands of "experts." In this respect, his vision for regulating the presentation of knowledge in print ran parallel to the efforts of Riaz Pasha, his colleagues, and European diplomats who sought to shift the locus of power in the government itself.[52] Proposing a board of experts to oversee publishing in all areas of knowledge reflected a fear that, in public controversies, the new periodical press might empower the wrong people. But it also testified to a faith, characteristic of the new bureaucratic elite, that the solution to this problem lay in the institutionalization of scientific expertise within the state.

[48] Hunter, *Egypt Under the Khedives*.
[49] A poem that Fikri wrote in the Khedive's honor helped secure his release, and Fikri resumed his career at the pinnacle of the Egyptian intelligentsia for the last years of his life. In 1889, he was a member of the Egyptian delegation to the International Congress of Orientalists in Stockholm. His son Amin published a memoir of the trip, which included a visit to Paris that Mitchell memorably analyzes in his account of "Egypt at the Exhibition," in *Colonising Egypt*, 1–33.
[50] Ahmad 'Urabi, *Mudhakkirat al-za'im Ahmad 'Urabi*, vol. 3, ed. 'Abd al-Mun'im Ibrahim al-Jumay'i (Cairo: Matba'at Dar al-Kutub, 2005), 1199.
[51] Ayalon, *The Press*, 42.
[52] On the rise of "expertise" in Egyptian governance, see Mitchell, *Rule of Experts*.

The *faqīh* and the Man of Astronomy:
Setting the Terms of Debate

While Fikri did not specify how his proposed "great scientific society" would determine the criteria of "acceptable" scholarship, the *Treatise* made clear that such scholarship should respect a particular arrangement of expertise. The bulk of the text comprises a detailed and multipronged defense of the Islamic legitimacy of the controversial astronomical claims that the editor of *Wadi al-Nil* had published. While this discussion noted certain points of agreement between post-Copernican and classical Islamic cosmology, ultimately it relied on a broader claim that interpreters of the Qur'an should defer, in matters of astronomy, to the current knowledge of astronomers. It was no accident that Fikri penned his treatise as a dialogue between a *faqīh* (traditional Muslim scholar) and a man of astronomy (*ṣāḥib al-hay'a*). The reconciliation (*tawfīq*) of "Sharia" with "astronomy" depended on setting the terms of engagement between newly distinct social groups.

As Fikri noted, a few of the astronomical ideas that had attracted controversy in *Wadi al-Nil* were, in fact, not new to Islam at all. Regarding the sphericity of the earth, for example, Fikri quoted extensively from the thirteenth-century Qur'an commentary of the Persian exegete, theologian, and philosopher Fakhr al-Din al-Razi, who referred to the earth's spherical shape as a matter "established by proofs."[53] However, not every aspect of the post-Copernican cosmos enjoyed such explicit support in Islamic exegesis. While this was most famously true of the axial and orbital motions of the earth, it was also true of less notorious ideas: for example, that the heavens do not comprise physical spheres, but rather a single, mostly empty, space. In fact, judging by the amount of attention it received in the *Treatise*, the broader conflict between new astronomy and Aristotelian metaphysics was just as controversial as the specific issue of the earth's motion.[54]

Fikri's *Treatise* dealt with such conflicts by quoting classical commentaries to show that Qur'anic verses could be interpreted to agree with new astronomy's perspective. In doing so, however, Fikri put himself in the awkward position of quoting figures from the exegetical tradition in

---

[53] Fikri, *Risala*, 5. Quoting Fikri on Q 13:3.
[54] Cf. Saliba, "Copernican Astronomy," which views the debate as having revolved around the issue of the earth's motion. Understanding the debate, instead, as a broader crisis of cosmology or "world system" suggests a parallel with the Copernican debates in Europe. See Thomas Kuhn, *The Copernican Revolution* (Cambridge, MA: Harvard University Press, 1957), ch. 3.

defense of ideas that were quite far from, or even opposed to, their own.[55] For example, Fikri explained Qur'an 36:40, "they [the sun and moon] float each in a *falak* (sphere/orbit)," by quoting Razi's comment on a similar phrase in Qur'an 21:33:

"Falak" in the speech of the Arabs is anything rotating, and its plural is *aflāk*. Scholars have disagreed regarding it. Some of them say that the *falak* is not a body, but merely the orbit of the stars... Most have said, rather, that they are bodies, on which the stars rotate. ... The first opinion agrees with new astronomy, and the second agrees with the old.[56]

Razi clearly favored the second opinion, that the stars are situated on physical spheres; further in the same passage, he says that this opinion is "closer to the apparent meaning of the Qur'an."[57] But Razi's preference evidently did not bother Fikri, who knew that he was quoting his sources selectively. Although he certainly invested a great deal of effort into arguing that the classical exegetes offer support for new astronomy, ultimately his argument, and the importance it attributes to the Islamic exegetical tradition, rested on a more fundamental point:

You find that many exegetes understood some of the apparent meanings (*zawāhir*) of the Qur'an in a way that agreed with what was said in old astronomy. The fact is that both the masters of old astronomy, and those of new astronomy, are on the same level relative to us, so is it not permissible for me then to resort to interpreting (*ta'wīl*) those apparent meanings (*zawāhir*) in order to agree with something based on decisive proofs in new astronomy, which the apparent meanings' words accept and their expressions can mean?[58]

In this version of the argument, the exegetical authorities do not offer support for new astronomy *per se*. Rather, what they authorize is the interpretation of the Qur'an according to astronomy. Therefore, when astronomy changes, it is permitted to change one's interpretation of the Qur'an.

What was new about this principle was not, by itself, the notion that scriptural interpretation should be informed by natural knowledge. In

[55] In addition to continued use of Razi, Fikri relied most often on a group of closely related *tafsīr*s by Nasir al-Din al-Baydawi, Abu al-Suʿud al-ʿImadi, and Shaykhzade: a highly specific vein of the Islamic exegetical tradition, rooted either in Fakhr al-Din al-Razi's *Mafatih al-Ghayb* or Zamakhshari's *Kashshaf ʿan haqaʾiq al-tanzil*. See J. Robson, "al-Baydawi," *Encyclopaedia of Islam*, 2nd edition (Brill Online, [2012]); J. Schacht, "Abū 'l-Suʿūd," *Encyclopaedia of Islam*, 2nd edition (Brill Online, 2012).

[56] Fikri, *Risala*, 11–12. This is an instance in which Fikri drew almost verbatim from the Turkish text of Elbistani's commentary on Bakülü. See *Efkâr ül-Ceberût*, 186.

[57] Fakhr al-Din Muhammad ibn ʿUmar al-Razi, *Al-Tafsir al-Kabir li-l-Fakhr al-Razi*, vol. 22 (Tehran: Shirkat Sahafi Nawin, 198-), 167.

[58] Fikri, *Risala*, 9.

addition to Razi's commentary, Fikri also quoted from the introduction
to Ghazali's *Incoherence of the Philosophers* to establish precedent for this
principle.[59] The departure from this precedent was the extent to which,
for Fikri, astronomers and exegetes constituted particularly distinct social
groups. In the context of scholarly astronomy, one could hardly think of
practitioners as "astronomers" any more than they were also grammar-
ians or jurists. Recall Muhammad al-Khudari, who studied and wrote
across the disciplines of Islamic scholarship; even his work on astronomy
is filled with literary and legal discussions. He did not write a *tafsīr*, but
we can imagine that when he read the Qur'an, he did so through the lens
of his own astronomical knowledge. Similarly, the cosmological passages
in Razi's *tafsīr* that Fikri excerpted are part of lengthy philosophical discussions that Razi himself had clearly mastered.[60] Fikri articulated a different set of relations of authority, in which those who studied the stars
and those who studied the scripture were newly remote from each other.

The link between the acceptance of new astronomy and the cultural
authority of specific social groups becomes explicit in Fikri's *Treatise*
when the *faqīh* pushes the man of astronomy to defend the scientific certainty that leads him to reinterpret the Qur'an and hadith:

> The gist of what you are saying is that you only move to metaphorical interpre-
> tation (*ta'wīl*) of anything in the Qur'an or *sunna* when it contradicts definitive
> evidence and powerful proofs that leave you no room for doubt. If, among the
> questions of your science that you speak of, there is something for which decisive
> evidence and definitive proofs exist, then bring your proofs so we may examine
> them! If they are as you describe and of the type that you claim, then we will
> examine the metaphorical interpretations toward which you have moved.[61]

Fikri's man of astronomy answers this call for "definitive evidence" in a
curious way. He tells the *faqīh* to go read. "What you have asked for is laid
out and settled in the books composed on astronomy according to the
new way" ('*alā al-ṭarīqa al-jadīda*).[62] Furthermore, the man of astronomy
emphasizes, "People's discourse on this way, the proof of it, their objections to its opposite, is found in their books, which are many, circulating,

---

59 See Michael Marmura, trans., *The Incoherence of the Philosophers: a Parallel English-Arabic
    Text* (Provo, UT: BYU Press, 1997), 7.
60 Razi apparently did not write on astronomy, but he had clearly mastered all of the argu-
    ments pertaining to the structure of the cosmos. He also wrote on geometry, and a
    refutation of astrology. See G.C. Anawati, "Fakhr al-Din al-Razi," in *Encyclopaedia of
    Islam,* 2nd edition, ed. P. Bearman et al. (Brill Online). For an analysis of a fourteenth-
    century astronomer and exegete who drew heavily on Razi's work, see Morrison, *Islam
    and science.*
61 Fikri, *Risala*, 18.
62 Fikri, *Risala*, 18.

and printed." Fikri did not see it as his job, or that of his protagonist, to explain the scientific proofs of new astronomy. This point is more striking if we consider that Fikri authored both sides of this conversation. We can assume that he only had the *faqīh* request proofs specifically so that the man of astronomy could reject the notion that such proofs belonged in the discussion. Proof belonged, instead, to "*their* books" (my emphasis) – the "circulating, and printed" books of astronomers.

The line from Muhammad al-Khudari's *zīj* commentary in the 1820s to ʿAbd Allah Fikri's *Treatise* in the 1870s traces a shift in astronomy's relationship to other kinds of knowledge in Egyptian society. Understanding celestial motion was no longer thought of as a field of Islamic scholarship, but as a science created in an independent space. While I have focused thus far on the content of Fikri's *Treatise*, not the least remarkable of its features is something it does not contain, namely reference to an astronomer between the time of Ptolemy and the age of Copernicus, or to any Muslim astronomer of more recent times. For the purposes of legitimation, Fikri did give heliocentric astronomy an older pedigree, but it was the Pythagorean pedigree that Copernicus himself had claimed for his theory in *De Revolutionibus*.[63] Fikri passed over the entire Ptolemaic tradition as an unmitigated error, a move we can also see in Qustantin Yusuf's article in *al-Jinan* of the same year.[64]

This act of historical erasure was hardly inevitable. Recall, by way of contrast, the history of astronomy according to Lalande's translators in Istanbul and Aleppo, for whom the body of valuable knowledge formed a continuum comprising the ancient Greeks, earlier and more recent Muslims, and Europeans. For Fikri, by contrast, astronomy was not an active field of Islamic scholarship, but something that scholars of Islam were to receive in the printed books of "the select people of astronomy in this recent age,"[65] which offered "true observations,"[66] and "decisive evidence and accepted proofs that leave no room for doubt."[67] It is not that these phrases could not have appeared in Islamic discourse of an earlier age. For Fikri, however, they referred to something new: people and spaces at a certain distance from Islamic scholarly discourse.

The mirror image of Fikri's conception of astronomy was a conception of Islam. In the absence of a living tradition of scholarly astronomy, Islamic authority existed only within what Fikri called, in his title,

[63] Fikri, *Risala*, 18. See Kuhn, *Copernican Revolution*, 142.
[64] Fikri, *Risala*, 25. Yusuf, "Radd ʿala jumlat Nasir Effendi Khuri," 809–10.
[65] Fikri, *Risala*, 7.
[66] Fikri, *Risala*, 15.
[67] Fikri, *Risala*, 17.

"the Sharia texts": essentially the Qur'an and the exegetical tradition, in this context, with occasional reference to classic theological and philosophical texts. Whereas 'ulama' like Khalil al-'Azzazi and Husayn Zayid sanctioned new astronomy by translating it into the norms of scholarly astronomy, Fikri's approach to legitimizing science drew on a more circumscribed, "religious" tradition. But the relationship between these areas of knowledge was not symmetrical. Whereas the *faqīh*, knowing nothing of modern science, must take his cues from the man of astronomy, the latter is entirely at ease quoting Qur'an commentaries, hadith, and generally outmaneuvering the *faqīh* even in the interpretation of religious texts. Of course, the 1876 *Treatise* was still some distance from a world in which any layperson could claim to understand such texts. For its author, after all, facility with Qur'an commentaries was a product of a youth spent learning in the scholarly circles of Mecca and al-Azhar. Nevertheless, in a dialogue in which a *faqīh* was barred from speaking about astronomy while a man of astronomy could speak fluently about the Qur'an, scientific knowledge appeared to be technical and exclusive in a way that religious knowledge was not. From here, the twentieth-century embrace of lay hermeneutical authority was not far off. Put another way, the distinction between *faqīh* and man of astronomy can only be seen as a differentiation of science from religion in the negative, one-sided sense that the religious scholar was not to speak authoritatively about science. Meanwhile, science acquired an unprecedented authority over the meaning of revelation. The point was not to separate science from religion, but to arrange them according to a particular relation of authority.

Seeing the Skies in Print: Images and Histories of Astronomy in *al-Muqtataf*

The word of one bureaucrat was hardly enough to set the terms of debate in the Arabic press. Fikri's *Treatise*, and the debate in which it intervened, open a window onto a contest for authority that transpired in the way that science was represented in the press. However, while Fikri's "great scientific society" did not come to be,[68] the print journalism through

---

[68] Fikri gave his proposed society not only the power of censorship, but also the power to determine the proper translation of European terms. Many years later, the latter power did become the function of the Royal Academy of Cairo (later the Arabic Language Academy), founded in 1932. Fikri's proposal appears to be one of the earliest calls for such an academy, adding a particularly authoritarian vision to its history. For the "fundamental anxieties over textual or discursive authenticity and authority" reflected in the debates leading to the establishment of the Royal Academy, see Marwa Elshakry,

which science became increasingly accessible to Arabic readers of the late nineteenth century largely adhered to the principles that he had advocated. Indeed, it was *al-Muqtataf*, the journal in which much of the 1876 controversy had unfolded, that did more than any other institution to popularize the notion that astronomical knowledge, like all modern science, was the product of a new class of actors whose methods and instruments were uniquely privileged. In 1884, the journal's editors and publishers, Ya'qub Sarruf and Faris Nimr, emigrated from Beirut to Cairo, where *al-Muqtataf* became the most widely read journal of science in the Arabophone world until it closed in 1952. Intellectuals of the early twentieth century often referred to it not only as a reliable source of information, but as having shaped their worldview and even literary style. This was true of a broad range of readers, from the secularist and early Egyptian socialist, Salama Musa,[69] to Muslim reformists like Rashid Rida, who cited it frequently in his writing and enjoyed conversing with its publishers. As a prestigious, widely circulated journal, *al-Muqtataf* played an important role in resolving controversies and, in the process, setting the terms of debate.[70]

For Sarruf and Nimr, science journalism was a powerful tool in a project of social, cultural, and political reform. Like their mentors at the Syrian Protestant College, Sarruf and Nimr believed that as people came to understand the methods and findings of modern science, they would adopt more rational and industrious behaviors – ultimately producing a stronger, more prosperous society.[71] Thus, for example, the journal played a leading role in popularizing Darwin's theory of evolution in Arabic, introducing it alongside the gospel of "self-help" and individualism promulgated by Herbert Spencer, Edmond Demolins, and Samuel Smiles.[72]

Astronomy, too, had a role to play in cultivating the virtues of a modern society. Sarruf and Nimr's closest mentor at the Syrian Protestant College, Cornelius Van Dyck, was the author of one of the first Arabic

---

"Knowledge in Motion: The Cultural Politics of Modern Science Translations in Arabic," *Isis* 99 (2008), 726.

[69] Salama Musa, *The Education of Salama Musa*, trans. L.O. Schuman (Leiden: Brill, 1961), 37.

[70] Dagmar Glass, *Der Muqtataf und seine Öffentlichkeit: Aufklaerung, Raesonnement und Meinungsstreit in der fruehen arabischen Zeitschriftenkommunikation* (Würzburg: Ergon Verlag, 2004); Nadia Farag, "Al-Muqtataf 1876–1900: A Study of the Influence of Victorian Thought on Modern Arabic Thought" (Ph.D. diss., Oxford University, 1969).

[71] M. Elshakry, "The Gospel of Science and American Evangelism in Late Ottoman Beirut," *Past and Present* 196 (2007), 214.

[72] M. Elshakry, *Reading Darwin in Arabic*; Olivier Meier, *Al-Muqtataf et le débat sur le Darwinisme: Beyrouth, 1876–1885* (Cairo: CEDEJ, 1996).

textbooks of modern astronomy, *Usul 'ilm al-hay'a* (1874),[73] and *al-Muqtataf* published regularly on astronomy over the decades, long after the debate of the 1870s had ended. One purpose of such articles was to serve an ongoing polemic against astrology, which represented the antithesis of certain virtues that the publishers saw as essential to the future of Ottoman society. Whereas science would promote rational religion, astrology was mere superstition; whereas science required trained, critical thinking, astrology's popularity testified to the lamentable credulousness of an uneducated population; whereas science enjoyed increasingly well-defined professional borders, astrology was a morass of charlatanism.

Much of this polemic transpired in direct exchanges between *al-Muqtataf* and its readers, who frequently asked the publishers to comment on apparently efficacious instances of astrology and related arts. In one characteristic response, Sarruf and Nimr remarked:

> Our view of this issue is well-known. ... Magic and astrology are false, and the abundance of their books does not establish their validity. ... Careful examination has demonstrated the falsity of astrology, too, for a scientist recently examined the circumstances of many people in relation to their horoscopes, and found that those with the same horoscope did not resemble each other in their circumstances more than they resembled people with contrasting horoscopes.[74]

In addition to reporting that critical methodology had contradicted the claims of astrologers, *al-Muqtataf* encouraged readers to adopt such a methodology in their own manner of responding to extraordinary claims. Asked by a reader in Damascus to comment on the case of a man who was gaining fame in the city for predicting the future using physiognomy (*'ilm al-firāsa*), *al-Muqtataf* suggested that the man be asked to predict a number of events in the near future, and that these predictions should be recorded in a book to be sealed until they came true – or not.[75]

As the latter example suggests, "astrology" sometimes served as a catchall for what was actually an eclectic set of fashionable practices that Sarruf, Nimr, and likeminded contributors and readers believed had no place in the current age of "civilization and progress" (*al-ḥaḍāra wa-l-taqaddum*).[76] Some of these practices, like geomancy, had deep roots

---

[73] Van Dyck's book was read widely, not only among missionary students. It can be found, for example, in the library of the early twentieth-century Grand Mufti Muhammad Bakhit al-Muti'i. Syed Junaid Quadri, "Transformations of Tradition: modernity in the thought of Muhammad Bakhit al-Muti'i" (Ph.D. Diss., McGill University, 2014).

[74] "Fasad al-sihr wa-l-tanjim," *al-Muqtataf* 19 (1895), 470.

[75] "Al-Munazara wa-l-murasala," *al-Muqtataf* 14 (1889): 37–39.

[76] Qasim Effendi Hilali, "Kadhaba al-munajjimun wa-law sadaqu," *al-Muqtataf* 12 (1888): 399–400, with quotation on p. 400; see also "Jawabuna 'ala al-sihr," *al-Muqtataf* 2 (1877): 28–31; and "Ghara'ib al-ittifaq," *al-Muqtataf* 2 (1877), 156.

in Ottoman-Egyptian society; others, like mesmerism, spiritism, and *zar*, were more recently assimilated as a result of increasing population movements across both the Mediterranean and Nilotic regions in the nineteenth century. Critics and defenders alike associated such practices with socially "other" populations, including Jews, Indians, Italians (particularly associated with mesmerism), and women – especially black African women, in the case of *zar*.[77] They also associated these practices with stargazing. "What is your view of the fortune-teller in the area of Matbuli?" Asked one Kamil Effendi Haqqi in 1905. "Is her knowledge and palm-reading a part of astronomy?"[78] Such a connection was, in part, an ironic result of astronomers' success in drawing attention to their ability to make astonishing predictions. As another reader complained to *al-Muqtataf* in 1915, "Most [people] think that astronomy is the very same thing as astrology, seeing as astronomers convey news of the times of eclipses and comets before these events happen."[79] Inasmuch as astrology served as a catchall for various forms of prognostication (as well as healing), insisting on the distinction between astrology and science served to debunk a host of widespread practices – from European fashions to East African rituals – that represented the antithesis of modern, rational ways of knowing. The fact that the purveyors of these practices were typically foreign, female, and/or black African underscored the link between the scientific and the social "other."

But defining astronomy in the Arabic press entailed more than a critique of astrology or any particular set of beliefs and practices. As *al-Muqtataf* introduced its readers to modern observatories, recent astronomical discoveries, and the new technologies on which they relied, astronomy came to provide an exemplary narrative about the conceptual and material gulf between premodern and modern knowledge, and the distinctive virtues of the people and tools who produce the latter. In this narrative, the tradition of astronomy practiced among 'ulama' did not disappear, but it was transformed into an object of historical significance rather than a living practice.

---

[77] On a Jewish spirit medium in Damascus, see "al-Sihr," *al-Muqtataf* 2 (1877): 28–31; on the association of mesmerism with Italians in Egypt, see "Ghara'ib al-ittifaq," *al-Muqtataf* 2 (1877), 156; for an Indian fortune-teller, see "al-'Arraf al-Misri," *al-Muqtataf* 50 (1917), 292; on astrology and *zar*, see Hilali, "Kadhaba al-munajjimun wa-law sadaqu," 400. On the coming of *zar* to Egypt among black African, particularly female, slaves in the nineteenth century, see Hager El Hadidi, "Survivals and Surviving: Belonging to Zar in Cairo" (Ph.D. Diss., University of North Carolina, 2006), 57.

[78] "Al-Masa'il: al-Mar'a al-'Arrafa," *al-Muqtataf* 30 (1905), 1033.

[79] Ahmad al-Sayyid, "Kidhb al-Munajjimin," *al-Muqtataf* 47 (1915), 589.

The astronomical and meteorological observatory was among the first scientific institutions to which *al-Muqtataf* introduced its readers, publishing regular updates on the activities of the Beirut Observatory starting in 1876.[80] Occasional articles also reported on the ʿAbbasiyya Observatory,[81] but less often – partly because of the relative inactivity at ʿAbbasiyya in the 1870s and 1880s, but mainly because of the editors' connections to the observatory in Beirut. Located in the Syrian Protestant College, where the editors studied and then taught until 1884, the observatory was intended by the college's founder to "prove useful in the direct education of students and in attracting the attention of natives to the superiority of Western knowledge."[82] The reports on the observatory in *al-Muqtataf* included precise meteorological data, as well as predictions of eclipses and the lunar phases. Articles also emphasized the way in which observatories increasingly formed an international network – highlighting, for example, contacts between the French, Ottoman, and Beirut observatories.[83] Readers were invited to acquire new kinds of instruments and contribute to the collection of meteorological data, adding themselves as spokes in the wheel of scientific production.[84] At the same time, *al-Muqtataf* told readers that certain kinds of scientific activity were unnecessary for them, given the existence of observatory publications. For example, when in 1889 a reader asked for information on how to predict eclipses, the editors noted that such calculations were superfluous given the publication of annual ephemerides. However, for those readers who wanted to attempt the calculations themselves, *al-Muqtataf* recommended Souchon's 1883 *Traité d'astronomie pratique*.[85]

Al-Muqtataf particularly enjoyed deploying the power of prediction as evidence of modern astronomy's credibility. "The meteor shower that we carried information about in the third issue happened at the time specified there," they reported with satisfaction.[86] Not only meteor showers but weather, lunar phases, and eclipses were all predicted in the pages of *al-Muqtataf*, sometimes for a variety of locations.[87] As Marwa Elshakry

[80] See, e.g., "Min al-marsad al-suri al-falaki wa-l-mitiyyuruluji," *al-Muqtataf* 1 (1876): 46, 69, 213, 237.
[81] "Al-arsad al-jawiyya fi al-marsad al-khidiwi," *al-Muqtataf* 3 (1878): 277; "Al-Marsad al-Misri," *al-Muqtataf* 22 (1898): 799; "Taqrir al-ahdath al-jawiyya," *al-Muqtataf* 28 (1903): 519.
[82] Quoted in M. Elshakry, "The Gospel of Science," 196.
[83] "Min al-marsad al-suri al-falaki wa-l-mitiyyuruluji," *al-Muqtataf* 1 (1876): 46.
[84] "Min al-marsad al-suri al-falaki," 47.
[85] "Masaʾil wa-ajwibatuha," *al-Muqtataf* 13 (1889): 849.
[86] "Min al-marsad al-falaki wa-l-mitiyyuruluji," *al-Muqtataf* 1 (1876): 94.
[87] See, for example, the prediction of lunar eclipse times for Beirut, Damascus, Jerusalem, Alexandria, and Cairo, in "Min al-marsad al-falaki wa-l-mitiyyuruluji fi Bayrut," *al-Muqtataf* 1 (1876): 213.

has noted, the predictive power of modern science was a favorite resource for missionary educators at the Syrian Protestant College, who wanted to demonstrate the rationality of their religion to students and potential converts.[88]

Another way in which astronomy served as an exemplar of modern science's unique credibility was its capacity for impressive feats of mathematical precision:

> Revealing the precision of astronomy, and the difficulty of achieving it and practicing it, they counted their calculations of a small part of the Transit of Venus that happened toward the end of 1874, and it was three million numbers. And they have estimated that its calculation will require millions of numbers and not be finished before two or three years. Obviously the goal of all these calculations is that they determine a very small quantity, not exceeding one third of a second of arc. It is clear to all the esteemed readers of al-Muqtataf that the scholars of this science only render their judgments after close examination and long research, and that contradicting them without deliberation and proof is unreliable and untrustworthy (lā yuʿtamad ʿalayhā wa-lā yurkan ilayhā).[89]

Such celebratory accounts established an ideal of science that was defined by certain types of places and people. The 1874 transit of Venus observations were a massive undertaking of international science, employing observatories and deploying observation expeditions around the globe.[90] The main observing parties in Egypt had been organized by the Royal Observatory at Greenwich, which, despite the presence of an observatory with trained astronomers in Cairo, had sent its own teams, which set up their own sites for the transit observations. The model of scientific authority that appeared in al-Muqtataf may have involved international networks, and it may have been open to certain kinds of reader participation, but its center was in Europe – as the recommendation of Souchon's textbook also suggests. Even the article about cooperation between the French and Ottoman observatories suggests a certain hierarchy: the French had contacted their Ottoman colleagues not to consult on a question of celestial observation, but because they were interested in obtaining the works of Abu al-Wafaʾ, a tenth-century mathematician and astronomer who made observations in Baghdad.[91]

As this incident and al-Muqtataf's reporting of it suggest, the emerging conception of astronomy was one that offered Ottomans an Arab-Islamic history of science in which they could take pride. In the very

---

[88] M. Elshakry, "Gospel of Science."
[89] "Akhbar wa-iktishafat wa-ikhtiraʿat," al-Muqtataf 1 (1876): 263.
[90] On the 1874 transit, including an incisive account of the Cairo station, see Ratcliff, Transit of Venus Enterprise.
[91] "Min al-marsad al-suri al-falaki wa-l-mitiyyuruluji," al-Muqtataf 1 (1876): 46.

first issue of the journal, an article entitled "Astronomers among the Arabs" celebrated the flourishing of science under the Abbasids: the patronage of the Caliph al-Ma'mun, the translation of Ptolemy's *Almagest,* and the towering achievements of ninth- and tenth-century scholars such as Thabit ibn Qura and al-Battani.[92] Such narratives echoed the history of Arabic astronomy that was just beginning to take shape among Orientalists. Indeed, a major way in which the historical significance of Arabic astronomy first came to the attention of a broad Ottoman audience was the delivery of Carlo Nallino's lectures on "Astronomy among the Arabs" at the Egyptian University in 1909–10. The prominent Italian Orientalist gave his lectures in Arabic; the Arabic text became available in print in 1911, and a substantial excerpt appeared in *al-Muqtataf* in 1913. So extensive was the journal's reliance on European sources for interpreting the history of science that the publishers embarrassed themselves at least once by misspelling an Arabic name, having mistaken the French transliteration of *al-Khāzin* for *al-Ḥasan.*[93]

Amid the celebration of "astronomy among the Arabs" as a historical phenomenon, Muslim scholarly participation in science came to occupy a different cultural place from the one it held for the students of Muhammad al-Khudari, for example. *Al-Muqtataf*'s readers encountered Muslim participation in astronomy as an object of memory: a source of cultural pride, but not a living culture. Even the historical significance of "astronomy among the Arabs" was circumscribed: the achievements of the Abbasids, for example, while impressive by the standards of the time, were said to bear little relation to modern science. When, in 1917, a reader wrote to ask whether medicine and astronomy are essentially modern fields of knowledge, *al-Muqtataf* answered that both are overwhelmingly products of the last three centuries.[94]

Those who articulated this history through the writing of astronomy in the press were hardly lazy or arbitrary in their thought. They had reasons for drawing a line between premodern and modern astronomical knowledge, and they explained these reasons in detail. Chief among them was the fact that modern astronomers could see the sky in ways that their predecessors could not. In their 1917 comment on the modernity of astronomy, the editors of *al-Muqtataf* remarked that "the most important discovery in astronomy is the discovery of the telescope and spectroscope, for they have brought astronomers to knowledge of the

[92] "'Ulama' al-hay'a 'ind al-'Arab," *al-Muqtataf* 1 (1876): 16–18.
[93] "'Ulama' al-hay'a 'ind al-'Arab," *al-Muqtataf* 1 (1876): 18n1.
[94] "Al-Masa'il," *al-Muqtataf* 51 (1917): 513.

size of the celestial bodies, their motions, their distances, and the internal elements of their composition."[95] An 1890 piece, which the editors thought was significant enough to reprint in 1905, celebrated the telescope as the "scientists' eye" that opened up new astronomical vistas.[96] The development of astrophotography was particularly important in this narrative, because it allowed for a new kind of archiving of observations, as well as the sharing of observations among observatories.[97] These instruments brought with them their own heroes, from Galileo to the more recent William Pickering, who discovered one of Saturn's satellites in photographic plates taken at the Harvard Observatory in Peru in 1898.[98]

Most of all, the emphasis on the telescope and other technologies of viewing reveals that the cultural ascendance of new astronomy entailed more than replacing one scientific theory or set of facts with others. Fundamentally, it meant training laypeople to envision the skies differently – to look, that is, from the perspective of specific kinds of people and instruments. Of course, no one suggested that all astronomical knowledge came from modern technology. As *al-Muqtataf* itself reminded readers in a 1930 article, Copernicus had not used a telescope, the ancients had identified the naked-eye planets and named the constellations, and Hipparchus had measured the solar year. The same article, however, introduced readers to the history and achievements of observatory instruments such as the telescope, spectroscope, and spectroheliograph, and their inventors.[99] Based on an article by the curator of astronomy at the American Museum of Natural History, it was entitled, "The Astronomer's Factory and Its Instruments: The Wondrous Machines that Scientists Have Designed to Explore Space between the Earth and the Stars." While few late Ottomans owned astronomical instruments or ever set foot inside an observatory, these kinds of articles enabled readers to look to the heavens from the perspective of such "wondrous machines."

[95] "Al-Masa'il," *al-Muqtataf* 51 (1917): 513.
[96] "'Ayn al-'ulama' wa-kawakib al-sama'," *al-Muqtataf* 30 (1905): 417–22.
[97] "'Ayn al-'ulama' wa-kawakib al-sama'," *al-Muqtataf* 30 (1905): 419.
[98] "'Ayn al-'ulama' wa-kawakib al-sama'," 420. Other representative articles in which the telescope occupied a key place in the narrative of astronomy included: "Al-Sayyarat wa-aqdaruha," *al-Muqtataf* 24 (1900): 273–76 (see p. 276 especially); "'Ilm al-falak fi al-khamsin sana al-akhira," *al-Muqtataf* 68 (1929): 18–24; Faris Nimr, "Ba'd 'ahdi bi-'ilm al-falak," *al-Muqtataf* 93 (1938): 51–58.
[99] "Ma'mal a-falaki wa-adawatuhu: ghara'ib al-alat allati istanbataha al-'ulama' li-riyadat al-fada' bayn al-ard wa-l-nujum," *al-Muqtataf* 77 (1930): 7–10.

Al-Manar: Modern Science for a Modern Umma

Early in 1904, the Cairo-based monthly *al-Manar* published a question from an anonymous reader: "Is there anything in the noble Qur'an that supports those who say that the Earth is round and revolves?"[100] The journal's publisher, Muhammad Rashid Rida, answered with reference to Qur'an 7:54: "He draweth the night as a veil o'er the day, each seeking the other in rapid succession." In case anyone might doubt that these words affirmed both the sphericity and revolution of the earth, Rida cited as his authority for this interpretation Gazi Muhtar Pasha, the Ottoman military officer, astronomer, and partisan of calendar reform. Muhtar Pasha was far from an exegete (*mufassir*), but he was, in Rida's words, "Someone known for skill in the astronomical sciences" (*man ta'arrafa bi-l-barā'a fī al-'ulūm al-falakiyya*).[101]

The importance of *al-Manar* to the shaping of a new, transnational Islamic conversation in the early twentieth century can hardly be overstated.[102] Founded in 1898, the monthly periodical quickly garnered loyal readers across the increasingly literate Islamic world, despite a relatively limited print run in its early years.[103] From Southeast Asia to Latin America, people not only subscribed, but actively sought Rida's guidance on a range of issues, submitting their questions to a section that he styled "The fatwas of *al-Manar*." In another section of the magazine, he serialized a commentary on the Qur'an, based in part on lessons that 'Abduh gave between 1899 and 1905. Although it only reached the twelfth sura, the "Manar Commentary" (*tafsīr al-Manar*), as it was popularly known, quickly became – and remains – one of the most widely read exegeses of the Qur'an, particularly outside of scholarly circles.[104]

Al-Manar can be read as a window onto the social and intellectual formation of the emergent Salafi movement. The legacy of the early Salafis, and of Rashid Rida in particular, is as ambiguous as it is vast.

---

[100] *Al-Manar* 7, 1 Rabi' al-Akhir 1322 (15 June 1904), 260.

[101] *Al-Manar* 7, 1 Rabi' al-Akhir 1322 (15 June 1904), 260.

[102] On the founding of *al-Manar* and its audience, see Umar Ryad, "A Printed Muslim 'Lighthouse' in Cairo," *Arabica* 56 (2009): 27–60. On the transnational reach of the Arabic Salafi press, see Ghazal, *Islamic Reform and Arab Nationalism*, 91–108.

[103] *Al-Manar*'s print run was 300–400 copies in 1901, compared with several thousand for the science monthly *al-Muqtataf* around the same time. Ayalon, *The Press*, 148. The number was surely higher in later years.

[104] Rida entitled his commentary *Tafsir al-Qur'an al-Hakim*, but such was the strength of its association with the magazine in which it appeared that it has almost always been called *Tafsir al-Manar*. See Jacques Jomier, *Le commentaire coranique du Manâr; tendances modernes de l'exégèse coranique en égypte* (Paris: Maisonneuve, 1954); J.J.G. Jansen, *The Interpretation of the Koran in Modern Egypt* (Leiden: Brill, 1974), 18–34. On the complex question of the commentary's authorship, see Zaman, *Modern Islamic Thought*, 47.

Today, "*salafi*" typically denotes a markedly conservative interpretation of Islamic norms, and is often associated with the historically distinct movement of Wahhabism (which Rida defended especially toward the end of his career). While today's Salafis may claim Rida as an intellectual progenitor, however, more liberal reformers arguably have at least as much in common with Rida's use of foundational texts to address contemporary problems.[105] The meaning of his appeal to the example of the "*salaf*," or forebears of Islam, shifted over time and in response to specific issues – as can only be expected, in light of the dramatically changing circumstances to which Rida was asked to respond over the course of nearly forty years in publishing.[106]

One theme that remained consistent in *al-Manar*, both in Rida's own work and in the letters and essays that he published by others, was a certain approach to writing about the sciences. In this approach, which strongly resembled the presentation of science in the work of 'Abd Allah Fikri Pasha or the journalists of *al-Muqtataf* (from whom Rida sometimes borrowed directly), the sources of scientific knowledge were not 'ulama', but rather the new scientific elite who had trained in European institutions or the late Ottoman civil and military schools. Far from being opposed to modern science, the emergent Salafi understanding of Islam in terms of clarity, accessibility, and uniformity, went hand in hand with a commitment to what they understood as modern ways of knowing.

The 1904 exchange over the motion of the earth was a case in point. The anonymous reader did not request information regarding the view of Muslim scholars of astronomy (whether historical or contemporary). Instead, the legitimacy of scientific knowledge depended directly upon the interpretation of the Qur'an – regarding which, in a matter of natural phenomena, Rida deferred to one of the Ottoman state's leading military officers and science popularizers. This approach, which would

---

[105] Zaman, *Modern Islamic Thought*, 7. A self-identifying "*salafiyya*" movement only began in the 1920s. On the complex relationship between the "modernist" and "puristic" strains of Salafism, see Henri Lauzière, "The Reconstruction of *Salafiyya*: Reconsidering Salafism from the Perspective of Conceptual History," *IJMES* 42 (2010): 369–89; and Henri Lauzière, *The Making of Salafism: Islamic Reform in the Twentieth Century* (New York: Columbia University Press, 2016).

[106] Dyala Hamza, "From 'Ilm to Sihafa or the politics of the public interest (maslaha): Muhammad Rashid Rida and his journal al-Manar (1898–1935)," in *The Making of the Arab Intellectual*, ed. Dyala Hamza (New York: Routledge, 2013), 90–127. For an interpretation of Rida that highlights his consistent emphasis on the need for Muslim political independence, see Haddad, "Arab Religious Nationalism." Much of the recent literature on Rida seeks to revise earlier critiques of his work as intellectually inconsistent or deficient: cf. Kerr, *Islamic Reform*.

have delighted 'Abd Allah Fikri, was consistent with al-Manar's tendency to celebrate the achievements of modern men of science in much the way that al-Muqtataf did. In a later issue of the magazine in 1904, for example, in the section devoted to the commentary on the Qur'an, Rida remarked upon the astonishing astronomical discoveries of recent scholars, who have found that other celestial bodies are composed of their own solar systems like our own. Rida presented such discoveries as part of a natural theology: thus, the fact that no one solar system interferes with another, but all exist in balance with each other, "proves that they issue from one God, with no partner in his creation."[107] In such discussions, however, Rida appears to have had some anxiety about the full-fledged "scientific exegesis" (tafsīr 'ilmī) that some of his contemporaries popularized. In that genre, current science became the definitive lens through which to interpret the meaning of the Qur'an's words. Rida, by contrast, cautioned that the purpose of revelation was not to teach the details of science; the Qur'an's references to natural phenomena were simply evidence of God's wisdom.[108]

Nevertheless, for al-Manar's most extensive discussion of the modern sciences, Rida ceded the page precisely to someone who would provide a more thorough reading of science in the Qur'an: Muhammad Tawfiq Sidqi. Sidqi was not someone who carried an argument halfway. In 1905 he had written, also for al-Manar, a strident defense of Islam against Christian as well as materialist criticisms; his anti-Christian polemics eventually went so far as to provoke the government's censorship.[109] Between 1906 and 1910 he sparked a fierce debate in the magazine by critiquing the authority of the entire hadith corpus and advocating reliance upon the Qur'an alone.[110] Arguably, however, it was his specific way of reading the Qur'an itself that proved to be most influential. As the hadith debate simmered, Rida published two additional, lengthy essays

---

[107] Al-Manar 7 (16 Rabi' al-Akhir 1322/30 June 1904), 285.

[108] See also "Shubahat al-nasara wa-hujaj al-muslimin," al-Manar 6, no. 9 (1 Jumada al-Uwla 1321/26 July 1903): 333–35, which defends the consistency of Qur'anic cosmogony with modern science, while maintaining that the point of the relevant verses of the Qur'an was not scientific detail but rather the greatness and wisdom of God.

[109] Jansen, Interpretation of the Koran, 43–44. Rida republished these essays as a booklet in 1927. Muhammad Tawfiq Sidqi, Al-Din fi nazar al-'aql al-sahih (Cairo: Matba'at al-Manar, 1346 [1927]).

[110] Sidqi's position on the hadith resembled that of the Ahl-i Qur'an in India, and was far more radical than Rida's own view. Perhaps the case of scientific exegesis is similar: Sidqi's way of reading the Qur'an in light of modern science was more extreme than Rida's own, but it was an extreme that lay in the direction in which Rida sought to prod his readers. Daniel Brown, Rethinking Tradition in Modern Islamic Thought (Cambridge: Cambridge University Press, 1996), 40–41.

by Sidqi: the first, serialized in 1908, on "The Qur'an and Science,"[111] and the second, in 1911, on "Astronomy and the Qur'an."[112]

Perhaps Rida turned to Sidqi on scientific matters for the same reason that led him to cite Muhtar Pasha as his authority on astronomy. Sidqi, whatever his views on the hadith may have been, was a doctor at the Tura Prison, where the British Occupation refashioned Egyptian incarceration practices along more European lines. Medicine occupied a central place in this project, which began under the direction of another doctor, Harry Crookshank.[113] Like Muhtar Pasha, Sidqi represented a conception of science that was linked with a self-consciously modernizing state.

Sidqi's approach to science, astronomy, and the Qur'an also calls to mind the kind of harmonization between science and scripture that Fikri Pasha had crafted in the 1870s. One goal was to reveal the agreement between the Qur'an and modern astronomy. Thus, Sidqi argued that the Qur'an spoke of light years ("A day for your Lord is like a thousand years as you count")[114] and described gravity ("By the heavens, containing pathways...");[115] that the heavenly "throne" (al-'arsh) referred to a central sun;[116] and that the description of God as "Lord of the worlds" revealed the existence of multiple solar systems.[117] Through such arguments, a reader also received a surprisingly thorough introduction to modern astronomical terms, technologies, concepts, and narratives: heliocentricity, elliptical orbits, asteroids, the ether, Halley's Comet, and the telescope, for example (see Figure 5.1). While sharing the general approach of Fikri Pasha's *Treatise,* however, Sidqi's series of essays significantly expanded the scope of scientific knowledge that the Qur'an was understood to include, and spoke to a much larger readership than Fikri had been able to reach in the 1870s – especially when the series was reprinted in a textbook for Egyptian schools.[118] Together, these differences in text and context made Sidqi's essays in *al-Manar* a major development in the growth of "scientific exegesis" as a genre in its own right.

---

[111] Muhammad Tawfiq Sidqi, "Al-Qur'an wa-l-'ilm," *al-Manar* 11, no. 3 (30 Rabi' al-Awwal 1326/1 May 1908), 208f.; no. 4 (29 Rabi' al-Awwal 1326/30 May 1908), 281f.; no. 5 (30 Jumada al-Uwla 1326/29 June 1908), 361f.; no. 6 (29 Jumada al-Akhira 1326/28 July 1908), 441f.

[112] Muhammad Tawfiq Sidqi, "'Ilm al-falak wa-l-Qur'an," *al-Manar* 14, no. 8 (End of Sha'ban 1329/24 August 1911): 577–600.

[113] Frank Dikötter and Ian Brown, eds., *Cultures of Confinement: A History of the Prison in Africa, Asia, and Latin America* (Ithaca, NY: Cornell University Press, 2007).

[114] Sidqi, "'Ilm al-falak wa-l-Qur'an," 579; see Qur'an 22:47.

[115] Sidqi, "'Ilm al-falak wa-l-Qur'an," 580; see Qur'an 51:7.

[116] Sidqi, "'Ilm al-falak wa-l-Qur'an," 590; see e.g., Qur'an 9:129

[117] Sidqi, "'Ilm al-falak wa-l-Qur'an," 593; see e.g., Qur'an 1:2.

[118] Jansen, *Interpretation of the Koran,* 44.

أما السياران اللذان في داخـل فلك الارض فهما عطارد (Mercury)

والزهرة (Venus) ويسميهما الفلكيون السيارين الداخلين أما السيارات الخمس

‏(١) يطلق لفظ مجموعة في هـذه المقالة على معنيين مختلفين (١) على المنظومة
المكونة من شمس وسيارات حولها كنظو متنا الشمسية (System) (٢) وعلى
مجموعة الكوا كب الثابتة كالدب الا كبر المركب من عدة شموس (Constellation)
والمجموعة بالمعنى الثاني مركبة من عـدة مجاميع بالمعنى الأول والسياق هو الذي يبين
أحد المعنيين فيها يأتي

Figure 5.1 Explanation of the solar system from Muhammad Tawfiq
Sidqi's essay on astronomy and the Qur'an in *al-Manar*. Muhammad
Tawfiq Sidqi, "'Ilm al-Falak wa-l-Qur'an," *al-Manar* 8, no. 14
(1908), 549.

The similarity between the astronomy of *al-Manar* and 'Abd Allah
Fikri's astronomy, or even the astronomy of *al-Muqtataf*, can partially
be explained in terms of social background. As a state-employed doctor, Sidqi had more in common with a lifelong bureaucrat like Fikri
Pasha than with an *'alim* such as Khudari, for example. And Sidqi and
Rida both had more in common than meets the eye with men like the
Muqtataf editors Ya'qub Sarruf and Faris Nimr. Sarruf and Nimr, it is
true, were Christian, missionary-educated Anglophiles: three traits that
hardly bring to mind Rashid Rida. Yet all three men were natives of the
Syrian lands who had come to Cairo in the late nineteenth century to
enter the small but growing world of Arabic journalism. Sarruf and Nimr
were already quite successful when Rida arrived in Cairo in 1897; despite
the differences between them, it is unsurprising that Rida sought them
out to discuss his prospective magazine. In a recent study based in part
on Rida's diary, Umar Ryad has written that Rida described his journal
to Sarruf as aiming "to remove the idea in the minds of the majority of
Muslims that philosophy contradicts religion" – a project with which
Sarruf and Nimr would have been deeply sympathetic.[119] All three had
studied with teachers for whom the harmony of science and religion was
a major concern of their life's work: for the *Muqtataf* editors, this was

[119] Ryad, *Islamic Reformism and Christianity*, 84. The quotation is Ryad's paraphrase, not
Rida's own words.

Cornelius Van Dyck; in Rida's case, Muhammad 'Abduh and Husayn al-Jisr.[120] Over the years, Rida frequently cited his colleagues at *al-Muqtataf* on matters of science, particularly questions about science and religion. Not coincidentally, the relationship between *al-Manar* and *al-Muqtataf* extended to a shared readership. Sometimes one even finds contributors to *al-Muqtataf* asking for fatwas in *al-Manar*.[121]

The admiration for modern science in *al-Manar* can also be understood in light of the entanglement of Rashid Rida's religious project with the simultaneously emerging project of anticolonial nationalism.[122] The contours of Rida's political advocacy shifted over the course of his career, as he lived through the substantially different political circumstances of the Hamidian period, World War I, and the post-Ottoman upheavals of the 1920s and 1930s. Throughout, however, Rida consistently emphasized the importance of Muslim independence from European rule.[123] Moreover, even as the kind of state that Rida envisioned for the Ottoman/Arab/Muslim community changed, one idea remained constant in his thinking about political independence: the need for Muslims to study the modern sciences in order to maintain that independence. According to Rida, Muhammad 'Abduh himself had emphasized this lesson when explicating the verse (Qur'an 3:200), "O ye who believe! Persevere in patience and constancy; vie in such perseverance; strengthen each other; and fear Allah; that ye may prosper." 'Abduh read this verse (said Rida) to encourage Muslims in the study of modern sciences, so that they would match other societies in military power.[124]

The importance of a strong state and the value of modern sciences converged in the narrative of Arab-Islamic history that Rida promulgated

---

[120] On Sarruf, Nimr, and Van Dyck, see M. Elshakry, "Gospel of Science"; regarding Rida and Jisr, see Johannes Ebert, *Religion und Reform in der arabischen Provinz: Ḥusayn al-Ǧisr al-Ṭarâbulusî (1845–1909) – Ein islamischer Gelehrter zwischen Tradition und Reform* (Frankfurt am Main: Peter Lang, 1991).

[121] Daniel Stolz, "By Virtue of Your Knowledge: Scientific Materialism and the *Fatwas* of Rashid Rida," *BSOAS* 72 (2012), 224.

[122] On the relationship between the emergence of Salafism and Arab nationalism, see David Commins, "Religious Reformers and Arabists in Damascus, 1885–1914," *IJMES* 18, no. 4 (November 1986): 405–25; and Ghazal, *Islamic Reform and Arab Nationalism*.

[123] Haddad, "Arab Religious Nationalism in the Colonial Era."

[124] "Bab tafsir al-Qur'an al-Hakim," *al-Manar* 12, no. 6 (End of Jumada al-Akhira 1327 / 17 July 1909), 408–9. This connection between modern science and political power can be traced to the first year of *al-Manar's* publication. In a long essay on "The national schools in Egypt," Rida argued that education – both moral and scientific – was central to political independence, and criticized the prevalence of English in the school system. "Seeking knowledge is an obligation for every Muslim," Rida reminded his readers, quoting a well-known hadith. (*Ṭalab al-'ilm farīḍa 'alā kull Muslim*). "Al-madaris al-wataniyya fi al-diyar al-Misriyya," *al-Manar* 1, no. 15 (9 Safar 1316/29 June 1898), 260–61.

in the pages of *al-Manar*. Historiography was (and remains) an important locus of Salafi debate. The Salafi hermeneutical stance is, after all, in part an assessment of Islamic history: who exactly count as *salaf?* (Contemporary Salafi circles continue to argue over the breadth of Islamic history to be valued, particularly in its scholarly traditions.)[125] Of course, there was no prevailing Salafi view of history when Rida began to publish, because no coherent Salafi movement existed yet. Rida's own writings, however, frequently embraced an expansive Arab-Islamic heritage – perhaps nowhere more so than when it came to science and its connection with a strong state.[126] In a 1900 series on "The Civilization of the Arabs," Rida celebrated the achievements of the great Muslim astronomers, mathematicians, and doctors, from Ibn Sina to Ibn Yunus, and from Baghdad to Maragha. Rida did not state his sources explicitly, but there was undoubtedly a connection here with Gustave Le Bon's *La Civilisation des Arabes* (1884). The appeal of Le Bon's theories to other turn-of-the-century Arab intellectuals, including Muhammad 'Abduh, has been well remarked. Le Bon's understanding of collective psychology and civilizational difference, as well as his emphasis on the role of the elite, had much to say to Arab nationalists and Muslim reformers alike.[127] It is likely that Rida's knowledge of Le Bon's work came directly out of his conversations with 'Abduh, who corresponded with Le Bon and attempted to visit him while in Paris.[128] For Le Bon, the rise and fall of the great Arab empires was a case study in the importance of racial characteristics in history.[129] For Rida, the notion that Muslims had once been the foremost civilization on the world stage suggested a model to which the beleaguered Muslim (not only Arab) societies of the early twentieth century should turn.

[125] Roxanne Euben and Muhammad Qasim Zaman, eds., *Princeton Readings in Islamist Thought* (Princeton, NJ: Princeton University Press, 2009), 20–21.

[126] Zaman has also found that Rida's understanding of the term "*salaf*" was broader than Hourani judged it to be. Zaman, *Modern Islamic Thought*, 8; cf. Hourani, *Arabic Thought*, 230.

[127] On Le Bon's significance for 'Abduh and others, see Mitchell, *Colonising Egypt*, 124–25; Hourani, *Arabic Thought in the Liberal Age*, 173; O. El Shakry, *Great Social Laboratory*, 96–97.

[128] Rida's series appeared well before the Arabic translation of Le Bon's work between 1909 and 1913, and even before Jirji Zaydan's *Tarikh al-Tamaddun al-Islami* (Cairo: Matba'at al-Hilal, 1906), which was also indebted to Le Bon. For 'Abduh's relationship with Le Bon, see Anwar Luqa, *Voyageurs et écrivains égyptiens en France au XIXe siècle* (Paris: Didier, 1970). On the Arabic translation, see Rashid Khalidi, "'Abd al-Ghani al-'Uraisi and al-Mufid: the Press and Arab Nationalism before 1914," in *Intellectual Life*, ed. Buheiry, 41.

[129] See Gustave Le Bon, *La civilisation des arabes* (Syracuse [Italy]: IMAG, 1969), especially 472–83.

What is most surprising in this presentation of Arab civilization is not that Rida found specific admirable individuals from relatively late in Islamic history, but rather that he explicitly held up for emulation the states and societies in which these individuals had flourished. The 'Abbasid Empire, the Fatimids, and even the Mongol Ilkhanids appear in glowing terms. Expanding on his view that a strong state requires a strong culture of science, Rida argued that the advancement of science likewise depends on the strength of the state.[130] In the conclusion to the series, he observed that today's Muslims were returning to the sciences because the government required that they be taught in its schools, whereas, by contrast, "most of the people of the noble Azhar are still fighting these sciences in the name of religion." Rida expressed his wish that the 'ulama' instead would return to the example of "their ancestors" (aslāfuhum), by which he apparently meant the mathematicians and astronomers of the 'Abbasid and Ilkhanid courts (although it is possible he was referring to the jurists and other scholars of these periods who did not object to science).[131] The echo of the term salaf in this narrative is telling. For Rida, the value of Islamic history did not end in a specific generation. Different points in history provided models for different aspects of his vision for a strong, independent, Muslim society. Wherever these models lay in the Islamic past, however, the point for Rida was that there was a stark contrast, a nearly total rupture, between the model Islam and the behavior of most contemporary 'ulama'.

Conclusion

Commenting on the debate of 1876, the historian George Saliba has remarked that none of the participants recognized the relationship between Copernican astronomy and mathematical models developed by Muslim astronomers of the thirteenth and fourteenth centuries. As a result, there arose an unnecessary "schism between the modern scientific theories and the indigenous culture," which Saliba attributes in large measure to the role of missionaries in introducing modern science into Arabic.[132] Ahmad Dallal has gone further, pointing to the 1876 debate as evidence that "the modern Islamic discourses on science do

---

[130] "Madaniyyat al-'arab," al-Manar 3, no. 18 (1 Jumada al-Uwla 1318/27 August 1900), 412.

[131] "Madaniyyat al-'arab," al-Manar 3, no. 23 (21 Jumada al-Thaniyya 1318/15 October 1900), 533.

[132] Saliba, "Copernican Astronomy," 154.

not inform and are not informed by a living, productive Islamic culture of science."[133]

It is true that the Arab-Islamic history of Copernicanism did not figure in this debate. But uncovering that history required decades of work by academic researchers with a specific historicist agenda. One reason why this option was not available to late Ottomans is that it was only in the pages of the late nineteenth-century press that scholarly astronomy became, for them, a historical object. The "discourses on science" that Dallal laments were not symptomatic of a stable relationship between science and "Islamic culture"; they were precisely the site at which such a relationship was redefined, and the means by which the absence of a "living, productive" science among Muslim scholars became conventional wisdom. While it would be too great a generalization to say that all debates over science and religion unfolded according to the terms laid out by 'Abd Allah Fikri in 1876, the notion that scientific–religious disputes should be adjudicated by a new class of scientific experts, rather than by the old learned elite, was shared by a remarkably diverse group of actors – not only missionary students and other products of European (or American) education. The promotion of modern scientific authority served Egyptian bureaucrats, who sought to expand the institutionalization of science within the state. It served intellectual allies of the British Occupation, who saw modern science as the foundation for social progress. It served Islamic reformists, who saw the established 'ulama' as a hindrance to the revitalization of the Muslim community, and the adoption of modern science as crucial to Muslims' political independence.

These agendas became only more urgent in the late nineteenth and early twentieth centuries, as a consequence of another great debate of 1876: what to do about the Egyptian debt. Less than half a decade after 'Abd Allah Fikri published his *Treatise*, European financial oversight would be challenged by an open revolt against viceregal authority, which was both restored and circumscribed by the onset of the British Occupation in 1882. Under Britain's quasi-colonial rule, the rapid expansion of science within the state would produce new challenges, but also new possibilities, for the fulfillment of major Islamic duties.

---

[133] Dallal, *Islam, Science, and the Challenge of History*, 157.

6   The Measure of Piety: Making Prayer Times Uniform

In November 1904, Rashid Rida recounted the following anecdote for readers of *al-Manar*:

Once, we were in the countryside with some of the great scholars of al-Azhar, when the Mufti of Egypt saw that the twilight had vanished, so he got ready to pray the evening prayer. Some of the scholars said to him: "there are still five minutes left until the time of the evening prayer!" The mufti said: "the twilight has vanished; nothing is left!" After some discussion the others agreed, and we all prayed together. But after the end of the prayers, I saw them opening their watches, and some of them said: "*now* the time has come." They had prayed with the knowledge that their prayer was valid (*ʿalā ʿilm bi-anna ṣalātahum ṣaḥīḥa*), but with the impression of contravening the practice to which they were accustomed (*mukhālafat al-ʿāda allatī jaraw ʿalayhā*).[1]

This story was, to begin with, a polemical criticism directed toward the usual targets of Rida's ire, the Azhari scholars of his day. More specifically, it was a comment – skeptical, to be sure, but not unambiguous – on the reliability of watches and prayer tables as aides in the performance of prayer. As such, the anecdote opens a window onto the relationship between a conception of correct Islamic practice that was popularized in journals like *al-Manar*, and the technological culture of the literate classes in Egypt in the early twentieth century.

For the moment, however, let us place Rida's story next to another, contemporaneous anecdote, this one recorded in a private diary by Captain H.G. Lyons, director-general of the Survey Department of Egypt. On 1 October 1905, Lyons inspected the department's new observatory, which had opened in 1904 on a plateau above the town of Helwan, fifteen miles south of Cairo. Lyons summarized his findings with dismay:

Found Observatory organization and control most unsatisfactory. Magnetic observations meagre. ... Time observations show no control. General absence of supervision. Senior observer sent on leave and time & magnetic observations

---

[1] "Bab al-fiqh fi ahkam al-din: mawaqit al-ʿibada min al-salat wa-l-siyam wa-l-hajj," *al-Manar* 7, no. 18 (16 Ramadan 1322/24 November 1904), 699.

left to two employees who are incapable of doing anything but the simplest routine work.[2]

The Helwan Observatory was one of the crown jewels of British science in Egypt. Its staff performed crucial work on behalf of the cadastral survey (finished in 1907); maintained a new time service that distributed synchronized signals across much of the Nile Valley and Suez Canal region; and participated in international astronomical research.[3] Yet, as Lyons' journal testifies, the human work that animated the observatory was liable to fall short. It required "organization," "control," and "supervision," all of which he found to be honored in the breach.

Rashid Rida and Henry Lyons typically figure in separate histories. Rida, as we saw in the previous chapter, was a pioneer of Islamic press activism, an influential disciple and interpreter of the legacy of Muhammad 'Abduh, and a major contributor to the formation of diverse Islamic movements in the early twentieth century, from Salafism to "modernism." Lyons, meanwhile, played a leading role in the "technopolitics" of the British Occupation of Egypt, and of the British Empire more broadly: geographer, surveyor, later a fellow of the Royal Society and founder of the Science Museum of London.[4] The two men probably never met, and their personal distance has, if anything, been amplified by the distance of the historiography of modern Islam from the historiography of science and imperialism.

Yet, both Lyons and Rida were deeply concerned with the synchronization of certain people and machines in Egypt in the early twentieth century. For Rida and his interlocutors, the concern was how Muslims ought to fulfill the duty of prayer at specific times during the day, in an era when mechanical timepieces had become ubiquitous among

---

[2] Journal of Henry Lyons, Science Museum (Wroughton) MS 570, 1 October 1905.
[3] For a fuller discussion of the Helwan Observatory, see Daniel Stolz, "The Lighthouse and the Observatory: Islam, Authority, and Cultures of Astronomy in Late Ottoman Egypt" (Ph.D. Diss., Princeton University, 2013), ch. 5. On the technical achievements of British and Egyptian astronomers working with the equatorial telescope at Helwan in the early twentieth century, see Jeremy Shears and Ashraf Ahmed Shaker, "Harold Knox-Shaw and the Helwan Observatory," *Journal of the British Astronomical Association* 125 (2015): 80–93; and A.A. Shaker, "The Scientific Work Done by the 30 Inch Reynolds Reflector at Helwan," *NRIAG Journal of Astronomy and Astrophysics*, Special Issue, 2008. For an overview of the observatory's history, there are several brief but useful internal accounts. See Jami'at Fu'ad al-Awwal, *al-Marsad al-Maliki bi-Hilwan* (Cairo: Matba'at Jami'at Fu'ad al-Awwal, 1949); Muhammad Jamal al-Din al-Fandi et al., *Tarikh al-Haraka al-'Ilmiyya fi Misr al-Haditha: al-'ulum al-asasiyya: al-irsad al-jawiyya wa-l-falakiyya wa-l-jiyufiziqiyya* (n.p., 1990); A.M. Samaha, *The 50th Anniversary of Helwan Observatory* (Cairo: Cairo University Press, 1954); Jami'at al-Qahira, *Ma'had al-Irsad bi-Hilwan: nabdha tarikhiyya* (Cairo: Matba'at Jami'at al-Qahira, 1958); Cairo University, *Helwan Observatory* (Cairo: Cairo University Press, 1958).
[4] Mitchell, *Rule of Experts*, 105–15.

educated men. For Lyons, along with the British and Egyptian men who staffed the Helwan Observatory, the concern – or at least one problem that occupied a great deal of their time – was how to ensure that clocks and other signals in multiple Egyptian cities and ports would all indicate the mean time of the thirtieth meridian east of Greenwich.

These two problems of synchronization became increasingly entangled over the course of the first quarter of the twentieth century. This chapter follows a history from 1898, with the founding of the Survey Department of Egypt, to 1923, when mosques in Cairo became legally obligated to issue the call to prayer at the times published in Survey Department almanacs. During this period, the ways in which an increasing number of Muslims measured one of their central religious duties became enmeshed in the efforts of British surveyors and astronomers to measure land, synchronize clocks, regulate commerce, and facilitate maritime navigation. Since public timekeeping regimes changed partly in response to the priorities of those who administered them, the measurement of time must first be understood within the political history of the British Occupation and the post-Ottoman Egyptian state.

However, the linking of observatory and mosque was not simply a matter of subordinating a religious space to the regulatory power of a state institution. As these opening anecdotes illustrate, for both Lyons and Rida, synchronization was a problem without a conclusive solution. In fact, for both men, the attainment of synchrony lay at a problematic intersection of observed celestial phenomena, human capacity for measurement, and technology. From the "small technologies" of pocket watches and printed prayer tables, to the specialized observatory tools of transit instruments, sidereal clocks, and telegraphic connections, technologies facilitated new ideals of precision, but also opened up new spaces for error, frustration, and disagreement. And error, whether scientific or religious, actual or potential, required constant work to correct or avoid. Uniform prayer timing was more than an ideal inscribed in law; it emerged, on a daily basis, to the extent that both British and Egyptian men successfully performed specific practices of measurement. These practices of measurement transcend the analytical categories of "science" and "religion," revealing historically specific anxieties and dispositions that they shared in the early twentieth century. For Cambridge mathematicians regulating clocks in the observatory, as well as for middle-class Muslims praying in mosques, the pursuit of synchrony in Egypt depended not only on the declared priorities of political elites, but also on less visible – yet socially pervasive – concerns regarding the ability of specific kinds of bodies to perform specific kinds of measurement.

The making of science and the making of piety were thereby linked, in particular, with the making of class and gender in a colonial context.

Inasmuch as these processes ultimately enabled a uniform, precise calibration of prayer times, they shed light on a global trend toward uniformity that characterized the long nineteenth century. As C.A. Bayly remarked, this was an era when elites and middling people in much of the world became more similar in behaviors ranging from dress to sexuality.[5] Time was only one of many standards of measure that became increasingly subject to institutionalization and international agreement, both facilitating and facilitated by European imperialism, technologies of rapid transit and communication, and scientific internationalism.[6] As Vanessa Ogle has pointed out, those who shaped the transformation of time in this context tended to speak broadly (and idealistically) of "time reform" and the attainment of "uniform" time. "Standardization" became the common term relatively late in this process, and adopting it uncritically may color our understanding of the late nineteenth and early twentieth centuries with Fordist overtones particular to the inter-war United States.[7] As an analytical term, however, "standardization" distinguishes a crucial way in which scientific and religious practices both achieved greater uniformity in this period, which was, of course, the creation of standards: texts, instruments, or institutions whose authority was agreed upon by a community of practitioners, yet which somehow came to stand outside the community as a fixed object. In the production of such standards, however, and in their adoption as a basis for uniform practice, the commission of error played a crucial role. It is not merely that standardization and uniformity were idealized outcomes. They entailed the recognition of new kinds of mistakes, and of specific kinds of people who were prone to commit them.

Approaching the practice of measurement at the intersection of science, religion, and empire requires a juxtaposition of sources and perspectives. Whereas disagreements about the correct way to use a pocket-watch or the time at which to pray appeared in newspapers and magazines, the history of error and disagreement inside scientific institutions is visible thanks to less widely circulated documents, such as personal diaries, internal correspondence, and technical papers. Following the practice of

---

[5] Bayly, *Birth of the Modern World*.
[6] Simon Schaffer, "Metrology, Metrication, and Victorian Values," in *Victorian Science in Context*, ed. Bernard Lightman (Chicago: University of Chicago Press, 1997), 438–74; Simon Schaffer, "Late Victorian Metrology and Its Instrumentation: A Manufactory of Ohms," in *Invisible Connections: Instruments, Institutions, and Science*, ed. Robert Bud and Susan Cozzens (Bellingham, WA: SPIE, 1992).
[7] Ogle, *Global Transformation of Time*, 18–19.

measurement places familiar voices from journals like *al-Manar* and *al-Muqtataf* in conversation with the papers of British surveyors and astronomers, publications of the Survey Department and related offices, and bureaucratic developments recorded in the Egyptian National Archives. The rise of uniform prayer timing links the institutional history of British science in Egypt with the politics of religion and state in the post-Ottoman 1920s, and points to a material and physical culture shared by Egypt's aspiring middle class and its British technical elite.

Borders, Ships, and Prayers: The Survey Department and the Reorganization of Measurement in Egypt

New practices of measurement in Egypt depended, in part, upon a new institutionalization of astronomy particular to the politics of the British Occupation that began in 1882. This was not a linear development, however. The state astronomy established under Mehmed Ali Pasha and cultivated by his successors atrophied, indeed almost perished, during the initial years of the British Occupation. Mahmud al-Falaki died in 1885, and Isma'il al-Falaki retired in 1887.[8] The job of running the 'Abbasiyya Observatory fell to Ibrahim 'Ismat, who had trained abroad in Paris, Berlin, and Washington, DC, but who was unable to maintain much more than meteorological work at 'Abbasiyya.[9] When a reader asked *al-Muqtataf* in 1895, "Is it true that there is an astronomical observatory in Cairo for observing the stars?" the publishers responded, "Yes, it exists, but it is effectively non-existent (*fi hukm al-'adam*)."[10] The only documented instance of British involvement in the observatory prior to 1898 was an 1886 attempt to shutter the place.[11]

The contrast between this initial neglect of observatory affairs, and the serious attention that the British began to give the observatory at the turn of the century, was consistent with the broader trajectory of the

---

[8] On Isma'il's retirement due to advanced age and inability to perform duties, see the exchange between the Ministry of Education, the Council of Ministers, and the Financial Commission, between 16 March and 20 March 1887, DWQ 75-043252.

[9] On 'Ismat's almanacs and meteorological reports, see Survey Department (Egypt), *A Report on the Meteorological Observations Made at the Abbassia Observatory, Cairo, during the years 1898 and 1899* (Cairo: National Printing Department, 1900), 8; and 'Ismat, *Mawaqi'*.

[10] "Marsad Misr al-falaki," *al-Muqtataf* 19 (1895), 778. Seconding this impression, an Egyptian government document from 1886 refers to 'Abbbasiyya as a "station météorologique." See DWQ 75-003142, 24 January 1886.

[11] When the matter was raised in the Council of Ministers, the cabinet rejected the British proposal and decided to maintain the observatory under the control of the Department of War. The episode is a small example of the limitations on British influence, at least during the first decade of the occupation. See DWQ 75-003142, 24 January 1886.

Occupation. British policy (at least in some quarters) initially viewed the intervention in Egypt as a short-term action to reinstate a friendly government, bring the country's finances into line, and direct its revenues toward the repayment of Egypt's European bondholders.[12] The assumption of increasingly broad and direct responsibility for the country's administration was a gradual process; the consolidation of British control over the Interior Ministry, a key moment, occurred only in 1894.[13] Four years later, acting on the advice of Captain Henry G. Lyons, the Egyptian government merged several formerly independent survey offices, as well as the 'Abbasiyya Observatory, under the aegis of a single Survey Department. Lyons, formerly head of the geological survey, served as the first director-general.[14] By coincidence, Ibrahim 'Ismat retired due to illness in the same year, but it was no coincidence that Lyons installed a British surveyor, E.B.H. Wade, to replace him as Observatory Superintendent.[15] Five years later, the department built a new observatory at Helwan, leaving 'Abbasiyya behind to remain as a meteorological station.

The British assumption of control over surveying and the observatory flowed from the occupying power's interest in putting the Egyptian government on an improved financial footing. Between 1898 and 1907, the central occupation of the department was the cadastral survey: a detailed mapping and registration of all land holdings, for the purpose of tax assessment.[16] By 1905, the relationship between the survey and the country's financial administration was formalized when the entire department moved from the Ministry of Public Works into the Ministry of Finance, where it remained.[17] In

---

[12] The classic account of this period is Robert L. Tignor, *Modernization and British Colonial Rule, 1882–1914* (Princeton, NJ: Princeton University Press, 1966). For a more recent consideration, see Owen, *Lord Cromer*.

[13] Harold H. Tollefson, Jr., "The 1894 British Takeover of the Egyptian Ministry of Interior," *Middle East Studies* 26, no. 4 (October, 1990), 547–60.

[14] "Mudhakkirat al-ashgal bi-sha'n tawhid maslahat al-misaha wa-ta'yin al-Kabtan Liyuns mudir^{an} 'umumiy^{yan} laha," 8 June 1898, DWQ 75-017240. Public Works Ministry (Egypt), *Report upon the Administration of the Public Works Department for 1899 by Sir W. E. Garstin, K.C.M.G., under secretary of state for public works department, with reports by the officers in charge of the several branches of the administration* (Cairo: National Printing Department, 1900).

[15] Survey Department, *Report on the Meteorological Observations... during the years 1898 and 1899*, 9.

[16] Survey of Egypt, "The Survey of Egypt, 1898–1948," ed. G.W. Murray, technical expert, topographical survey, *Survey Department Paper* no. 50 (1950), 9. See Mitchell, *Rule of Experts*.

[17] Murray, "Survey of Egypt," 2. The Observatory came under the Ministry of Public Works only as part of the 1898 reorganization that established the Survey Department. Formerly, it had been run by the Ministry of Education and funded by the Ministry of War, as discussed in Chapter 2. See "Talab al-sirdariyya tatabbu' al-rasadkhana

general, the department's early work prioritized the Nile Valley and Delta: the slim area within which most people lived, and the heart of the tax base. A telling exception was the 1908 surveying of sections of the Eastern Desert that contained mining claims.[18]

Though rooted in the measurement of land, the Egyptian Survey Department, and particularly its observatory at Helwan, became a center for a great deal of science – especially British science – in Egypt and throughout the Middle East in the early twentieth century.[19] The Observatory was home to an equatorial telescope, where the astronomer H. Knox-Shaw took advantage of Egypt's clear skies to participate in international astronomical projects, such as the observation of southern nebulae and the tracking of Halley's Comet in 1909–10.[20] Helwan also housed one of Isma'il al-Falaki's old transit instruments, which Knox-Shaw, Wade, and others now used to regulate the observatory's clocks. Beyond such instruments of celestial observation, however, the observatory also housed a new weights and measures service, for the storage and correction of standards of measure; a facility for studying terrestrial magnetism; a meteorological facility; and a workshop for the repair of instruments. An internal history of the Survey later wrote that "almost all scientific work in Egypt connected with measurement was either initiated or re-organized during the first decade of the present century by Captain Lyons."[21] The audacity of this claim offers a revealing self-image of the department: a place for "all scientific work... connected with measurement" to be not only performed, but "re-organized."

In part, the centralization of measurement under the Survey Department grew directly out of the technical needs of surveying. The Survey Department controlled the accuracy of the cadastral survey by means of geodetic triangulation, the classic chains-on-the-ground method of measuring angles and distances; hence the need to maintain

al-harbiyya," 24 Safar 1303/1 December 1885 (approved in the Council of Ministers on 10 December), DWQ 75-014381.

[18] Captain H.G. Lyons, Director-General, *A Report on the Work of the Survey Department in 1906* (Cairo: al-Mokattam Printing Office, 1907), 11.

[19] During World War I, the Survey Department met the cartographical needs of the British Army not only in Egypt, but also in Palestine, Iraq, and Turkey. Murray, "Survey of Egypt," 2, 4. Dov Gavish, *A Survey of Palestine under the British Mandate* (London: Routledge Curzon, 2005).

[20] H. Knox-Shaw, "Observations of Halley's Comet made at the Khedivial Observatory, Helwan," *Survey Department Paper* no. 23 (Cairo: National Printing Department, 1911), 1–4; Shears and Shaker, "Harold Knox-Shaw and the Helwan Observatory"; Shaker, "Scientific Work Done by the 30 inch Reynolds Reflector."

[21] Murray, "Survey of Egypt," 1.

standards of length.[22] Triangulation itself, however, introduced a certain degree of error, which the department sought to control by establishing a baseline of astronomically determined points. Survey Department officers produced these points on the map either by carrying chronometers, set at the Observatory,[23] into the field, or by performing transit observations in tandem with the observatory superintendent back in Cairo.[24] As had been the case in the days of Mahmud al-Falaki and Isma'il al-Falaki, the unique precision of observatory techniques for the measurement of time, and hence for the measurement of space, invested state astronomy with immediate political significance.

However, much as Mahmud al-Falaki and Isma'il al-Falaki had begun measuring time in order to measure space, yet soon assumed responsibility for public timekeeping itself – establishing the Citadel cannon as a signal of local noon in 1874 – British surveyors quickly made their own, distinctive intervention in social routines of timekeeping in Egypt, and not only in Cairo. One of the first decisions of the new Survey Department was to implement uniform mean time across Egypt – or rather, more specifically, throughout the Nile Valley and along the Suez Canal. When the department was founded at the end of the nineteenth century, public time signals in Egypt included the Citadel cannon in Cairo, but also time balls at Alexandria, Port Said, and Suez, which ships relied on to regulate their chronometers. Each of these time signals, however, was determined locally. Thus, according to a report by Lyons, the mean time at Cairo was 2 hours, 5 minutes, and 8.9 seconds ahead of Greenwich; at Port Said it was 2 hours, 9 minutes, and 15 seconds; at Suez, 2 hours, 10 minutes, 15.5 seconds; and at Alexandria, where they used the meridian of the Great Pyramid at Giza, mean noon was 2 hours, 4 minutes, and 31.1 seconds ahead of Greenwich. This pastiche of local arrangements, Lyons noted, "was especially inconvenient for shipping passing through the canal."[25] (A minute of error in the setting of a chronometer yields a quarter-degree of error in the measurement of longitude at sea.)

[22] The objective of controlling the accuracy of the cadastral survey is stated in Captain H.G. Lyons, *A Report on the Work of the Survey Department in 1905* (Cairo: al-Mokattam Printing Office, 1906), 19. On the difference between the plodding work of triangulation and the romanticized job of the traverse surveyor, see Burnett, *Masters of All They Surveyed*, 10.

[23] Survey Department (Cairo), *Meteorological Report for the Year 1903* (Cairo: National Printing Department, 1905), 5.

[24] For example, Lyons exchanged signals with Col. M.G. Talbot (also of the Royal Engineers) in the Sudan in 1900. See Survey Department (Cairo), *A Report on the Meteorological Observations... during the year 1900*, 8.

[25] Survey Department (Cairo), *A Report on the Meteorological Observations... during the year 1900*, 2.

The transition to a uniform mean time required Lyons to arrange for an unusually long morning in Cairo on 18 September 1900. On that day, the observatory announced noon 24 hours and just over 5 minutes later than it had the previous day, a one-time correction rendering Cairo mean time exactly 2 hours ahead of Greenwich. By October, the port cities had been brought into line as well, so that all time signals in Egypt now announced a single, synchronized noon, based on the thirtieth meridian east of Greenwich. The Survey Department Almanac for 1902 remarked, "The convenience of this for the purpose of shipping, international telegraphy and meteorology hardly requires more than mention to be appreciated."[26]

It is noteworthy that Egypt first adopted a uniform mean time for reasons of international commerce and science. Accounts of late nineteenth- and early twentieth-century time reform tend to emphasize its origins in the American railroad, which crossed vast longitudinal distances within a single nation-state.[27] By contrast, as Lyons himself noted, the Nile Valley (where most Egyptians lived, and where most Egyptian trains ran in the late nineteenth century) is a relatively narrow area with a roughly meridional orientation. The differences in local time that uniform time erased were not very large. In fact, Lyons argued, "Egypt is particularly well adapted for a single system of time."[28] Another way of looking at the same geography, however, would be to say that Egypt did not particularly need a single system of time. This had presumably been the view of the viceregal astronomers at 'Abbasiyya, and of the local authorities at Suez and Alexandria who had administered the old time signals. Even the ships passing through the canal apparently had managed. What changed in 1900 were the people inside the observatory, who weighed the value of local time against the convenience of international shipping according to a different calculus.

One factor that Lyons and Wade appear not to have considered was the role that the Citadel cannon played, for some people, in the performance of prayer. Recall that, since 1874, many residents of the city had been setting their watches to 12:00 at the firing of the cannon (a modification of the Ottoman, *gurubi saat* convention of setting watches to 12:00 at sunset).[29] With the switch to uniform mean time, this "12:00"

[26] Survey Department (Egypt), *An Almanac for the Year 1902: Compiled at the Offices of the Survey Department, Public Works Ministry* (Cairo: National Printing Department, 1902).
[27] Wolfgang Schivelbusch, *The Railway Journey: The Industrialization of Time and Space in the 19th Century* (Berkeley, CA: University of California Press, 1977), 43–44 and *passim*. See also Galison, *Einstein's Clocks*, 122–28.
[28] Captain H.G. Lyons, "Civil Time in Egypt," *Cairo Scientific Journal* 2 (1908), 54.
[29] See Chapter 3.

was now roughly five minutes later than it had been. The correct times of prayer – defined, as always, according to the actual position of the sun in the sky – now fell roughly five minutes "earlier" according to the watch.

Yet many people, including many publishers of prayer tables and almanacs (*arbāb al-natā'ij*), neglected to make this adjustment. Mahmud Effendi Naji, who produced the Survey Department almanacs from their inauguration until at least the 1930s,[30] later explained:

12 + 5 − 5 = exactly 12: this is what the government did regarding the regulation of clocks and determining prayer times in its almanacs, and this is correct, without any doubt or ambiguity.

12 + 5 = exactly 12: this is what the masses did regarding the regulation of clocks, and the almanac-makers did regarding the determination of prayer times. This is incorrect, causing tumult and confusion (*laghaṭ wa-tashwīsh*) among the masses, and doubt and error in the performance of the devotional practices ('*ibādāt*).[31]

The logic of this explanation obscured a history in which "error" had resulted from the Survey Department's modification of time signals in order to accommodate certain people (steamship navigators and meteorologists) rather than others (e.g., muezzins). As this history was translated into the first two terms of an equation (12 + 5), "error" (= 12) became the result of popular ignorance, and the obvious solution became a further act of standardization: reliance on government almanacs, rather than others.

Indeed, Naji's analysis appeared in print as the justification for a 1923 ordinance, issued by the Ministry of Endowments (Wizarat al-Awqaf), that required mosques in Cairo to keep time and issue the call to prayer according to a single set of practices. Noting that the disagreement over five minutes had led to "agitation among the masses and confusion in the determination of the prayer times," the Ministry issued the following rule intended to facilitate "unity in the order of practice":

It is obligatory for the heads of the mosques of Cairo, and those who give the call to prayer and their timekeepers, to set their watches to exactly twelve at the firing of the noon cannon, and to observe the [prayer] times according to the calculation of the government almanac.[32]

[30] See Mahmud Nagi, "The Mohammedan Hours of Prayer," *Survey Notes* no. 4 (January, 1907): 151–53; "Al-Natija al-qada'iyya," *al-Muqtataf* 59 (1921): 502; and Naji, *Natijat al-dawla al-misriyya*.

[31] Naji, *Natijat al-dawla al-misriyya*, 11.

[32] Mosques Administration, Ministry of Endowments (Egypt), Ordinance # 8, 1923, quoted in Mahmud Naji, *Natijat al-dawla al-misriyya li-sanat 1354 hijriyya* (Cairo: al-Matba'a al-Amiriyya, 1935), 13.

This unusual ordinance, on its face a matter of religious practice, can only be understood as a piece of the broader history of measurement particular to Egypt under British occupation.

For Muhammad al-Nabuli and his fellow 'ulama' in the 1860s, even the problem that the ordinance addressed would have been unthinkable. In their Cairo, the basic time signal by which people set their watches was the call to prayer itself. Tables such as Nabuli's "almanac of the position of watch-hands at the times of prayer," and technologies such as Mahmud al-Falaki's original design of the Citadel cannon, facilitated the synchronization of mechanical timekeeping with the real solar day of Islamic prayer (see Chapter 3). This tightly woven fabric of celestial motion, prayer, and the practices of timekeeping unraveled, however, as the vicissitudes of viceregal state-building, British colonial administration, and even the maritime contingencies of the Suez Canal, made themselves heard in the shifting echo of the Citadel cannon. The 1923 ordinance was an attempt to stitch the fabric back together by drawing upon the credibility – as a center of measurement – of the very institution that had helped set the problem in motion to begin with.

Post-Ottoman Calculations: Religion and State in Interwar Egypt

From Mahmud Effendi Naji's perspective, the need for the 1923 ordinance was a matter of arithmetic. "Doubt and error in the performance of the devotional practices" had arisen because other almanacs rested on the absurd premise of $12 + 5 = 12$. While Naji's math was indisputable, however, his explanation was incomplete. Almost a quarter-century passed between 1900, when the Citadel cannon's "noon" shifted to uniform mean time, and 1923, when the Ministry of Endowments noted that this discrepancy had "caused agitation among the masses and confusion in the determination of the prayer times" (*ahdath tahwīshan fī nufūs al-'āmma wa-irtibākan fī tahdīd mawāqīt al-ṣalāt*).[33] The gap between the creation of the technical problem and the issuing of the ordinance suggests that the Ministry of Endowments was not responding to a new phenomenon, but interpreting an old phenomenon as a problem newly requiring state intervention.

Nineteen twenty-three was a post-revolutionary moment in Cairo, in more ways than one. The monarchy had only acquired its claim to sovereignty in 1914, when the United Kingdom, responding to the Ottomans'

---

[33] Naji, *Natijat al-dawla al-misriyya*, 13.

entry into the Great War, ended Cairo's 400-year suzerainty to Istanbul and declared Egypt its own "sultanate" and protectorate of the British Empire. In the War's aftermath, the 1919 Revolution prompted a formal end to the protectorate in 1922, even as the British military presence continued to hold *de facto* sway in many arenas. Less often remembered in the context of Egyptian history is that 1922 was also the year that the Turkish nationalists formally renounced any claim to sovereignty over the old Ottoman provinces, signing the death certificate of the Ottoman Empire.

The 1920s were thus a decade when many people in Egypt, as in much of the Middle East, urgently debated the basis and boundaries of their political community. The 1919 Revolution, a confluence of middle-class activism, urban labor organizing, and peasant protest, ultimately brought to power the Wafd party, led by Sa'd Zaghlul and other members of Egypt's landholding elite. At the same time, Fu'ad I (who adopted the title of "King," rather than "Sultan," in 1922) also sought to assert his leadership on both a national and regional level. As Gershoni and Jankowski showed in their classic study of this period, the early interwar years were, on a certain level, the heyday of territorial nationalism. The "teacher of the generation" (*ustādh al-jīl*) was Ahmad Lutfi al-Sayyid, founder of the *Umma* Party and its widely read newspaper, *al-Jarida*, who had argued since the beginning of the century for Egypt's independence from a greater Ottoman or Islamic community.[34] (Lutfi famously broke ranks with many anticolonial activists in 1906 by taking the British side in a dispute with the Ottomans over the border between Egypt and the Hijaz province – a boundary ultimately drawn by the Survey Department.)[35]

Such heady political trends notwithstanding, when it came to the quotidian practices of timekeeping, Egypt in the 1920s was not one country, but several. As the Survey Department almanacs – the very instruments of synchronization – disclose, even as the use of uniform mean time was legally mandated in Cairo's mosques, the times of prayer still had to be published in two columns: one according to uniform mean time, the other according to the old Ottoman time. Moreover, within the columns of prayer times – sunset, dawn, sunrise, noon, afternoon, evening – an additional, mediating time of day was inserted: "cannon" time (*midfa'*), expressed according to Ottoman convention in terms of hours since sunset. A preface to the tables instructed readers on how to use this

---

[34] Gershoni and Jankowski, *Egypt, Islam, and the Arabs*, 8–9.
[35] Survey Department (Egypt), "A Report on the Delimitation of the Turco-Egyptian Boundary, between the Vilayet of the Hejaz and the Peninsula of Sinai (June–September, 1906), by E.B.H. Wade, together with additions by B.F.E. Keeling and J.I. Craig," *Survey Department Paper* no. 4 (Cairo: National Printing Department, 1907).

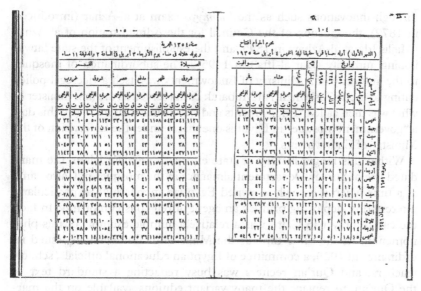

Figure 6.1 Table for the first half of Muharram 1354 (mid-April, 1935), from the Almanac of the Egyptian State, as computed by the Survey Department. Columns, from right to left: day of the week; calendar dates (*hijrī*, Gregorian, Coptic/Diocletian, Seleucid/Alexandrian, Jewish); degree of the sun in the ecliptic; times of prayer (sunset, evening, dawn, sunrise, noon, cannon (!), afternoon); lunar rising and setting. Mahmud Naji, *Natijat al-dawla al-misriyya li-sanat 1354 hijriyya* (Cairo: al-Matba'a al-Amiriyya, 1935).

correspondence to convert times between the two systems on a given day.[36] Similarly, each month's table of prayer times was itself embedded in a larger table that enabled the conversion of dates between the *hijrī*, Gregorian, Coptic, Jewish, and Greek Orthodox calendars. From the hours of the day to the months of the year, the Survey Department almanacs were an attempt to manage, rather than to eliminate, the multiplicity of "Egyptian" identity in the interwar period (see Figure 6.1).

Assertions of state control over religious practices in Egypt had been expanding since the viceregal period, and extended far beyond prayer and timekeeping practice. Mehmed Ali Pasha's seizure of *waqf* properties dealt lasting damage to the economic independence of al-Azhar and, more broadly, of the 'ulama' in the nineteenth century. Subsequent viceroys sought to regulate Islamic learning and jurisprudence more directly,

---

[36] Naji, *Natijat al-dawla al-misriyya*, 100.

through innovations such as the *'ālimiyya* exam at al-Azhar (introduced in 1872), the creation of the Council for the Administration of al-Azhar (Majlis Idarat al-Azhar) in 1895, and the establishment of the state fatwa-issuing authority (Dar al-Ifta') in 1895.[37] The subordination of mosques to the authority of a Ministry of Endowments was itself a viceregal policy dating to the 1870s, when 'Ali Mubarak Pasha served as the first minister of what was then called "Endowments and Public Instruction."[38] (To this day, a "government mosque" in Egypt is one that is under the supervision of the Ministry of Endowments.)[39]

Within this longer history of state efforts to regulate Islam, the mandated use of Survey Department almanacs in the mosques of Cairo came at a time when the newly crowned Egyptian monarchy was particularly interested in deploying print as an agent of religious conformity. In fact, the call to prayer in Cairo was a relatively humble example of this phenomenon. At the same time as the Ministry of Endowments issued its ordinance in 1923, a committee of Egyptian educational officials, school-teachers, and Qur'an reciters was busy redacting a standard text of the Qur'an, to replace the many variant editions available on the market. Under the patronage of King Fu'ad, the "royal" Qur'an had its first printing in 1924 and quickly became the standard edition across most of the Islamic world. The former abundance of variant Qur'an readings (*qirā'āt*) began to dwindle, while editorial choices of the Egyptian committee – the labeling of chapters as Meccan or Medinan, the numbering of verses – became normative.[40] In this context, it is not enough to say that the 1923 ordinance represented the state's attempt to use law to rectify a

---

[37] Eccel, *Egypt, Islam and Social Change*, 198–99. According to Skovgaard-Petersen, although documentary evidence of the Dar al-Ifta' dates to 1895, the institution may have existed earlier. Skovgaard-Petersen, *Defining Islam*, 36.

[38] The original title suggests the close relationship between *waqf*s, mosque space, and learning at that time. Eccel, *Egypt, Islam and Social Change*, 74. Khafaji suggests a remarkably different history, tracing the institution to 118/736, when the qadi of Egypt assumed oversight of the *waqf*s. Khafaji provides no evidence for the continuity of this practice, however, and in any case, the authority of a qadi in the eighth century differed from the role of a late nineteenth-century bureaucracy. Muhammad 'Abd al-Mun'im Khafaji, *Al-Azhar fi alf 'am*, vol. 3 (Cairo: al-Matba'a al-Muniriyya bi-l-Azhar, 1374 [1954–55]), 96. Regarding the relationship between qadis, the Sultan, and *waqf* properties in the thirteenth-fourteenth century, for example, see Joseph H. Escovitz, *The Office of Qadi al-Qudat in Cairo under the Bahri Mamluks* (Berlin: Klaus Schwarz, 1984), 148–54.

[39] Gaffney, *The Prophet's Pulpit*, 45–47.

[40] The decision of the Royal Committee to adopt the reading (*qirā'a*) of Hafs '*an* 'Asim was likely its most consequential choice; the use of other *qirā'a*s has declined markedly as a result. Fred Leemhuis, "From Palm Leaves to the Internet," in *The Cambridge Companion to the Qur'ān*, ed. Jane Dammen McAuliffe (Cambridge: Cambridge University Press, 2006), 159; Michael W. Ibin, "Printing of the Qur'an," in *Encyclopaedia of the Qur'an*, ed. Jane Dammen McAuliffe (Brill Online).

technical and religious problem of its own making. The very recognition of the problem as requiring correction was particular to the politics of the moment. In the post-Ottoman and semi-postcolonial decade of the 1920s, creating "unity in the order of practice" took on new importance.

Measurement, Resistance, and Authority: Towards a Relational Understanding

The Survey Department's effort to regulate the timing of prayer appears to have succeeded without much controversy. In a 1939 essay advocating a new, astronomically determined Islamic calendar, the distinguished Muslim scholar and jurist Ahmad Shakir wrote:

Our country, Egypt, has an observatory considered to be one of the best observatories, staffed with experts in astronomy. ... They can, according to the scientists, issue a conclusive and authoritative ruling based on compelling knowledge. What harm would it do us if we accepted their word and knowledge and relied on their calculations in this matter [of the calendar], as we already do on their calculations for prayer times and other acts of devotion?[41]

Shakir's essay is typically read for its contribution to a debate about the determination of the month of Ramadan (which was his intention). Shakir took for granted, however, that such a calendar would be produced by the observatory, and that everyone already agreed on the authority of the observatory to regulate prayer time. Shakir's premise indicates that the Survey Department prayer tables were relevant beyond the idealized world of bureaucratic ordinances, and in fact achieved broad social credibility. Although Shakir's essay reflects the practice of the following decade, there is no evidence that the 1923 ordinance provoked significant debate in its immediate context.

In fact, well before 1923, Egyptians already looked to the Survey Department in order to measure their religious practices, even beyond those duties (such as prayer and Ramadan) linked specifically with timekeeping. Just as the department distributed standard scales for the weighing of cotton, its laboratory verified the quality of the silk that went into the *kiswa*, the black fabric that adorns the Kaʻba during the annual hajj. Meanwhile, the Assay Office tested the gold and silver with which the fabric was embroidered.[42] The Egyptian pilgrimage caravan (*maḥmal*

---

[41] I use the translation of Ebrahim Moosa, "Shaykh Ahmad Shakir and the Adoption of a Scientifically-Based Lunar Calendar," *Islamic Law and Society* 5, no. 1 (1998), 57–89; the quotation here is on p. 77.

[42] A. Lucas and B.F.E. Keeling, "The Manufacture of the Holy Carpet," *Cairo Scientific Journal* 7, no. 81 (June, 1913), 130. On the standard cotton scales, see Survey Department

al-ḥajj) had been responsible for providing the *kiswa* since the thirteenth century; its manufacture and travel were steeped in significance. The route of the *kiswa* – which, since the 1880s, made its way from Cairo to Jeddah by rail and steamer – constituted its own site of pilgrimage for crowds who sought the blessing of the fabric's presence.[43] As of 1912, even this unique object felt the calibrating touch of the new center of measurement in Egypt.

The Survey Department's successful measurement of religious practice stands in contrast to the fierce resistance that the department encountered in its nominal area of expertise, the measurement of land. As Timothy Mitchell has shown in his account of the cadastral survey, survey markers intended to be permanent frequently shifted or simply disappeared in the hands of property-holders.[44] In the province of Fayyum, this problem was so rampant that it required the repetition of enough work "to survey a province half as large again."[45] Since the work of the Survey Department facilitated taxation, it is not surprising that the department's officers encountered such resistance. Moving survey markers, however, was an act of protest that subverted, but did not reject, the logic of the department's approach to measuring land. A more unusual form of protest came from property-holders whose lands, by the department's reckoning, were less than they had held according to the old registers. This was a common occurrence, since Egyptian surveying had customarily measured the area of triangular plots by multiplying half the base by the mean of the two sides, while taking the area of quadrilaterals as the product of the mean of the two sides. A prominent British surveyor reported:

> As a result of this (imaginary) decrease in the area of the land they owned and paid tax upon, many people became dissatisfied with the results of the new survey. An immense number of complaints was received and many owners went on paying tax on the old areas of their land instead of the new reduced ones in a vain hope that, at some future period, they would be found entitled to the larger area.[46]

The extraordinary occurrence of land-holders paying more in taxes than the state required of them suggests that, if we are to understand the new regulation of prayer times in Egypt, we must do more than account for the motives and practices of the state. We must also look to the motives

---

(Egypt), *Report on the Meteorological and Physical Services for 1912* (Cairo: Government Press, 1913), 8.

43   Barak, *On Time*.
44   Mitchell, *Rule of Experts*, 104–7, 114–15.
45   Murray, "The Survey of Egypt," 10.
46   Murray, "The Survey of Egypt," 13.

and practices of the diverse people who adopted the Survey Department times as authoritative. It was not necessary that they should have done so. Given the "immense number of complaints" regarding lower taxes, it is quite possible that some of the people who referred to Survey Department almanacs for the times of prayer were among those who rejected the department's definition of the area of a triangle.[47]

It would therefore be insufficient to view the standardization of prayer times simply as a centralization of power, a relocation of knowledge from diverse actors to a single agency.[48] Having traced the history of how the Ministry of Endowments came to issue its 1923 ordinance, the remainder of this chapter will explore the ways in which people actually worked inside the network – observatory, clocks, signals, muezzins, prayers – that the ordinance simply posits. From this perspective, the standardization of prayer times comes into view as a relational construction of authority, in which multiple parties performed related kinds of work for different reasons.[49]

Gears, Papers, and Prayers: Precision Measurement as Effendi Practice

The "tumult and confusion" to which the Ministry of Endowments responded has not only a political history, but also a material and cultural history. The 1923 ordinance did not conceive of people, like the mufti in Rashid Rida's 1904 anecdote, who prayed according to their personal observation of the sun. Rather, it assumed that muezzins already relied upon two specific objects when issuing the public call to prayer: a table, written by an "almanac-maker," and a mechanical watch. The ordinance sought to mend a specific flaw that had arisen within this technological system of muezzins, watches, and tables, by linking the muezzin's watch to a particular time signal and specifying the one "almanac-maker" (the Survey Department) on which muezzins should rely. Such involvement of papers and machines in the performance of prayer is now a matter of course (except, of course, where they have been replaced by websites and smartphones), but in 1923 its normativity had a more recent history. As Chapter 3 showed, prayer tables began to acquire new uses and a broader audience in the manuscript culture of Cairo in the eighteenth and nineteenth centuries, as part of the longstanding Ottoman trade in

---

[47] In 1904–05, for example, the Department registered 16,172 complaints. Of the 14,322 that they deemed to be "without basis," most were cases in which the Department's measurement had reduced the registered size of the plot.

[48] Cf. Mitchell's interpretation of the cadastral survey: Mitchell, *Rule of Experts*, 115.

[49] I borrow the concept of relational authority from religious and legal studies: see Krämer and Schmidtke, *Speaking for Islam;* and Zaman, *Modern Islamic Thought.*

mechanical watches and the increasing literacy of Cairene society. As of the late nineteenth century, the printing of prayer tables, as well as the proliferation of cheaper and more reliable watches, facilitated broader participation in this phenomenon. However, the increasing use of watches and prayer tables in late nineteenth- and early twentieth-century Egypt was shaped not only by material availability, but by specific articulations of technology with problems of class, gender, and religious reformism.

The period that saw the standardization of prayer times in Cairo was also the period that saw the emergence of the *efendiyya*, the "middle-class" stratum of men (in reality a broad spectrum, from the middling to the quite wealthy) who came to dominate Egyptian politics and cultural production from the 1920s until 1952. As Lucie Ryzova has shown, the process of "becoming effendi," the ideal to which generations of young Egyptian men began to aspire in the late nineteenth century, had many valences.[50] It typically entailed a certain amount of "modern" education, meaning schooling outside of al-Azhar and its affiliated religious institutes. A government position was ideal. Along with such formal signifiers, however, being an effendi meant dressing in a certain way. Instead of the *galabiya* or gown that their fathers had worn, effendis adopted European fashion. As early as 1896, a journalist in Egypt parodied the pretension to "civilization" (*tamaddun*) of those who wore "a jacket, pants, and watch and chain or perhaps a black pair of socks and shiny shoes."[51] The presence of the watch in this satirical portrait suggests its ubiquity, as well as its role in the performance of a middle-class identity.

Similarly, when Ahmad Amin (1886–1954), the pioneering Egyptian historian, scholar of Islam, and litterateur, wanted to rebel against his father's decision to move him from a state school to al-Azhar, he modi-fied his newly required gown, cloak, and turban, and sought "to be chic in that, like elegant men, I left the end of my waist belt loose and car-ried a watch in my right pocket."[52] Amin's sartorial statement provoked a beating from his father, who accused him of acting like "the son of

[50] Lucie Ryzova, *The Age of the Efendiyya: Passages to Modernity in National-Colonial Egypt* (Oxford: Oxford University Press, 2014). Ryzova points out that the common English spelling, "effendi," originates in British usage; a correct transliteration of the Arabic term has only one "f." However, given the broad acceptance of "effendi" in English, and the fact that the concept of the "effendi" in Egypt is imbued with the history of the British Occupation regardless of how one spells the word, I use the standard English spelling. By contrast, *efendiyya* – the class – transliterates an Arabic term for which no English equivalent exists.

[51] Quoted in John Baskerville, Jr., "From Tahdhiib al-Amma to Tahmiish al-Ammiyya: In Search of Social and Literary Roles for Standard and Colloquial Arabic in late 19th Century Egypt" (Ph.D. Diss., University of Texas, Austin, 2009), 9; quoted also in Barak, *On Time*, 241.

[52] Ahmad Amin, *My Life*, trans. Issa J. Boullata (Leiden: Brill, 1978), 47.

al-Suyufi," a merchant family prominent in the Indian Ocean and British trade.[53]

In effendi fashion, the watch was not merely a symbol to display one's status or aspirations, but an instrument that one regulated with care. In 1922, an unsigned piece in *al-Muqtataf* (probably by the editor, Ya'qub Sarruf) advised readers to follow a strict daily routine:

Beneficial Instructions on the Use of Watches

To preserve the regularity (*dabt*) of small watches that are carried, their ambient temperature must be approximately the same at night and during the day. If it is kept in a pocket by day, let it be kept in a warm place at night. Its position must be the same, so if it is kept in a pocket by day, let it be hung on the wall by day [*sic*, read "by night"], with a piece of soft fabric behind it. If the watch is a chronometer or one of the precise watches of the type called "Duplex," it is absolutely forbidden to move its hands backward, whereas for simple watches, there is no problem with that. One thing it is necessary to pay attention to is that, if the watch is early or late, its hand should not be moved a lot with one motion. Rather, it should be moved forward or backward little by little, until the watch reaches the most precise time of which it is capable. Otherwise, moving the hand a great distance suddenly toward this direction or that has no benefit, but rather damages it greatly.[54]

As this advice suggests, the watch occupied a distinctive place on the body as object that was not only a piece of dress, but also a tool requiring precise, manual work. The accuracy of the watch hands rested ultimately on the motion of human hands, for which there was a proper direction, a proper speed, even a kind of bedtime routine to ensure that the watch rested in its proper position. Such attitudes were a prerequisite for the kind of formal standardization inscribed in the 1923 ordinance, which characterized a disagreement of five minutes as a matter of "tumult and confusion."

Inasmuch as the watch was an object of manual attention and self-discipline, we may also link the standardization of prayer timing – as a continually performed practice – with the role of "physical culture" (*al-riyāḍa al-badaniyya*) in the formation of effendi identity in the early twentieth century.[55] In their performances of gender and sexuality, many effendis adopted conceptions of masculinity popularized through global

---

[53] Amin, *My Life*, 47. On the Suyufi family, see Gad G. Gilbar, "Muslim tujjar of the Middle East and Their Networks in the Long Nineteenth Century," *Studia Islamica* 100/101 (2005), 186. Ryzova discusses Amin as a classic example of the role of dress as a signifier of effendi identity. Ryzova, *Age of the Efendiyya*, 57.

[54] "Fawa'id mufida fi isti'mal al-sa'at," *al-Muqtataf* 60 (1922), 70.

[55] Wilson Chacko Jacob, *Working Out Egypt: Effendi Masculinity and Subject Formation in Colonial Modernity*, 1870–1940 (Durham, NC: Duke University Press, 2011).

phenomena such as Edmond Demolins' bestseller *À Quoi tient la supéri-
orité des Anglo-Saxons?* and the spread of Robert Baden-Powell's scouting
movement. By emphasizing physical strength, independence, and discipline, Egyptian men refuted British claims that Egypt was unready for
self-governance – claims that, since the days of Lord Cromer, had specifically portrayed effendi men as effeminate.[56] Thus scouting, for example,
flourished in Egypt specifically amid the unsettled politics of the 1920s.
Although the National Scouting Society and the Royal Egyptian Boy
Scouts Association competed for troops, their activities rested on the
same premise: "dressing in uniform, chanting in unison, and so forth,
was meant to create *within* the Scout a point from which his duties to the
nation and the monarch would flow naturally."[57] In other words, although
the *efendiyya* is often identified with the cause of Egyptian nationalism,
effendi masculinity transcended the increasingly fractured Egyptian political scene of the interwar period, forming a common set of dispositions
among nationalists, royalists, fascists, and Islamists (see Figure 6.2).[58]

One such common disposition was an increased attention to the precise measurement of time. This concern is evident not only in the prominent place that the (precisely regulated) watch occupied on the effendi's
body, but also in the printed prayer tables that effendis read.

The records of two of the major sites of book collection in Cairo in this
period, the al-Azhar mosque and the newly founded Khedivial Library
(est. 1870), suggest a society awash in printed sets of annual tables. In
addition to the astronomers Mahmud al-Falaki and Isma'il al-Falaki,
authors of late nineteenth-century Egyptian print almanacs included a
captain in the viceregal navy, Sulayman Halawa (who computed his for
Alexandria); the Zagazig schoolteacher Mustafa Muhammad al-Shafi'i
(whose almanac disputed Rudolf Falb's doomsday prediction in 1899),
and lesser-known figures such as al-Sayyid Muhammad Salih al-Baqli
and Muhammad 'Izz al-Sabbagh. For the 1880s alone, the catalogue of
the Khedivial Library records at least seven different authors of printed,
annual almanacs.[59]

Authors of such printed tables were increasingly concerned to refine
their accuracy and precision, for which they looked to state observatories.
A striking example is the set of tables that Idris Bey Raghib published for

---

[56] See, for example, Jacob's discussion of Lord Milner's *England in Egypt*, in Jacob, *Working Out Egypt*, 46.
[57] Jacob, *Working Out Egypt*, 94.
[58] Jacob, *Working Out Egypt*, 109.
[59] *Fihrist al-kutub al-'arabiyya*. Most of these almanacs could no longer be found in the Egyptian National Library in 2010–12. However, almanacs are notoriously ephemeral objects; for most users, their value expires with the year for which they were written.

Figure 6.2 The effendi athlete, victorious and timely: when 'Abd al-Fattah 'Amru Bey defeated English athletes in a squash-racquets championship in 1931, the Egyptian press depicted the triumphant effendi in formal attire, with watch-chain hanging from his pocket. "Egyptian Athlete triumphs over English Champions," *Majallat al-'Arusa*, 30 December 1931.

1312 (1895).[60] Born into one of the city's elite Ottoman-Egyptian households, Idris Bey was educated in Turkish, Arabic, French, and English.[61] His prayer tables corrected the observed time of sunset based on air pressure and temperature data from the meteorological observations of Mahmud al-Falaki.[62] To ensure the most precise measurement of the sun's position, he also employed the laborious procedure of using data from the Observatoire de Paris, which he converted for the location of Cairo.[63]

The audience for this exceptionally rigorous approach to the timing of prayer was much the same new class of men who carefully displayed and attended to their watches in the late nineteenth and early twentieth centuries. In addition to being a scion of Cairo's old Ottoman elite, Idris Bey was also one of the founders of the National Sporting Club (al-Nadi al-Ahli), a key site for the cultivation of effendi masculinity, in 1907.[64] The intended readership for Idris Bey's prayer tables can also be inferred from the *taqariz*, or prefatory commendations, that accompanied them.

[60] Idris Bey Raghib, *Kitab Tib al-Nafs bi-Ma'rifat al-Awqat al-Khams* (Cairo: Matba'at al-Muqtataf, 1312 [1894–95]).
[61] 'Umar Rida Kahhala, *Mu'jam al-mu'allifin* (Damascus: al-Maktaba al-'Arabiyya, 1957–62), I:32. Idris was the son of Isma'il Pasha Raghib.
[62] Raghib, *Kitab Tib al-Nafs*, 63.
[63] Raghib, *Kitab Tib al-Nafs*, 42.
[64] Jacob, *Working Out Egypt*, 85.

A venerable genre in Islamic literature, *taqārīz* attest to the trustworthiness of a text, and thus identify the kinds of people who could perform such a validating function for the work's audience.[65] In this case, one such person was Isma'il al-Falaki, who vouched for the tables in his capacity as "former director of the Engineering School and Observatory."[66] Another was Faris Nimr, who signed as "founder of *al-Muqtataf*," whose press published the tables.[67] A third was Gazi Ahmed Muhtar Pasha (Mukhtar Basha al-Ghazi), the senior Ottoman officer in Egypt, who had a reputation for scientific publishing and had participated in a British solar eclipse expedition in Egypt in 1882, just months before the bombardment of Alexandria (see Figure 6.3).[68] These men were not 'ulama'. (Nimr, in fact, was not a Muslim.) If they could serve to endorse the reliability of prayer tables, it was because such tables rested ultimately on correct measurement, and the users of these tables defined correct measurement by looking to state observatories, military schools, technocrats, and the press – specifically, a prestigious scientific periodical like *al-Muqtataf*. The fact that Nimr's press published the tables, along with Nimr's endorsement in the front matter, strongly suggests that the audience for such tables comprised not only the dwindling Ottoman elite from which Idris Bey haled, but a broader swath of Cairo's literate classes.[69]

It is easy enough to see the increasing, and increasingly precise, use of watches and prayer tables as elements of effendi class and gender performance, because that is how effendis themselves portrayed them. But this confident self-portrayal concealed an unstated tension. While the particular significance that watches and prayer tables acquired in the context

[65] On the premodern genre of the *taqrīz*, see Rosenthal, "'Blurbs' (*taqrīz*)." The significance of this genre in modern print culture – where it has flourished – warrants more attention.

[66] Raghib, *Kitab Tib al-Nafs*, 68.

[67] Raghib, *Kitab Tib al-Nafs*, 72.

[68] Muhtar Pasha attended military preparatory school in Prussia, graduated from the Imperial War College in Istanbul, and rose to prominence through service on various fronts, particularly in the Balkans and Yemen. Dispatched to Egypt in 1882 to help suppress the 'Urabi Revolution, he stayed until 1908 and developed a reputation for scientific expertise by publishing books on math, timekeeping, and the calendar. See Yücel Dağlı and Hamit Pehlivanlı, "Gâzî Ahmed Muhtar Paşa'nın Hayatı," in Gâzî Ahmed Muhtar Paşa, *Takvimü's-Sinîn* (Ankara: Genelkurmay Basımevi, 1993), xiv–xvii; and 'Abbas 'Azzawi, *Tarikh 'ilm al-falak fi al-'Iraq wa-'alaqatihi bi-l-aqtar al-Islamiyya wa-l-'Arabiyya fi al-'uhud al-taliyah li-ayyam al-'Abbasiyyin* (Baghdad: Matba'at al-Majma' al-'Ilmi al-'Iraqi, 1958), 541. See also Ahmad Mukhtar Basha, *Riyad al-Mukhtar* (Cairo: Matba'at Bulaq, 1885); Gazi Ahmet Muhtar Paşa, *Islah-ü Takvim* (Cairo: Matba'at Muhammad Effendi Mustafa, 1307 [1889–90]). Muhtar Pasha wrote in Turkish, but Arabic translations of his work appeared promptly or even alongside the original, as in the latter case.

[69] Muhtar Pasha's *taqrīz* appeared in Turkish, suggesting that the Ottoman elite was certainly part of the book's audience. The other prefatory commendations were in Arabic.

Figure 6.3 Photograph, dated 17 May, of participants in an expedition to observe the solar eclipse of 1882 in Egypt. The formidable figure at the center is Gazi Ahmed Muhtar Pasha, senior Ottoman officer in Egypt and scientific authority, who wrote a prefatory commendation (*taqrīẓ*) for Idris Bey Raghib's prayer tables in 1896. (Immediately to Muhtar Pasha's right is Norman Lockyer, founder of the journal *Nature*.) RAS ADD MS 261. Courtesy of the Royal Astronomical Society.

of effendi usage was new, watches and prayer tables were longstanding features of urban Muslim society in much of the Ottoman Empire, certainly including Cairo. As Chapter 3 demonstrated, in the eighteenth and early nineteenth centuries, the watch became a specifically Ottoman and Islamic instrument of measurement in Cairo through the work that ʿulamaʾ performed in linking the mechanical timepiece with the tradition of *mīqāt* (astronomical timekeeping), and particularly with the timing of prayer. Increasing precision, both in the computation of prayer tables and the regulation of timepieces, was a part of the material and religious culture of ʿulamaʾ in Cairo long before the rise of print, the British Occupation, anticolonialism, or the *efendiyya*.

In fact, even as watches came to be identified with effendi fashion in the late nineteenth and early twentieth centuries, they also complicated the stark opposition that nominally defined effendi versus Azhari self-presentation. Thus Ahmad Amin, who wore a pocket-watch to try to escape the embarrassment of his Azhari garb, noted that one of his teachers at al-Azhar also carried such a watch – and what seems unusual, in his recollection, is not that an Azhari shaykh had a pocket-watch, but that it was "one of two ever made," with the other belonging to the German emperor.[70] The ubiquity of less remarkable watches at al-Azhar in the early twentieth century is confirmed by Rashid Rida's 1904 anecdote with which this chapter began. Given Rida's typical view of Azhari 'ulama' as hidebound, stuck in the ways of scholasticism, and politically passive, it would seem that the use of watches and tables in the performance of prayer was so well-established by this time that it could be caricatured as a kind of fusty traditionalism.

It is telling that such a caricature comes from precisely the same period as Jufani al-Zananiri's satirical depiction of the "watch and chain" as the pretension of a modernizing middle class. Effendis may have defined their "modern" identity in opposition to the "traditional" figure cut by 'ulama', yet this very modernity was, at times, rooted in the effendi's appropriation of material habits, technical knowledge, and forms of piety that had emerged specifically among the 'ulama'. This process of borrowing appeared as differentiation, rather than mimicry, because it was precisely in this period that the role of 'ulama' as competent guardians of knowledge, particularly scientific knowledge, came under widespread criticism. As the previous chapter demonstrated, an important aspect of Rida's critique of the 'ulama' – and, more broadly, a theme of Islamic press activism in the late nineteenth and early twentieth centuries – was the notion that 'ulama' were ignorant of science, and were weakening the Muslim community by refusing to permit the teaching of science in places like al-Azhar.

The 1904 dispute in the countryside produced an odd but illustrative variation on this theme. In this case, Rida appeared to heap scorn on a group of Azhari 'ulama' for being slaves not of scholastic books, but of modern machines. In fact, however, these two critiques had much in common. Thus, Rida provided a scientific explanation for why the observed sunset varied from the time at which the 'ulama' wanted to pray according to their watches and prayer tables. It was not that their watches were wrong, or even that the tables had been calculated incorrectly.

---

[70] Amin, *My Life*, 46.

Rather, he noted, the published times were calculated conservatively, to allow for a certain range in which sunset would be observed under different conditions of humidity. In other words, the 'ulama' were not at fault for relying on technology instead of personal observation of the sky; they were at fault for adopting, in their use of technology, the same blinkered and timid hermeneutics that they displayed in their jurisprudence and theology. Thus watches and prayer tables became, in the hands of the effendi, tools of discipline, precision, and modernity, while in the hands of 'ulama' they became symbols of ignorance and error.

The standardization of prayer times in Cairo was more than a top-down political intervention in popular religious practice, resulting from the timekeeping conventions of British surveyors. It was also an outgrowth of a material and cultural history of watch-use, literacy, class, gender, and religious polemics among Muslims in Egypt. Formal standardization as codified in the 1923 ordinance assumed the normativity of relying upon watches and prayer tables to determine when to pray. This practice was established in Cairo well before 1923, and it was established through notably heterogeneous means. It was the work of a rising middle class of effendis, for whom the watch was both a showpiece and a tool of measurement and discipline. But it was also the work of 'ulama', who had linked the use of watches with prayer tables in Cairo since the eighteenth century. From "beneficial instructions" in al-Muqtataf to polemics in al-Manar, the key concern that people across this spectrum harbored was not *whether* one should use a watch and table, but *how* to use them correctly, and *who* was particularly able (or unable) to do so. In this respect, the work of performing uniform prayer timing in mosques and debating it in newspapers had much in common with the related work of coordinating uniform mean time inside the Helwan Observatory.

Manual Apprenticeship: The Measuring Body in Colonial Labor Practice

For Henry Lyons and his subordinates, maintaining uniform mean time in Egypt required more than the temporal hiccup of adding five minutes to a single morning. The synchronization of time signals in multiple cities required a center – an observatory – where the mean time of the 30th meridian could be measured with the utmost precision and accuracy; and it needed a reliable network, which could distribute this time instantaneously across significant distances. Maintaining both center and network required constant, intensive labor, which also had a material and cultural history – much of it, a history of error. Like the *efendiyya*, British surveyors understood precision measurement as an aptitude of certain kinds of bodies rather than others. Whereas effendis used precision measurement to fashion themselves in opposition to an old learned elite, however, the British understood the capacity for measurement in national and racial terms.

While overseeing a decade of work that measured the length of the Nile Valley from Rosetta to Wadi Halfa, Lyons found it difficult to recruit the Egyptian labor that the survey required. The root of the problem, Lyons remarked in 1907, was social:

The execution of the work of the Department necessitates the employment of a considerable number of employees, which varies according to the amount of work in hand, while they are mostly to be classified as skilled artisans who are employed at their trades. Some of these trades are but little followed by Egyptians at present, and consequently other nationalities have to be employed; among these are superior topographical draughtsmen, engravers, lithographers, lithographic printers, senior computers and observers – in short, callings in which considerable intellectual ability has to be combined with technical apprenticeship or training. The reason probably is that at present those of requisite intelligence find more lucrative and attractive employment in other professions which do not necessitate the same manual apprenticeship.[71]

For Lyons, scientific work was well within the intellectual grasp of at least some Egyptians. The problem was that those who had the intellectual capacity were put off by the manual aspect of scientific work. Working implicitly within a Cartesian dualism, Lyons conceived of measurement practices as distinctive because they demanded both cognitive and physical capacity. And such practices became problematic specifically in Egypt, where mind and body (in his view) were not properly cultivated in tandem.

Lyons' concern for the capacity of Egyptian bodies to perform precision measurement went hand in hand with the Survey Department's reliance on an increasingly large staff of Egyptians to perform a range of crucial tasks. Even as the department excluded Egyptians from the most elite roles, it employed, as of 1906, 1,520 Egyptians next to only 112 staff of other nationalities. The Observatory alone employed Egyptian observers, technical assistants, and computers.[72] The very availability of such information reflects the department's protocol of classifying – and compensating – labor according to nationality. The majority of Egyptians were employed under the category of "inferior Egyptian" labor, a classification that referred to "men trained in special work, but not to be

[71] Lyons, *A Report on the Work of the Survey Department in 1906*, 6–7.
[72] See the Ministry of Public Works (Egypt), "1917–1918 call for budget," DWQ 4003-033363, and the 1919–20 budget papers of the Physical Department in DWQ 4003-02146.

classed as skilled artisans," including chainmen and the staff of the printing and photographic offices.[73] Based on such hierarchies of skill, of the Egyptians working in the Survey Department in 1906, less than 3 percent (43) had pensioned status, whereas over 23 percent (26) of the non-Egyptian staff were pensioned. In other words, at the same time as Egyptian bodies were excluded from certain scientific roles, the roles in which most Egyptian bodies did serve were devalued – and, in published work, usually erased.[74]

The subordination of people like observers, computers, and technical assistants to a new class of scientific "managers" is a common theme in the history of the sciences in the nineteenth century, from the disciplining of observers at Greenwich under George Biddell Airy, to the industrialization of scientific labor in Pavlov's laboratory.[75] In an era of industrialization, as Simon Schaffer has remarked, "'Mere' observers were relegated to the base of a hierarchy of management and vigilance, inspected by their superiors with as much concern as were the stars themselves. Observation was mechanized, and observers transformed into machine minders."[76] In this context, a disciplined body was in some ways more important than a keen mind for the performance of science.

In the colonial context of Anglo-Egyptian relations, however, while bodies certainly had to be disciplined in order to produce knowledge, discipline alone could never be fully successful – at least, not from a British perspective. In much the same way that the continuity of the Occupation rested on deferring *ad infinitum* the achievement of its stated objectives, British science in Egypt, while dependent on huge numbers of Egyptian bodies, was committed to their imperfections. In this context, a strategy emerged that compensated for indiscipline through instrumentation. The standardization of time at its center, inside the observatory at Helwan, relied upon a particular material and physical culture, in which a specific hierarchy of bodies corresponded to a specific hierarchy of instruments.

---

[73] Lyons, *A Report on the Work of the Survey Department in 1906.*

[74] For example, the first volume of the Helwan Observatory *Bulletins*, published between 1911 and 1923, contained the following names: B.F.E. Keeling, E.B.H. White, H. Knox-Shaw, P.A. Curry, and Walter S. Adams (a lone American, from the Mount Wilson Observatory), T.L. Eckersley, and C.C.L. Gregory. One would never know anyone else worked there. Survey Department and Physical Service (Egypt), *Helwan Observatory Bulletins*, vol. 1 (Cairo, 1923).

[75] Schaffer, "Astronomers Mark Time"; Daniel Todes, *Pavlov's Physiology Factory: Experiment, Interpretation, Laboratory Enterprise* (Baltimore: Johns Hopkins University Press, 2002).

[76] Schaffer, "Astronomers Mark Time," 119.

The "master clock" that sat at the heart of the system for distributing mean time in Egypt was not in England, but in Cairo.[77] In 1874 – by coincidence, the very year that Mahmud established the Citadel cannon as a signal of apparent, local noon – the ʿAbbasiyya Observatory had received a mean time clock as a gift from the Royal Observatory at Greenwich. The clock, number 1914 by the precision instrument-maker Dent, was a token of gratitude for the viceroy's hospitality toward a British expedition that observed the 1874 transit of Venus from the Muqattam hills outside of Cairo. The gift may also have been a suggestion. Charles Orde Browne, the leader of the 1874 expedition, commented that Mahmud's cannon, and the practice that some Cairenes had adopted of adjusting their timepieces according to its shifting pronouncement of noon, constituted an "ill use of the watch."[78] But Browne could do little more about it. He spent hardly two months in Egypt, eight years prior to the British Occupation.

In 1900, British astronomers in Egypt were in a position of greater power. Lyons and Wade had an instrument-maker rig Dent 1914 with electric contacts on the hour and second hands.[79] Returned to the observatory in ʿAbbasiyya, and subsequently moved to Helwan, the clock could now distribute Greenwich mean time across Egypt almost without the intervention of a human hand. Wade explained:

The hour signal passes through two relays in series. The first brings in a battery on a private line to the citadel of Cairo. The second completes a circuit between the Observatory and the central Telegraph Station in Cairo. Here the current actuates a distributing relay capable of simultaneously completing the circuits of several lines. At Port Said the Administration of Ports and Lighthouses has availed itself of this current from the beginning of September to drop a time-ball daily, and a similar ball will shortly be dropped at Alexandria. The current to the citadel of Cairo deflects a galvanometer there, at which signal a time gun is fired by hand, but early in 1901 the current will be used to fire the gun automatically.[80]

Telegraphic connections linked the internal Observatory switchboard to the Citadel and the Central Telegraph Office, and the latter to local telegraph offices (of which Egypt had over 200 in 1902, and 300 in 1908) (see Figure 6.4).[81] In all of these locations, "noon" now meant two hours later than mean noon in Greenwich.

[77] Cf. Barak, On Time, 240.
[78] See Chapter 3.
[79] Papers of William Christie: Correspondence on Clocks, RGO 7/252/1.
[80] E.B.H. Wade, "Report of the Superintendent, Abbassia Observatory," in Survey Department, A Report on the Meteorological Observations made at the Abbassia Observatory, Cairo, during the year 1900, 8.
[81] Karl Baedeker, ed., Egypt: Handbook for Travellers, 5th edition (Leipsic: Karl Baedeker, 1902), xix; Karl Baedeker, ed., Egypt and the Sudan: Handbook for Travellers, 6th edition (Leipzig: Karl Baedeker, 1908), xviii.

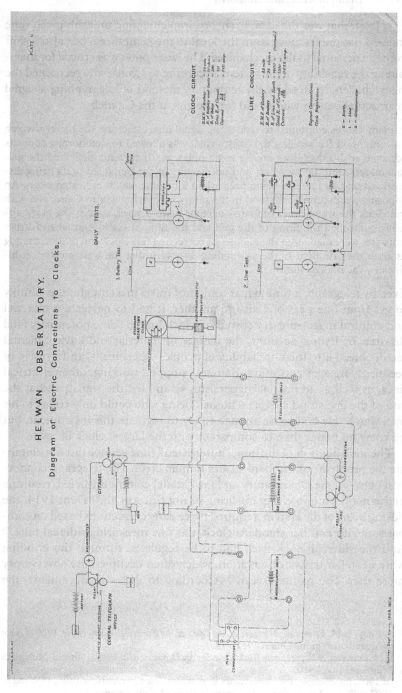

Figure 6.4 Telegraphic distribution of time signals from the Observatory to the Citadel cannon and the Central Telegraph Station. "A Report on the Work of the Survey Department in 1907, By Captain H.G. Lyons" (Cairo: Al-Mokattam Printing Office, 1908), Plate V. © The British Library Board. Cartographic Items 8766.ee.

At least, that was the idea. In theory, implementing uniform mean time entailed not only a move from the local to the centralized, but also a shift from the personal to the mechanized. The latter point was crucial for maintaining the reliability of the system, according to Lyons, who recounted the many challenges faced by the temporary method of telegraphing a signal to an actual person who would fire the gun at the Citadel:

> ... many causes combined to render the signal unsatisfactory: the battery, which was only used for sending the daily signal, was allowed to deteriorate; contacts occurred with other lines on the same poles; earth currents deflected the galvanometer, until the weakness and uncertainty of the signal led to its being disregarded and to the gun being fired when a reliable watch indicated noon. An imposing pendulum clock stood in the shop of a watchmaker in Cairo, and with this clock the watch was compared once or twice a week, so that the clock virtually controlled the firing of the gun and the time of Cairo; but when further enquiries were made into the manner of correcting this clock and of determining its error they resulted in the astounding discovery that the watchmaker set his clock by the midday gun![82]

Even by telegraph, a time signal could not travel in a straight line. Simply to get from one part of Cairo to another, it had to navigate a material and cultural maze, in every corner of which lurked some potentially fatal disaster. In Lyons' account, the danger of material and environmental factors bleed into the unreliability of people – specifically, an Egyptian of seemingly meager education. Batteries stopped working; other electrical lines, as well as terrestrial magnetism, scrambled the signal; and at the end of the line stood a hapless human being who could only resort to his own watch. He had sense enough to try to regulate this instrument, but he knew no better than to compare it with the largest clock in town.

The introduction of the new, "automated" time service did not eliminate the problems of mechanical and human error. The telegraph connection between the Observatory and the Citadel commonly failed, resulting in the noon cannon either misfiring or not firing at all.[83] Dent 1914, the ticking heart of the system, required twice daily corrections based on comparison with another standard clock, this one measuring sidereal time.[84] And the sidereal clock itself had to be regulated through the grinding work of stellar transit observation, which often occupied the observatory more than 100 nights a year.[85] According to these measurements, the

---

[82] Lyons, "Civil Time in Egypt," 55–56.
[83] See e.g., E.M. Dowson, *A Report on the Work of the Survey Department in 1910* (Cairo: al-Mokattam Printing Office, 1911), 12.
[84] E.M. Dowson, *A Report on the Work of the Survey Department in 1909* (Cairo: al-Mokattam Printing Office, 1910), 7.
[85] In 1905, for example, the Observatory took transits on 129 nights; in 1906, 155; and in 1907, 127. See Lyons' Survey Department *Reports* for these years.

standard clocks were frequently off by half a second, with greater errors – two, four, even six seconds – not unheard of.[86] A new sidereal clock, from the German instrument-maker Riefler, promised a lighter schedule of maintenance in 1913, but promptly stopped working in 1914.[87] Even had it continued to work, however, it would not have eliminated human labor from the circuit of the "automatic" time service.

Moving out from the center of the system to its tentacles, railway station-masters were required to correct their station clocks according to the telegraph signal on a daily basis, to keep a register of these corrections next to the clock, and, if the error should ever exceed one minute, to submit the clock for repairs.[88] Lyons concluded with satisfaction: "Thus the clocks of the Telegraph offices and Railway stations throughout the country are daily corrected by reference to the signal sent automatically by the Observatory clock."[89] As the regulation of station-masters illustrates, however, Lyons' certainty ultimately rested more on the disciplinary power of human record-keeping than on the electrified hands of a clock – which itself was frequently in error.

Where human labor was involved, however, its performance was often shockingly undisciplined, as Lyons discovered in his disastrous 1905 visit to the Observatory, when he found that "time observations show no control." Such observations had stakes quite apart from the implications they ultimately held for the timely fulfillment of Muslim prayer. In addition to guaranteeing the accuracy of the Citadel cannon, of Egyptian railway clocks, and of time balls on the Mediterranean and Suez Canal, maintaining time at the Observatory was a crucial link in the ability of British surveyors to map Egypt and the Sudan onto longitudes measured ultimately from Greenwich. Given these stakes, it is perhaps not surprising that the British men who were responsible for maintaining the "control" of observatory work in Egypt in the first decade of the twentieth century found the task exceedingly stressful. In an earlier journal entry, Lyons reported that Wade "had boxed the ears of an observer at the Observatory; had been sleeping badly and was in a nervous excited state."[90] This behavior troubled Lyons, who felt that such a blunt approach to discipline suggested a lack of self-discipline. Lyons soon concluded that the observatory needed a new director:

---

[86] Dowson, *A Report on the Work of the Survey Department in 1908*, 52; Dowson, *A Report on the Work of the Survey Department in 1910*, 12.
[87] See B.F.E. Keeling, "Khedivial Observatory, Helwan," *MNRAS* 73 no. 4 (February, 1913), 255; and H. Knox-Shaw, "Khedivial Observatory, Helwan," *MNRAS* 74, no. 4 (February, 1914), 311.
[88] Lyons, "Civil Time in Egypt," 58.
[89] Lyons, "Civil Time in Egypt," 59.
[90] Journal of Henry Lyons, 13 July 1905.

Wrote to Wade telling him that... I shall relieve him of superintendence, & that in view of his inability to control his temper toward subordinates, as well as a tendency to the immoderate use of stimulants he would be wise to consider whether he is sufficiently confident of his power of self control to resume work here.[91]

Wade's conduct was particularly disturbing at a time when scientific work had increasingly come to depend on a hierarchy of correctly disciplined bodies. His erratic behavior, his "immoderate" use of drugs as well as his proclivity for physical violence, called into question his own capacity for scientific work.

Insofar as the correct performance of precision measurement in the observatory was a distinctively embodied act, it was also – as much as the watch-use of effendi men – part of a distinctive physical culture. Whereas the old observatory at 'Abbasiyya had offered quarters for the superintendent and a single observer,[92] the Helwan Observatory, fifteen miles south of Cairo, was a residential community – equal parts desert encampment, English college, colonial outpost, and country club.[93] The Oxford professor of astronomy H.H. Turner, returning from a visit to Helwan in 1909, wrote fondly of

a little community of University men – Cambridge mathematicians for the most part – who have their headquarters there. Since my last visit in 1905 a new dwelling-house has been built... with general dining-room and drawing-room and a few bedrooms. Those for whom there is no bedroom in the house sleep in tents outside, as is befitting a nomad people who are here to-day testing a level or a magnetometer, and gone to-morrow to use the rectified instruments perhaps in Syria or in the Soudan. The community is thus stable but not stagnant; it is a college with a mobile personnel; the inmates come and go as the exigencies of their field-work dictate, but are united by their common interests and their common difficulties in dealing with the gentle Arab and his language – shall I say also by their common subjection to the Head of the Observatory?[94]

Turner was not exaggerating the prevalence of Cambridge men at Helwan. Lyons, Wade, Knox-Shaw, B.F. Keeling (Wade's successor as observatory superintendent), and Philip Curry (the founder of the Observatory's weights and measures service, later another Observatory director) had all taken degrees at Trinity College. The sense of Spartan

---

[91] Journal of Henry Lyons, 1 October 1905.
[92] Survey Department, *A Report on the Meteorological Observations... during the year 1900*, 3.
[93] The rapid move to Helwan, just five years after the new department assumed control over the old viceregal facility at 'Abbasiya, partly reflected longstanding concerns about 'Abbasiya's proximity to sea level, where the mist frequently obstructed astronomical work, as well as a new concern that the electric lines of the Cairo tramway would interfere with efforts to use the observatory as a site for studying terrestrial magnetism.
[94] Turner, "From an Oxford Note-Book," 111.

ruggedness that pervades Turner's description of life at Helwan is consistent with the distinctive masculinity of the training that the observatory's staff had endured at Cambridge, which, as the premier site of British mathematics in the early twentieth century, emphasized competition in everything from written exams to physical fitness.[95] Translated into the exotic aesthetic of colonial territory, these qualities appeared vividly in Turner's description of the observatory superintendent (Keeling, at the time) overseeing a caravan of mules on its way up to the plateau. Seeing that the hindmost mule required encouragement, Keeling gave it "a flying kick that would have done credit to a harlequin." Turner concluded, "I had not realized until that moment how varied might be the functions and the accomplishments of the director of a great Mountain Observatory."[96]

Notwithstanding the stress that the rigorous work of observatory science placed upon its practitioners, British surveyors and astronomers at Helwan enjoyed particular privileges and pleasures available to Britons in Egypt in the early twentieth century. It is difficult to imagine now, but Helwan was best known in this period as a site of elite leisure, a place where affluent Europeans came to enjoy the town's sulfur springs and salubrious desert climate. At Helwan, the astronomer Knox-Shaw played tennis with Ethel Smyth, the noted composer and suffragist.[97] Smyth wrote to Emmeline Pankhurst that Knox-Shaw, a "sport companion," was "such a dear clever boy (about 26 I daresay he is)," who allowed her to look through the Observatory's telescope at night, and had even given her a star atlas to take home.[98] In addition to tennis, Observatory staff also enjoyed an eighteen-hole golf course in the desert beneath their facility. "Indeed," Turner observed, "the big dome is a very good line to drive on for one of the holes, which has been called the Khyber Pass. (For another hole you drive a little bit to the right of a Pyramid on the horizon.)" And where there was golf, of course, there were also caddies: "Arab boys, with red jerseys added to their usual Arab dress, and a few English golf terms added to their vernacular."[99]

In the physical culture of measurement in a colonial context, bodies were differentiated, in part, according to the categories of colonial rule. The first decade and a half of the Helwan Observatory's existence coincided with the height of British power in Egypt; even before

---

[95] Warwick, *Masters of Theory*, especially chs. 3 and 4.
[96] Turner, "From an Oxford Note-Book," 114.
[97] Personal communication from Peter Knox-Shaw, 11 April 2011.
[98] Ethel Smyth to Emmeline Pankhurst, 7 March 1914, Ethel Mary Smyth Letters, Jackson Library, University of North Carolina, Greensboro.
[99] Turner, "From an Oxford Note-Book," 111.

the establishment of the formal protectorate in 1914, the 1904 Anglo-French entente recognized British priority in Egyptian affairs.[100] When, for example, it was suggested that an Egyptian be trained for the position of astronomer at Helwan, Lyons argued strenuously that such a step would create a "principle of suppressing highgrade scientific work," with the consequence that "all scientific work would fall in standards." He pointed out that the Observatory's new equatorial telescope was a gift from a British industrialist, J.H. Reynolds, who had stipulated the hiring of a "competent observer," and would probably want the telescope back if this condition were not met.[101] Within two weeks, Knox-Shaw had the job.[102]

While no Egyptian would qualify as a "competent observer" for the equatorial telescope, there was another role, crucial to its operation, that an Egyptian did perform – and in a specifically Egyptian way, according to Turner:

... at the Yerkes Observatory [of the University of Chicago] there is an Engineer-in-Charge of the 40-inch refractor, who sets the telescope, turns the dome, and raises the floor, all by means of elaborate electric machinery. For the 30-inch reflector at Helwan the corresponding functionary is a bare-footed Arab. He pulls round the dome by hand, as is only befitting in a country where the merry-go-rounds are still worked in that way not to mention the lifting of water. He also winds up the clock stealing silently into the proper position whenever the alarm bell rings which gives warning that the weight is nearly down. ... But after all he is, perhaps, most interesting in his repose, standing in the still, patient attitude of the East through the watches of the long night, while western civilization is performing strange rites with the telescope.[103]

Turner perceived a cultural specificity to the type of labor used to position the telescope. The *division* of labor, however, befit not the state of

[100] Owen, *Lord Cromer*, 324.
[101] Journal of Henry Lyons, 1 April 1907. Lyons' argument was with the government financial adviser, which suggests that the idea of hiring a native came not from some Egyptian nationalist, but from one of Lyons' frugal superiors in the British administration of Egypt, for whom the work of the Reynolds reflector was evidently not as high a priority as the survey work of the transit instrument.
[102] By contrast, after the 1919 revolution and the formal end of the British protectorate in 1922, Knox-Shaw, trying to hire an additional astronomer from England, was unable to replicate Lyons' success. He explained to Reynolds: "I put Hurst to try & see whether the Undersecretary of State [of the Public Works Ministry], now an Egyptian, would consent to the appointment of an astronomer for a period of two years. We pointed out that there were no Egyptians qualified at the moment & none likely to want to be astronomers. The latter point did not worry him. He said a man was to be chosen & sent to England to be trained as an astronomer whether he wanted to be one or not & he absolutely refused to allow an Englishman to be appointed in the meantime. So there we are." Knox-Shaw to Reynolds, 13 August 1922, RAS Reynolds Papers 1/4/7.
[103] Turner, "From an Oxford Note-book," 112–13.

technology in Egypt, but the relationship between British and Egyptian men, particularly at Helwan. When Knox-Shaw and the (tellingly anonymous) "bare-footed Arab" left the domed equatorial house, after all, they looked down on British golfers and Arab caddies.

Given the imbrication of scientific discipline with colonial power relations in British-occupied Egypt, one suspects it was an Egyptian observer whom Wade physically abused in 1905. Such an eruption of violence inside the observatory was itself a violation of discipline. But it gave voice to the tensions that lay beneath the daily routines of precision measurement, routines that sought to manage – but could never eliminate – the threat of error. These tensions constituted a set of anxieties analogous to the concerns that animated discussions of prayer timing among Muslims in Egypt at the same time. In both contexts, the attainment of precision was part of a material and physical culture in which the ability to use specific instruments distinguished certain kinds of bodies from others. In the observatory, precision was a British virtue, and imprecision an Egyptian vice; in the mosque, precision was the virtue of a new middle class that distinguished it from the old learned elite.

Conclusion

From a certain perspective, it seems odd that the Survey Department of Egypt, an institution founded in 1898 and most closely associated with colonial land and tax policy, quickly came to define the normative performance of a duty that Muslims had been practicing in Cairo for well over a thousand years. Yet, few attempts at state standardization of a religious practice have been so successful, or so enduring. In 2014, the published prayer times of the Egyptian General Authority of Survey are the standard not only in Cairo, but in most parts of Egypt and – via the internet – for large parts of the Middle East and Africa. (I return to the relationship between contemporary *adhān* websites and the colonial history of the Survey Department in the conclusion.)

One way of understanding the origins of this new institutionalization of religious authority is to place it at the intersection of global trends toward uniformity in scientific and religious practice. To a considerable degree, the standardization of prayer timing in Cairo flowed from the distinctive power and ambition of British surveyors to implement new systems of measurement in Egypt at the end of the nineteenth century and the beginning of the twentieth, as well as from the interest of the post-Ottoman Egyptian state in maintaining "unity in the order of practice." As the question of prayer timing became inseparable from imperial practices of coordination and national projects of unification, the duties of piety became linked with the technical requirements of surveying, finance, and navigation.

At the same time, however, standardization also emerged from longer, and socially broader, histories of measurement. The late nineteenth and early twentieth centuries were the period when watches and prayer tables became normative in the urban, literate performance of prayer. Whereas once such technologies had been the province of 'ulama', now they became markers of a new, particularly masculine, middle-class identity, which linked precision measurement with ideals of self-discipline and physical culture. Absent the ubiquity of such objects and attitudes, there would have been no "tumult and confusion" for the state to rectify, any more than if Henry Lyons and E.B.H. Wade had never attempted to implement uniform mean time throughout the Nile Valley.

When viewed as an ongoing process of daily, error-prone routines, the standardization of prayer timing emerged from shared concerns and practices that characterized the work of measurement inside the observatory as well as the mosque. The performance of stellar transit observations at Helwan was as much a part of a specifically colonial physical culture as the self-fashioning of effendis with their watch-chains and duplex escapements. For technocrats like Henry Lyons, as for religious activists like Rashid Rida, technologies of measurement were linked with views about the sort of people who could use them correctly. As a continually performed practice, the pursuit of synchrony relied upon hierarchy and work that addressed such anxieties, but did not eliminate them. Standardization was an ongoing relationship of authority: a relationship in which power was distributed unequally, but in which diverse actors joined in the use of coordinated precision to enact their communal identities.

7 Different Standards: The Ramadan Debates and the Establishment of Lunar Crescent Observation

In October 1922, *al-Manar* published a question from one of its readers in Bangkok. 'Abd Allah ibn Muhammad al-Mas'udi, director of a madrasa in the Siamese capital, described a long-running dispute among the imams of the city's mosques over the correct way to determine the beginning of the Islamic months, particularly Ramadan. Mas'udi identified four distinct methods currently in use:

1) Determining the day of the week on which Ramadan will start by using a table found in the thirteenth-century cosmography, *Wonders of Creation and Oddities of Existing Things*, by the Persian scholar Zakariyya ibn Muhammad al-Qazwini.[1]

2) Determining the day of the week on which Ramadan will start by counting five days from the day of the week on which it began in the previous year, according to the procedure that Qazwini, again in the *Wonders of Creation*, attributed to the Shiite Imam Ja'far al-Sadiq.[2]

3) Beginning Ramadan on the third night after two nights when the moon rose after dawn: a practice that Mas'udi found authorized in a gloss by the eighteenth-century Egyptian scholar Sulayman ibn 'Umar al-Bujayrimi.[3]

4) Beginning Ramadan by looking for the new lunar crescent, in accordance with the hadith, "Fast when it is seen, and end the fast when it is seen."[4]

As a result of following these distinct methods, the mosques of Bangkok annually disagreed on the dates of the fast. Mas'udi, who identified

---

[1] See Zakariyya al-Qazwini, *'Aja'ib al-makhluqat wa-ghara'ib al-mawjudat*, ed. Faruq Sa'd (Beirut: Dar al-Afaq al-Haditha, 1973), 113–14.

[2] Qazwini, *'Aja'ib al-makhluqat*, 113.

[3] See Sulayman ibn 'Umar ibn Muhammad al-Bujayrimi, *Hashiyat al-Bujayrimi 'ala al-manhaj al-musammat al-tajrid li-naf' al-'abid 'ala sharh manhaj al-tullab li-Shaykh al-Islam Abi Yahya Zakariyya al-Ansari*, vol. 2 (Cairo: Mustafa al-Babi al-Halabi, 1369 [1950]), 67.

[4] *Al-Manar* 23, no. 8 (29 Safar 1341/20 October 1922), 584.

himself as a "student" of *al-Manar*'s publisher, Rashid Rida, requested Rida's guidance in this perplexing situation.[5]

Questions about the calendar formed one of the most public and enduring controversies at the intersection of Islam and science in the first half of the twentieth century. A growing number of Muslim jurists, drawing on previously marginalized voices within the history of Islamic jurisprudence, argued for the adoption of a calendar computed according to the mathematics of lunar motion, rather than determined on a monthly basis by observation of the lunar crescent – which historically had been the dominant position in all of the legal schools. In Egypt, the advocates of a computed calendar ranged from the Hanafi mufti Muhammad ibn Bakhit al-Muti'i, to the Salafi scholar Ahmad Shakir. Such scholars were responding to a confluence of trends, most of which had begun in the previous century: the integration of Muslim societies within a global economy that favored commensurability and predictability; the spread of European rule across North Africa, the Middle East, and South Asia; the rise of self-consciously modernizing nation-states in parts of the Islamic world that maintained political independence; an organized global movement that promoted time reform as a matter of social and economic rationalization; the rising influence of modern scientific epistemologies; and, perhaps above all, the spread of technologies of communication that revealed disagreement among Muslim communities in newly public and rapid ways.[6] Confusion over the beginning and end of Ramadan, the month of fasting, was especially distressing: it cast doubt on the fulfillment of a major religious duty, and it made Islam an object of mockery among non-Muslims – a particular concern in societies where Muslims were a minority.[7] For many, computation offered a way of celebrating the month that would be appealingly "modern" while also conducive to unification of the *umma*.

And yet, in contrast to the case of prayer timing, no consensus emerged on the standardization of the Islamic calendar. To this day, even as prominent scholars and activists continue to argue the merits of astronomical computation, the fall of Ramadan brings a moment of uncertainty to

---

[5] *Al-Manar* 23, no. 8 (29 Safar 1341/20 October 1922), 585. The salutation, "your student" (*talmīdhukum*) was not uncommon among the questions that Rida published in *al-Manar*. It indicated the way in which Rida's readers relied on him for answers; no personal relationship was necessarily implied.

[6] See especially Skovgaard-Petersen, *Defining Islam*; Ogle, *The Global Transformation of Time*, 149–76; Quadri, "Transformations of Tradition"; and Barak, *On Time*, 115–44.

[7] For example, see the complaint of a Russian qadi to Rashid Rida that disagreement on the timing of Ramadan had become a laughingstock among other religious communities (*adhūka 'ind ahl sā'ir al-milal*). *Al-Manar* 6, no. 18 (16 Ramadan 1321/December 1903), 706.

those who keep the fast, as authorities in major Muslim communities around the world search the skies for the sliver of light whose appearance will allow them to declare the commencement of the new month. This very public ritual and its controversies have long attracted academic as well as popular scrutiny. Recent scholarship has drawn attention to the epistemology of reformists like Muti'i, who argued that the *presence* of the new crescent was the definitive cause of Ramadan, whether it was seen or not.[8] Meanwhile, to explain the endurance of lunar crescent observation despite such arguments, historians have emphasized the way in which timekeeping conventions became markers of religious and national identity in the early twentieth century, especially in the context of anticolonial politics.[9]

Such accounts have tended to interpret the calendar debate in terms that equate the practice of looking for the new crescent with Islamic tradition, and to identify the use of astronomical computation as a response to the demands of modern science and technology. On this view, the search for the lunar crescent is a shining triumph of traditionalism; its endurance illuminates the boundaries that canonical texts, particularly as mobilized under anticolonial and postcolonial circumstances, have placed on Islam's adaptation to modernity. This interpretation rests, however, on the premise that lunar crescent observation was actually the practice of all Muslim communities until the twentieth century, rather than a norm articulated in *fiqh* but (like many norms of *fiqh*) observed differently in different places at different times. As 'Abd Allah al-Mas'udi's letter to *al-Manar* suggests, this premise obscures a complex history. At the beginning of the twentieth century, Muslim communities engaged in a variety of practices for determining the beginning and end of the month, which included computation as well as searching for the lunar crescent. Moreover, computation itself was a compound category, including sophisticated mathematical techniques of astronomical calculation alongside more vernacular methods for approximating the lunar cycle. This pastiche of techniques derived from an eclectic reading of Islamic scholarly sources.

In this context, reformists like Rashid Rida began to emphasize lunar crescent sighting as the authoritative Islamic practice, not in opposition

---

[8] Quadri shows that Muti'i's argument rests on an epistemology similar to what Charles Taylor termed the "representationalism" of modern science. See Quadri, *Transformations of Tradition*, and Charles Taylor, *Philosophical Arguments* (Cambridge, MA: Harvard University Press, 1995). For the broader history of this epistemology in Egypt, see Mitchell, *Colonising Egypt*, 7–10.

[9] Ogle, *Global Transformation of Time*, 149–76.

to science or standardization but rather – to the contrary – out of a belief that direct observation of the moon would serve as the most accurate and, crucially, most unifying way for Muslims to celebrate their holidays. This belief was contingent upon the specific institutions and technologies available to them, and the specific communities that they addressed. All four fatwas that Rida published on the question of the calendar were written to Muslim communities under non-Muslim rule: a 1903 fatwa to a qadi in the Russian Empire; a 1907 fatwa to a reader in Suakin, in the Anglo-Egyptian Sudan; the 1922 fatwa to the madrasa director in Bangkok; and a 1930 fatwa to a newspaper-owner in China. While this geographic range is not unusual for the readership of *al-Manar*, it helps make sense of two core themes that Rida's fatwas on the calendar emphasized: the importance of celebrating Ramadan in unison with the greater Islamic community (*umma*), and of not deputing religious authority to a potentially corrupt leadership. While these considerations generally led Rida to emphasize the importance of lunar crescent sighting, he was also interested in the possibility that, under the right circumstances, computation of the month by a central, credible institution – under Muslim rule – would serve the cause of uniformity most effectively.

Following a brief overview of the astronomical and juridical considerations that framed the terms of debate over the calendar, this chapter seeks to contextualize the debate in terms of on-the-ground computational practices – and their politics – in the late nineteenth and early twentieth centuries. This on-the-ground perspective is based on a variety of travel memoirs and newspaper sources from the period, in addition to requests for fatwas. Because debates over the calendar were debates about the geographic scale at which religious communities and technologies of coordination should function, they placed voices in Egypt in conversation with voices from across the Islamic world. While a global picture of Islamic calendrical practices would require its own monograph, in order to capture this transnational dimension of the debates, I have drawn on an extensive news archive from Southeast Asia, whose Muslim communities were geographically remote from Egypt, but were linked to the Middle East through the presence of a British colonial administration, an Arab diaspora, and the circulation of Arabic journals including *al-Manar* itself. Understanding the diverse and evolving means of determining the calendar in both the Middle East and Southeast Asia in the early twentieth century reveals how transnational circles of Islamic reformism began to place new emphasis on lunar crescent sighting as the best way to unite the *umma* in correct observance of the fast.

The Lunar Crescent: Astronomical and Juridical Considerations

Participants in the debates over the Islamic calendar often assumed a certain degree of knowledge of the lunar month, whose relevant characteristics are worth reviewing. A mean lunar cycle lasts approximately 29.5 days. This is the amount of time that elapses between successive "conjunctions," when the moon is in alignment with the sun and earth. For astronomers, this moment of conjunction, when no moon is visible at all, defines the "new moon." In Islamic law, however, the new *month* was generally considered to begin later, when the new lunar crescent becomes visible in the sky. The difference between the "new moon" and the rising of the new lunar crescent was one reason that some Muslim jurists suspected the reliability of almanacs for determining the months in the early twentieth century.

As in Muhammad al-Khudari's discussion, jurists typically held that the new month should be declared based upon the testimony of witnesses who have actually seen the lunar crescent. This practice is called "seeing the crescent," or *ru'yat al-hilāl*; the more specific term, "searching for the Ramadan crescent" (*istihlāl Ramaḍān*) may also be used. Of course, the practice of *ru'yat al-hilāl* did not imply that a month could end or begin at any time. Rather, the question was whether the old month would contain twenty-nine or thirty days. Those who practiced *ru'yat al-hilāl* would wait for reports of the new crescent's sighting on the eve of the old month's thirtieth day. This occasion was called the "night of seeing" or "night of doubt" (*laylat al-ru'ya, laylat al-shakk*). If no reliable report came, a thirtieth day would be ascribed to the old month, and the new month declared for the following day. Thus, juristic discourse paired "seeing the crescent" (*ru'yat al-hilāl*) with "completing the period" (*ikmāl al-'idda*), meaning counting one more day for the old month if no crescent was reported on the "night of seeing." The practice of *ru'yat al-hilāl* with *ikmāl al-'idda* drew support from numerous authoritative hadiths.[10]

Even within the normative discourse of *fiqh*, however, this framework contained significant ambiguities. In determining the appearance of the new crescent, what role should jurists accord to astronomical knowledge? Does the new month begin when one knows – by whatever means – that the crescent has risen, or must the crescent actually be seen? In evaluating the credibility of a witness, to what extent should a jurist

---

[10] E.g., Bukhari, *Sahih al-Bukhari*, Kitab al-Sawm, 31:124, 131, 133; Muslim ibn al-Hajjaj al-Qushayri, *Sahih Muslim* (Cairo: Dar Ihya' al-Kutub al-'Arabiyya, 1955), Kitab al-Sawm, 6:2365, 2378–81.

consider the astronomical parameters of lunar motion? Total reliance upon astronomical computation – with no need for physical observation – was unusual as a position within the schools of *fiqh*, but a number of intermediary positions were common.[11] Muhammad al-Khudari's view, that most people should begin the new month when there is testimony to the moon's appearance, but that one who is able to calculate (*al-ḥāsib*) should do so, was associated with the Shafiʿi school to which he belonged. Another position held that everyone should rely on testimony to the moon's appearance, but that such testimony should be evaluated in light of astronomical parameters.[12]

Neither the dominant *fiqh* position nor its alternatives can be described as the exclusively "scientific" practice, as has sometimes been done.[13] It is, at the least, unclear why the difference between calculating celestial phenomena and observing celestial phenomena is a difference between science and non-science. Rather, the difference is better understood according to the terms in which the jurists articulated it, as a matter of "seeing" (*ruʾya*) versus "calculating" (*ḥisāb*). Even among jurists, however, the meaning of *ḥisāb* was never monolithic. Scholars of astronomy computed the new months in different ways for different purposes. A conventional calendar, with a predetermined cycle of 29- and 30-day months composing predetermined 354- or 355-day years, served to convert dates between the *hijrī* and other calendars: to compute almanacs, for example. For religious (*sharʿī*) purposes, meanwhile, the month would be computed according to the rising of the new lunar crescent, or even (some held) according to the earlier time of conjunction.[14] Determining the new month by calculation (*bi-l-ḥisāb*) was already a heterogeneous category within Islamic jurisprudence, even before we account for the variety of practices authorized in other discourses.

Debates over *ruʾyat al-hilāl* and *ḥisāb* were part of a broader reevaluation of the means of determining the months among ʿulamaʾ in the late nineteenth and early twentieth centuries. It was even possible in this period for a teacher of astronomy at al-Azhar to argue in favor of standardizing the Islamic months by converting them to a solar calendar.[15] Perhaps the

---

[11] For a partisan but useful review of the various *madhhab* positions on this question, see Muhammad Bakhit al-Mutiʿi, *Kitab Irshad ahl al-milla ila ithbat al-ahilla* (Beirut: Dar Ibn Hazm, 1421 [2000]), 31ff.

[12] Khudari, *Sharh al-lumʿa*, 198–99. See Chapter 1.

[13] Cf. Moosa, "Shaykh Ahmad Shakir."

[14] See, for example, the discussion in Khudari, *Sharh al-Lumʿa*, Michigan Isl. Ms 722, pp. 197–99.

[15] See Samuel Zwemer, "The Clock, the Calendar, and the Koran," *The Moslem World* 3 (1913), 265. On Ahmad Musa al-Zarqawi's teaching career, see "Mukatabat ila diwan khidiwi bi-shaʾn talabat iʿanat li-l-ʿadid min al-ashkhas sanat 1913," DWQ 69–747.

most commonly debated question, however, was not the issue of seeing versus calculation, but rather the issue of whether telegraphic reports of the lunar crescent's appearance in one location should be accepted as evidence to declare the new month in another location. Requests for legal guidance on this question in Egypt date to at least the 1870s, and a consensus in favor of accepting telegraphic reports began to emerge by the 1910s.[16] Here, too, participants in the debate drew on a long history of juristic discussion. Muslim communities had exchanged news before the advent of telegraphy, after all, and premodern scholars were aware of the problem that the new crescent became visible at different times at different latitudes, even under ideal atmospheric conditions.

While the debate over accepting telegraphic reports turned on a different set of legal considerations than the debate over seeing versus calculating, the two issues were closely linked in practice. They converged, in particular, on the question of unity of the *umma*. Reformists like Rida, who worked ardently for pan-Islamic solidarity in the face of European imperialism, could comfortably argue against standardizing the months through *ḥisāb* in part because they foresaw that the telegraph had the potential to effect unity through *ru'yat al-hilāl*. This calculus made sense specifically at a time when Muslim communities around the world were in fact using a bewildering array of methods to determine the months.

Computing Ramadan: Diverse Practices, Eclectic Sources

The questions that readers submitted to *al-Manar* offer a window onto actual practices of determining the Islamic months in the early twentieth century. Historians of Islam have debated the degree to which requests for fatwas may shed light on social realities, since some jurists composed "fictitious fatwas," answers to questions of their own framing, as a way of articulating their positions.[17] There is, however, no evidence that Rashid Rida engaged in this practice in the pages of *al-Manar*; if anything, his fatwas on the calendar are particularly difficult to construe as fictitious, since their authors were identified by personal names, locations, and institutional affiliations. In any case, as the publisher of a periodical, Rida

---

[16] For a discussion from 1877, see Muhammad al-ʿAbbasi, *al-Fatawa al-Mahdiyya fi al-Waqaʾiʿ al-Misriyya*, vol. 1 (Cairo: al-Matbaʿa al-Azhariyya, 1301 [1883–84]), 14. On arguments in favor of the validity of telegraphic reports in the early twentieth century, see Skovgaard-Petersen, *Defining Islam*, 80–99.

[17] On the problem of the "fictitious fatwa," see the introduction to Khalid Masud, Brinkley Messick, and David Powers, eds., *Islamic Legal Interpretation: Muftis and Their Fatwas* (Cambridge, MA: Harvard University Press, 1996).

was unlikely to print a question – whether "invented" or not – unless he thought it would be relevant to his audience of readers.

In this light, the questions that Rida received about the timing of Ramadan complicate the dichotomy posed in *fiqh*, as well as in scholarly astronomical sources, between observation and sophisticated calculation. For example, according to a practice attributed to the Shiite Imam and popularly venerated figure Ja'far al-Sadiq, which appears in a question to *al-Manar* from the Sudan as well as the one from Bangkok, Ramadan will begin on the fifth day of the month in which it began in the previous year.[18] In other words, if Ramadan last year began on a Sunday, this year the fast will begin on a Thursday. Or, viewed from the perspective of a solar calendar, Ramadan will begin ten days earlier every year. This technique not only ignores the question of the lunar crescent's actual visibility in favor of a predetermined month; by assuming that the lunar year is a round 355 days (rather than between 354 and 355), it would quickly diverge even from the annual lunar cycle. Of course, the simplicity of the procedure is also its great advantage. Whereas other computational techniques require more complex arithmetic, as well as the use of tables, this procedure could have been performed by anyone who knew the days of the week. Such a potentially broad audience accords with the appearance of this technique, though attributed to Ja'far al-Sadiq, in Qazwini's *Wonders of Creation* – a text, unlike the technical manuals of scholarly astronomy, widely copied and illustrated since its original composition in the thirteenth century.[19]

In addition to this back-of-the-envelope technique, however, Qazwini's *Wonders of Creation* also contained a more sophisticated, but still relatively accessible, practice of computing Ramadan. This technique was based on an eight-year cycle of days of the week on which the fast could begin. The use of this table required the ability to perform arithmetic as well as to read. The technique appears to be a simplified version of a more elaborate procedure and set of tables that scholars of astronomy like Muhammad al-Khudari used to compute the *hijrī* calendar for the purposes of the *zij*. It would not, however, have corresponded (in any necessary way) to the visibility of the lunar crescent.

The only computational practice to surface in *al-Manar* that sought to determine the actual rising of the new lunar crescent was, unsurprisingly, based on a discussion in a work of *fiqh*. This case, too, however, shows that computation included a greater variety of practices, and among more kinds of people, than standard accounts suggest. When Mas'udi

---

[18] For the question from the Sudan, see *al-Manar* 10 (1907), 530–34.
[19] On the textual history of Qazwini's *Wonders*, see Persis Berlekamp, *Wonder, Image, and Cosmos in Medieval Islam* (New Haven, CT: Yale University Press, 2011).

cited Sulayman al-Bujayrimi in his letter to Rida, he was invoking an eighteenth-century gloss that was commonly used among Azhari scholars to teach one of the classic works of Shafi'i jurisprudence, Zakariyya' al-Ansari's commentary on al-Nawawi's *Minhaj al-Talibin*.[20] The relevant passage from Bujayrimi's gloss is worth quoting in full:

The moon is never hidden more than two nights at the end of the month. It is hidden for two nights if it is a complete [30-day] month, and one night if it is an incomplete [29-day] month. The meaning of "hidden for two nights" is that the moon is not visible during them, and becomes visible after dawn. According to some, if the moon is hidden for two nights and the sky is clear on both of them, then the third night is the beginning of the month without a doubt. Every Muslim should understand this, for one who understands it has no need to wait for the sighting of the Ramadan crescent (*ru'yat hilāl Ramaḍān*), and a day of the fast will not pass him by; it was a complete month. The hadith, "Fast upon seeing it..." is for one who does not understand this.[21]

The procedure that Bujayrimi described was a version of the Shafi'i practice of *ru'yat al-hilāl* and *ikmāl al-'idda* (completing the period) based on certain astronomical principles. Instead of searching for the new crescent on the eve of the old month's thirtieth day, followers of this procedure would take note of the new moon, and begin the fast two days later. In 1923 (a year after Mas'udi's letter was published in *al-Manar*), a similar procedure, fixing the beginning of the month as the third day after the new moon, became the basis for computing Ramadan among Muslims in England, due to the difficulty of observing the lunar crescent.[22]

Although the computational techniques attested in these letters to *al-Manar* were based on pre-nineteenth-century texts, the age of the source does not necessarily equal the age of any particular community's practice. Qazwini's thirteenth-century magnum opus, long in demand as a manuscript, was printed by several Egyptian presses in the 1880s and 1890s, and had even been the subject of a German edition (in Arabic) in 1848.[23] Bujayrimi's gloss was even more widely printed in Egypt, with at

[20] On the use of al-Nawawi's *Minhaj* at al-Azhar at the mid nineteenth century, see Mubarak, *al-Khitat*, 4:27.

[21] Bujayrimi, *Hashiyat al-Bujayrimi*, 67.

[22] "The Date of Ramadan," *The Straits Times*, 31 May 1923, p. 3.

[23] Zakariyya ibn Muhammad al-Qazwini, *'Aja'ib al-makhluqat wa-ghara'ib al-mawjudat* (Cairo: Matba'at al-Taqaddum, 1886); Muhammad ibn Musa al-Damiri and Zakariyya ibn Muhammad al-Qazwini, *Min hayat al-hayawan, wa-bi-hamishihi Kitab 'Aja'ib al-makhluqat wa-l-hayawanat wa-ghara'ib al-mawjudat* (Cairo: al-Matba'a al-'Amira al-Sharafiyya, 1306 [1888]); *Min hayat al-hayawan, wa-bi-hamishihi baqiyyat Kitab 'Aja'ib al-makhluqat wa-l-hayawanat wa-ghara'ib al-mawjudat* (Cairo: Matba'at al-Taqaddum, 1309 [1891]); Zakarija ben Muhammed ben Mahmud el-Cazwini, *Kosmographie*, ed. F. Würstenheld, 2 vols. (Göttingen, 1847–48).

least six Cairo printings between 1869 and 1926, reflecting the popularity of the text for students of the Shafi'i *madhhab*, as well as the increasing market for print in Azhari circles during the late Ottoman period.[24] When authorities in places as distant from each other as Suakin and Bangkok used these texts to compute Ramadan in the early twentieth century, they may have been perpetuating a longstanding practice in their communities, but they could also have been building on a recent resurgence of interest in specific scholarly works, as an "Islamic canon" began to take shape in print in the late nineteenth and early twentieth centuries.

Computing Ramadan: Almanacs in Egypt

In addition to using their own methods of computation, some communities in Egypt also began relying on the printed calendars that became increasingly available over the course of the nineteenth century. Lady Lucie Duff Gordon, who lived in upper Egypt from 1862 to 1869 (seeking relief from tuberculosis), wrote to her husband in 1868 that her "red pocketbook... determined the beginning of Ramadan at Luxor this year. They received a telegram fixing it for Thursday, but Sheykh Yussuf said that he was sure the astronomers in London knew best, and made it Friday."[25] Gordon's report is remarkable, first, because it suggests that religious authorities in Luxor readily accepted telegraphic communication of the beginning of Ramadan decades before the permissibility of this practice was debated by prominent Muslim legal authorities in more cosmopolitan locations. At least on a plain reading of this incident, the Luxor authorities rejected the telegram's advice not because they suspected the reliability of the telegraph, but because they had greater confidence in the astronomical computation recorded in an English pocket diary than they did in whatever method the Cairo authorities had used to fix the beginning of the month. While it is also conceivable that Lady Gordon's pocketbook appealed to Sheykh Yussuf specifically as a means of rejecting the telegraphic communication of Ramadan, this interpretation, too, would point to the cultural authority of the printed

---

[24] The first edition was Sulayman ibn Muhammad al-Bujayrimi, *Hashiyat al-'Alim... Sulayman al- Bujayrimi 'ala Sharh Manhaj al-Tullab* (Cairo: Dar al-Tiba'a, 1286 [1869]); subsequent Cairo editions appeared in 1875 (al-Matba'a al-'Amira), 1884 (publisher unknown), 1891 (al-Matba'a al-Amiriyya), 1910 (al-Maktaba al-Islamiyya), and 1926 (Mustafa al-Babi al-Halabi); an Istanbul edition by al-Maktaba al-Islamiyya appeared in 1910.

[25] Lucie Duff Gordon, *Letters from Egypt* (London: Virago, 1983), 373. Late eighteenth- and nineteenth-century Britons commonly carried pocket diaries, such as William Peacock's *Polite Repository*, in red-leather wallets. Sandro Jung, "Illustrated Pocket Diaries and the Commodification of Culture," *Eighteenth-Century Life* 37 (2013), 69.

calendar (and of the "astronomers of London" who were assumed to have produced it).

Even when religious authorities determined the beginning of the month themselves, they deliberated in the shadow of increasingly well-circulated calendars and almanacs. In 1875, the Khedivial court was frustrated by the inability of the city's shaykhs to agree on whether ʿId al-Adḥā would fall on the eighteenth or nineteenth of January. The shaykhs finally declared ʿId for the seventeenth – the date that the court had originally planned on, since it was the date printed in the almanac of the local publisher Kaufmann. Due to the back-and-forth, however, transportation for the court's festivities had to be re-arranged at the last minute.[26] Since ʿId al-Adḥā falls toward the middle of its lunar month (Dhu al-Hijja), such disagreements were less common than those concerning the beginning and end of Ramadan. Perhaps the shaykhs changed their calculations once they saw when the new month actually began. Whatever its cause, however, their change of heart must have lent credence to the notion that it would be easier simply to rely on almanacs to begin with.

A certain degree of comfort with regulating the calendar by means other than *ruʾyat al-hilāl* may have derived from the fact that rural communities had long relied on urban authorities to communicate the beginning and end of Ramadan. Even within the discourse of *fiqh*, it was well-established that villagers around Cairo could – indeed, should – consider the sounding of the Citadel cannon or the appearance of certain lights above the city's mosques on *laylat al-ruʾya* to be reliable indicators of the new month, since "the possibility of their being for something other than Ramadan is remote."[27] While this practice bore most directly on debates over the validity of telegraphic reports as testimony to the new moon's appearance, it also suggests that, even in times and places where people theoretically relied on lunar crescent sighting, actually going out to look for the moon was not necessarily a part of the community's experience.

Muslim communities at the dawn of the twentieth century were familiar with a range of techniques for computing, rather than observing, the rising of the new lunar crescent. The use of these techniques – in at least some communities, some of the time – recasts the calendar debates of the subsequent decades in at least two ways. First, it demonstrates that

---

[26] Ellen Chennells, *Recollections of an Egyptian Princess, by Her English Governess, being a record of five years' residence at the court of Ismael Pasha, Khedive*, vol. 2 (Edinburgh: William Blackwood and Sons, 1893), 219–20.

[27] Al-ʿAbbasi, *al-Fatawa al-Mahdiyya*, 1:14. The fatwa, dated 5 Dhu al-Hijja 1294 (1877), cites the opinion of the great early nineteenth-century Hanafi work of Ibn ʿAbidin, *Radd al-Muhtar ʿala al-durr al-mukhtar* (Bulaq: al-Matbaʿa al-Misriyya, 1286 [1869]).

advocates of *ru'yat al-hilāl* were not simply defenders of the status quo against innovation; they were making an intervention amid a complex field of practices. Second, it reminds us that the use of *ḥisāb*, rather than *ru'yat al-hilāl*, did not guarantee a more organized, let alone globally uniform, observance of the fast in the early twentieth century. Differences in computation methods as well as in the definition of the new month made computation a very uncertain tool.

Making Holidays Public: The Politics of Sighting and Calculating in the Straits Settlements

Something of the politics of this uncertainty, and of the shift toward *ru'yat al-hilāl*, can be understood from the experience of one of *al-Manar*'s non-Middle Eastern audiences: the Muslim communities in Britain's crown colony of the Straits Settlements and neighboring protectorate of the Federated Malay States. Although none of the questions that Rashid Rida published regarding the timing of Ramadan came from this region, its evolving practices over the course of the late nineteenth and early twentieth centuries shed light on the complex and changing circumstances in which Rida sought to intervene as a transnational mufti. The extensive news archive that exists for the Straits Settlements from this period suggests that, although *ru'yat al-hilāl* was always one way of determining the beginning and end of Ramadan in British Malaya, computation was also commonly used until the 1920s, when lunar crescent sighting became institutionalized as a result of the changing relationship between religion and the colonial state, and the growth of a more assertive, transnational Islamic activism.

Beginning in 1894, if not earlier, the official gazette *The Straits Times* regularly published the date of 'Id al-Fiṭr (Hari Raya Puasa) well in advance, usually for the first of the two possible days on which it could fall according to *fiqh*.[28] Although the newspapers occasionally reported on the practice of lunar crescent observation as well, until the 1920s, such reports – few and far between as they were – concerned instances of disagreement within the community, with some following the practice of observation while others held the festival on the computed date. Such

---

[28] See, e.g., "Arrangements," *The Straits Times*, 3 April 1894, p. 3; "The Holidays," *Singapore Free Press and Mercantile Advertiser*, 29 January 1897, p. 2; Untitled, *Straits Times*, 27 November 1902, p. 4; Untitled, *Eastern Daily Mail and Straits Morning Advertiser*, 16 November 1906, p. 2; Untitled, *The Straits Times*, 23 September 1908, p. 5. Based on the day of the astronomical new moon (i.e., conjunction) according to some of these columns (such as the last), it seems that the festival was predicted for the first date on which the new crescent was expected to be visible.

disagreements may have sprung from deeper cleavages in the community. Thus, in 1897, the Javanese Muslims of Singapore ended Ramadan a day earlier than the Malay Muslims, who waited because they had not seen the new moon.[29] As late as 1921, the *Singapore Free Press and Mercantile Advertiser* reported that the festival was "largely celebrated" in the city on the computed date, implying that only a minority chose to wait because the moon had not actually been seen.[30]

The extent to which computation was accepted in the early twentieth century is indicated by the fact that the computed dates were not only circulated in print, but also announced, at least in some communities, by means of gunfire. Thus, in 1919, the timetable for the Ramadan signals in the Malay state of Selangor was published on May 30 as follows:

At Klang [Kelang] and Kuala Lampur commencing from to-day for the Manmaybu Puasa [welcoming the fast], nine guns will be fired at 4 p.m. on every day until June 27, inclusive. At Klang only, at mid-night, one gun for Saher. At all stations from May 30 to June 28, one gun at 6:30 p.m. for Berbuka Puasa [iftar]. At Klang only, on June 29, nine guns at 4 p.m. for the Menyambut Hari Raya Puasa, and again at Klang, only, nine guns at 9 a.m. for the Menyambut Sambahyang Hari Raya Puasa.[31]

This timetable suggests not only that the signals for the beginning and end of Ramadan were scheduled in advance, but also that the daily signal for iftar was fired at a conventional time of 6:30 p.m., rather than at the moment of sunset (whether as observed or calculated). Indeed, this convention seems to have held no matter when Ramadan fell in the solar year. In 1911, when Ramadan ended in late September, a poet satirized the gusto with which Muslims ended the fast:

Bulan Puasa being ended and done with at last,
Mohamed bin Jait commenced breaking his fast,
Nor did he restrain his desires in the least;
But made Hari Raya a glorious feast,
Starting off at six-thirty with curry and rice,
Garnished freely with sambal, he ate in a trice...[32]

This was crude satire, to be sure, but the plausibility of ending the fast at 6:30 p.m., without regard for seasonal (let alone daily) variation in the length of the day, suggests that reliance on the computed dates of the month was less a matter of exactitude than of convenience.

---

[29] "The Hari Raya," *The Straits Times*, 5 March 1897, p. 3.
[30] "Hari Raya Puasa," *The Singapore Free Press and Mercantile Advertiser*, 9 June 1921, p. 6.
[31] "Bulan Puasa," *The Straits Times*, 30 May 1919, p. 15.
[32] "A Lay of Hari Raya," *The Singapore Free Press and Mercantile Advertiser*, 28 September 1911, p. 14.

In part, such practice was symptomatic of an era in which believers were still negotiating the proper usage of certain newly widespread technologies. In the Yemen, for example, a controversial practice arose in the 1920s of performing the evening prayer ('*ishā*') thirty minutes after the sunset prayer (*maghrib*), apparently using watches or clocks to fix the difference between the two times as a matter of convenience, rather than trying to determine the precise, variable duration between sunset and darkness on a daily basis.[33] In Alexandria, meanwhile, some broke their fast on the first day of Ramadan according to the signal broadcast from Cairo over Radio Egypt – only to hear the blast of Alexandria's "breakfast cannon" (*midfa' al-fuṭūr*) four minutes later, and wonder nervously whether they had voided their fast by eating too early.[34] It was far from obvious at what geographic scale, and with what degree of temporal precision, new technologies of coordination should serve to organize the performance of religious duties.

But prioritizing convenience over precision also made sense specifically in light of the politics of holidays in British Malaya at the beginning of the twentieth century. Particularly in the 1910s, public controversy over Ramadan centered not on intra-Muslim disputes about the correct way to determine the timing of the month, but rather on the interconfessional issue of the degree of recognition that Muslim holidays would enjoy from the non-Muslim state. Until 1914, although public institutions were sometimes closed in honor of 'Id al-Fiṭr on an *ad hoc* basis, the feast did not have the official status of a public holiday in the crown colony. Beginning in 1911, however, several public meetings of Muslim organizations discussed lobbying the governor to make the holiday official.[35] In 1913, the Chief Imam of the Sultan Mosque, Imam Yusof, requested the backing of the Moslem Union, the Moslem Association, the Indian Moslem Society, and "all other Moslem bodies in Singapore" to petition the governor to this effect, and in 1914 the colony's Holidays Ordinance was amended to include the first day after the end of Ramadan.[36] Tellingly, the Hindu holiday of Thaipusam was added at the same time, just as 'Id al-Aḍhā and Deepavali would be added together

---

[33] *Al-Manar* 29 (1928), 529–30.

[34] Muhammad ibn Bakhit al-Muti'i, *al-Fatawa lil-Imam al-'Allama Muhammad ibn Bakhit al-Muti'i*, ed. Muhammad Salim Abu 'Asi (Lebanon: al-Siddiq lil-'Ulum, Dar Nur al-Sabah, Dar al-'Ulum, 2012), 84.

[35] "Moslem Festivals," *The Straits Times*, 16 September 1911, p. 10; "The Hari Raya Question," *The Straits Times*, 4 September 1913, p. 8.

[36] See "Hari Raya Holidays," *The Straits Times*, 11 September 1913, p. 8, for quotation; see also "The Hari Raya Holidays," *The Straits Times*, 13 September 1913, p. 8; Untitled, *The Straits Times*, 16 September 1913, p. 8; "Hari Raya Holiday," *The Straits Times*, 10 January 1914, p. 8; "Legislative Council," *The Straits Times*, 31 January 1914, p. 10.

in the late 1920s.[37] The list of official holidays was a locus for negotiating the public visibility and privileges of one's religious community in a multiconfessional society under colonial rule.

In this context, the predictability of a calendar computed in advance must have been especially attractive. After all, in a multiconfessional, colonial society, it could become confusing – for everyone – if public holidays were not fixed in advance. Thus, in 1928, the government holiday of ʿId al-Fiṭr fell on Friday 23 March, and Muslims on the Peninsula indeed celebrated the feast on that day. In Singapore, however, the Muslim authorities held that ʿId did not begin until Saturday, since the new moon had not been visible (due to clouds) on Thursday night. As a result, "Muslims of Singapore fasted on Friday and were very sad about it, and it was a public holiday for nothing. ... Saturday, they are celebrating Hari Raya but without a public holiday."[38]

As this turn of events indicates, by the late 1920s, many Muslims in Singapore were particularly committed to the practice of observing, rather than computing, the visibility of the new lunar crescent. This was a recent development, which flowed from two broader trends in the city's community. First was the establishment of official representation of the community before the colonial state in 1915, when the Mohammedan Advisory Board was founded.[39] Although sometimes derided as an instrument of colonial rule, and lacking credibility as an Islamic legal authority, the board's existence made it possible to represent "Muslims of Singapore" as a community with a single voice, which could announce the beginning and end of Ramadan for all its members.[40] By itself, however, the board does not seem to have taken the lead in pushing for the primacy of lunar crescent observation. This particular impetus came, instead, from an overlapping group of communal leaders who formed the United Islamic Association in 1923.[41] Beginning in 1924, the UIA and Mohammedan Advisory Board cooperated in securing permission from the colonial port authorities to use the telescopes at Mount Faber and Fort Canning, as well as two boats, to look for the new moon of

---

[37] "A Holiday Problem," *The Singapore Free Press and Mercantile Advertiser*, 15 December 1927, p. 8.

[38] H. G. Sarwar, "Hari Raya," Letter to Editor, *The Straits Times*, 24 March 1928, p. 10.

[39] "Matters Muslim," *The Singapore Free Press and Mercantile Advertiser*, 12 December 1929, p. 13.

[40] On criticisms of the Mohammedan Advisory Board, see Nurfadzilah Yahya, "The Question of Animal Slaughter in the British Straits Settlements During the Early Twentieth Century," *Indonesia and the Malay World* 43 (2015), 185.

[41] "Moslems and Their Clubs," *The Straits Times*, 3 January 1924, p. 10.

Ramadan.[42] These official observation parties would testify to the crescent's appearance (or not) before the Chief Qadi, and the Advisory Board would declare the fast accordingly. This highly public performance of communal authority, staged within the structure of the colonial state, became a matter of routine press commentary in the 1920s.[43]

Testifying to the new moon was by no means a novel practice for Muslims in Singapore, but the decisive prioritization of this practice over others originated in the religious politics of the interwar period: the institutionalization of Islam within the colonial state, and the rise of a more assertive and religiously specific form of public advocacy from within the Muslim community itself. In contrast to the community's earlier focus on achieving recognition of Islam in civil terms (e.g., ʿId al-Fiṭr as a public holiday), the UIA promoted a specifically pious agenda of religious education and conformity, for example by working to have Islamic instruction included in the Malay Vernacular Schools.[44] In this sense, the establishment of *ruʾyat al-hilāl* was part of a history of Muslim elites' usage of the British colonial infrastructure to promote certain religious viewpoints over others.[45]

Whether in Cairo or Kuala Lumpur, *ruʾyat al-hilāl* was certainly a familiar and well-established practice at the dawn of the twentieth century – but it was one of several. In some places, old practices of non-astronomical computation, rooted in scholarly texts outside the dominant *fiqh* traditions, were widely credited. In villages, residents still turned their eyes (or ears) to the city for non-verbal communication of the month's beginning. Where newspapers, almanacs, and pocket diaries were available, relying on the astronomical computations disseminated in print was often accepted as unproblematic. When Muslim jurists and other activists insisted on the need for observation of the lunar crescent, they were not simply defending "tradition" against "modern science." They were seizing on a particular thread within a fabric of Islamic traditions and highlighting that thread at the expense of a pastiche of

---

[42] "Hari Raya Puasa," *The Straits Times*, 2 May 1924, p. 9; "Proving the Moon of Shawwal," *The Singapore Free Press and Mercantile Exchange*, 3 May 1924, p. 9.

[43] "The Moslem Community," *The Straits Times*, 8 March 1926, p. 10; Untitled, *The Singapore Free Press*, 13 March 1926, p. 8; "The New Moon of Ramadan," *The Straits Times*, 24 February 1927, p. 10; "Observation of Ramadan," *The Singapore Free Press and Mercantile Advertiser*, 21 February 1928, p. 8; "Town and Country," *The Straits Times*, 21 February 1928, p. 10.

[44] "Religious Teaching in Malay Vernacular Schools," *Singapore Free Press and Mercantile Advertiser*, 19 June 1924.

[45] For an example of this phenomenon from the late nineteenth century, see Nurfadzilah Yahya, "Craving Bureaucracy: Marriage, Islamic Law, and Arab Petitioners in the Straits Settlements," *The Muslim World* 105 (2015): 496–515.

alternatives. Some of these alternatives were new, while some were quite old; despite their mathematical nature, however, none had enjoyed much success in unifying Muslims around the timing of the fast. In this context, searching for the lunar crescent was not only a practice rooted in foundational texts; it also became the most plausible candidate for producing a uniform calendar. The fatwas of Rashid Rida shed further light on this process.

Accessibility and Unity: Rashid Rida on the Virtues of *ru'ya*

At least four times between 1903 and 1930, readers of *al-Manar* posed questions to Rashid Rida regarding the correct way to determine the new month, particularly Ramadan.[46] Without fail, Rida's responses adopted the dominant *fiqh* position: the new month begins either upon the sighting of the new lunar crescent (*ru'yat al-hilāl*), or, if the moon does not appear, the completion of thirty days of the previous month (*ikmāl al-'idda*). In support of this position, Rida usually cited one or more of several widely related hadiths, such as, "Fast when it is seen (*ṣūmū li-ru'yatihi*), and if it is hidden to you, then complete the number of Sha'ban as 30."[47]

In addition to articulating his commitment to the authority of the hadith, Rida's responses on the new moon consistently identified two desirable objectives that *ru'yat al-hilāl* achieved. The first was that the knowledge of how to perform Islamic duties should be accessible to all, not concentrated among a few. As Rida noted in 1903, if the "religious times" (*al-mawāqīt al-dīniyya*) are based on direct observation, rather than on calculation, they are equally knowable to "the masses and the elite" (*al-'āmma wa-l-khāṣṣa*).[48] The ability to know the correct timing of religious duties should be available to "the knowledgeable and the ignorant, the Bedouin and the urbanite,"[49] he elaborated in 1904. The alternative, he warned in 1922, was for people's "religious affairs" (*umūruhum al-dīniyya*) to be controlled by "individuals among the scholars of a special science (*fann*)," who would corrupt Islam in the same way

---

[46] See *al-Manar* 6, no. 18 (16 Ramadan 1321/5 December 1903), 705f.; *al-Manar* 10, no. 7 (Rajab 1325/September 1907), 530–34; *al-Manar* 23, no. 8 (29 Safar 1341/20 October 1922), 585f.; *al-Manar* 31 (Jumada al-Uwla 1349/22 October 1930), 270–78.

[47] Rida outlined the hadith evidence for *ru'yat al-hilāl* in a 1903 article that shortly preceded his first answer to a question on the topic. See "Fasl fima yuthbat bihi al-sawm wa-l-fitr," *al-Manar* 6, no. 17 (1 Ramadan 1321/20 November 1903), 654f.

[48] *Al-Manar* 6, no. 18 (16 Ramadan 1321/5 December 1903), 706.

[49] *Al-Manar* 7, no. 18 (16 Ramadan 1322/24 November 1904), 694.

that other religions had been corrupted by their leaders.[50] Emphasis on the simplicity of Islam was a theme of Rida's writing, as it would become for the Salafis generally. The ability of any person to know the beginning of Ramadan flowed from the same epistemic commitment that held the meaning of the Qur'an to be available to any reader. But this was a commitment conceived in opposition to the supposed intellectual decadence of the old scholarly elite (implicitly compared, here, with rabbis and priests); opposition to modern science had nothing to do with it.

In fact, in order to render the practice of ru'yat al-hilāl as an egalitarian alternative to elitist computation, Rida had to push the practice well beyond its boundaries in *fiqh*, which historically had placed strict criteria on the qualifications of a reliable witness ('adl). These criteria made it difficult for observing the new lunar crescent to be a practice shared by "the knowledgeable and the masses," in Rida's idealistic language. The question from the director of the Bangkok madrasa, for example, noted that one of the problems with ru'yat al-hilāl was that none of his community's members possessed the qualities of probity that the Shafi'i legal school required of a witness. For Rida, however, such qualifications were a matter of custom ('urf) – which could change according to time and place – rather than textual mandate (naṣṣ). He advised his follower in Bangkok that the key criterion for a witness to the new lunar crescent was simply the person's credibility among others (al-'ibra bi-taṣdīq al-nās lahu).[51] Minimizing the necessary qualifications of the witness was one strategy by which Muslim jurists also legitimized the use of telegraphic reports of the new moon in this period. Thus, in 1910, Muti'i had argued for the acceptability of telegraphic news by downgrading the status of the new month to a matter of "report" (khabar), rather than "testimony" (shahāda). A similar move allowed Rida to articulate a sharp contrast between the egalitarianism of ru'yat al-hilāl and the elitism of alternative practices.

The second virtue that Rida ascribed to ru'yat al-hilāl was communal unity: by looking for the lunar crescent, Muslims would all come to agree on when the month began, and thereby fast on the same days. One could well ask how Rida came to believe that the virtue of epistemological egalitarianism would be conducive to the virtue of communal unity. It is far from obvious why commencing the fast upon any reliable person's testimony of having seen the new moon should lead to less disagreement than if a group of specialists were to calculate the month in advance. Yet Rida explicitly linked the accessibility of religious knowledge to the unity

---

[50] *Al-Manar* 23, no. 8 (29 Safar 1341/20 October 1922), 587.
[51] *Al-Manar* 23, no. 8 (29 Safar 1341/20 October 1922), 586.

of the religious community, writing in a 1930 fatwa that both the months and the times of prayer depend on "sensorially observable things" (*umūr mashhūda bi-l-ḥiss*) so that "Muslims will neither disagree nor require leaders and scholars to know their religious times" (*la yakhtalif al-muslimūn wa-la yakūnū muḥtājīn fī mawāqīt dīnihim ilā al-ru'asā' wa-l-'ulamā'*).[52] It may be common sense that calculating the month would lead to a concentration of religious authority in the hands of a small group, but why would it also increase religious dissension?

Rida's fatwas responded to highly specific circumstances. All the questions that he published about the calendar originated in situations that caused Rida to suspect the ability of local figures to calculate the new month in such a way as to produce agreement in practice. In the 1922 Bangkok fatwa, for example, when Rida sided strongly with the practice of *ru'yat al-hilāl*, he did so in opposition to a specific set of alternatives and their sources: Qazwini's thirteenth-century cosmography; the back-of-the-envelope arithmetic technique attributed to Ja'far al-Sadiq; Bujayrimi's eighteenth-century gloss on a Shafi'i legal text of the fifteenth century. Similarly, in his 1907 fatwa requested from Suakin, Rida's defense of *ru'yat al-hilāl* was contrasted with the Ja'far al-Sadiq technique. Under such circumstances, it is unsurprising that Rida could think of *ru'yat al-hilāl* as the practice most likely to result in the correct and unified observance of Ramadan.

Put another way, Rida's endorsement of lunar crescent observation was hardly a stand against astronomy. Rather, it was part of his characteristic stance in favor of – as he framed it – the clear dictates of the Prophet's example, as opposed to the abstruse scholasticism of certain 'ulama'. In Rida's own words, addressed to the school director in Bangkok, "It is one of the marvelous errors of imitation (*taqlīd*) that one who knows the clear, correct practice of the Prophet (*sunna*) will abandon it and take up with the opinion of this or that guy."[53] Much as Rida faulted 'ulama' for unthinking use of prayer tables and watches (see Chapter 6), here he linked computation of Ramadan not to modern science but to *taqlīd*, or deference to an authority within one's legal school – a practice frequently caricatured, even made notorious, by reformist critiques in this period.[54] It was one of Rida's greatest aims to replace this scholarly mode

---

[52] "Mas'alat hilal Ramadan," *al-Manar* 31 (Jumada al-Uwla 1349/22 October 1930), 278.
[53] *Al-Manar* 23, no. 8 (29 Safar 1341/20 October 1922), 585.
[54] The term *taqlīd*, in the language of Islamic law, denoted a way of adhering to the authority of a specific legal tradition – not necessarily a renunciation of independent reasoning or of the possibility of new conclusions. See Wael Hallaq, *Authority, Continuity, and Change in Islamic Law* (Cambridge: Cambridge University Press, 2001), 86. It was in the polemical usage of critics of the legal tradition, such as Rida, that the word acquired the pejorative sense of "imitation," which is why I translate it that way here.

of authority with a practice more directly engaged with foundational texts. The clarity of these texts, he believed, was the ultimate guarantor of communal unity. Thus, when Rida argued that anyone could understand the Qur'an, he did not mean that the Qur'an was amenable to anyone's interpretation. Rather, the meaning of the Qur'an was so clear, in Rida's view, that all sensible people would arrive at the correct understanding. Only corrupt intellectual traditions stood in the way. Rida seems to have conceived of the observation of the new lunar crescent as a similarly straightforward exercise. Disunity was a result of the practice's corruption, whether because witnesses lied, honorable witnesses were not taken seriously, or people tried to force observation to agree with what was printed in calendars – which even disagreed among themselves.[55] Once these obstacles were cleared up, it would be as easy for people to agree on the visibility of the new moon as it was, in Rida's view, to agree on the plain meaning of a text. Foundational texts notwithstanding, however, *ru'yat al-hilāl* was not a universally preexisting custom of Muslim communities that had to be defended against modern science. Different communities determined the new month in various ways, including *ruy'at al-hilāl*, but also including computational practices both old and new.[56] It was Rida, with his emphasis on *ru'yat al-hilāl* as the exclusively legitimate, most egalitarian *and* most unifying practice, who sought to introduce a kind of uniformity.

In the Pasha's House: Reconsidering the Reliability of Calculation

The model of uniformity that Rida articulated in these fatwas differed from the technoscientific approach to time reform that was ascendant in the early twentieth century. The latter approach, of which our modern system of uniform mean time is a product, was an unabashed effort to promote regional and global coordination over local autonomy; authority flowed to a limited number of institutions, and ultimately to only one, at Greenwich. For Rida, however, as for the emergent Salafi movement, the

---

[55] *Al-Manar* 6, no. 18 (16 Ramadan 1321/5 December 1903), 707.
[56] The question from Bangkok provided the most detailed description of local practice. Rida also received a question from the Sudan that mentioned the calculation method ascribed to Ja'far al-Sadiq; a question from China that referenced local disagreements but did not specify what methods were in use; and a question from Russia, the only case in which *ru'ya* was explicitly contrasted with the use of calendars. This was an odd case, however: although Rida criticized the use of calendars for Ramadan, the question actually concerned the determination of the other months of the year, for which Rida said that *ru'ya* was not necessary. See *al-Manar* 6, no. 18 (16 Ramadan 1321/5 December 1903), 705.

certainty of the outcome in any encounter between an honest believer and an authoritative source made such a compromise unnecessary. Epistemological egalitarianism and regional, even global, coordination would go hand in hand.

Or rather, that was one possibility. Rashid Rida enjoyed a career of almost four decades in publishing. During these years, it became possible, at least in certain places, to conceive of the alternative to *ru'yat al-hilāl* not as an eclectic set of practices that generated disagreement, but rather as a centralized, reliable computation of the new month that all Muslims would follow. In fact, when controversy over the calendar arose in these terms, Rashid Rida, at least, adopted a position that was substantially different from the staunch advocacy of *ru'yat al-hilāl* that he articulated in his fatwas and closer to the centralized version of time reform modeled, notably, by British administrators in Egypt. This is not to say that Rida evinced a shift from one view to the other. However, his advocacy of centralized computation under specific circumstances suggests that the value of communal unity ultimately took precedence over the virtue of equal access to knowledge.

In 1904, the qadi of Cairo, members of the Sharia Court (al-Mahkama al-Shar'iyya), and a number of 'ulama' gathered to hear testimony from witnesses to the new moon on what would either be the thirtieth of Sha'ban or the first of Ramadan.[57] No one came to give testimony, because, according to Rida, it was widely known that "definitive astronomical calculation" (*al-ḥisāb al-falakī al-qaṭ'ī*) had established that the new crescent would not be visible that night, and anyone who claimed to see it would be lying.[58] Although Rida may have been poking fun at the court for convening under these circumstances, even such a polemical stance would imply that he could reasonably attribute awareness of the "definitive astronomical calculation" to a significant portion of the city's population. Nevertheless, in the morning, the qadi of Cairo received a telegram from Fayyum, informing him that the qadi of the oasis town about 100 km south of Cairo had heard testimony from two witnesses to the new crescent the previous night, and that the month of Ramadan had therefore commenced. The qadi of Cairo refused to credit a report by telegraph and asked his colleague to send the witnesses to him in the capital. After hearing their testimony himself, the qadi of Cairo declared that Ramadan had indeed begun according to their word. As a result, most people in Cairo had missed a day of the fast. According to Rida,

---

[57] "Ithbat Ramadanina hadha fi Misr," *al-Manar* 7, no. 18 (16 Ramadan 1322/24 November 1904), 697.
[58] "Ithbat Ramadanina hadha fi Misr," 697.

much debate ensued among "elite people of knowledge and understanding" (*ahl al-'ilm wa-l-fahm min al-khawāṣṣ*)[59] regarding the fact that the moon was not supposed to have been visible, but also questioning the way in which the qadi of Cairo seemed to place himself above the qadi of Fayyum, when their authority to validate testimony should have been equal.[60]

Rida considered the whole situation a travesty. On the one hand, he heaped scorn on the qadi of Cairo for refusing to accept news by telegraph: if he did not believe that the message had come from the qadi of Fayyum, then why would he reply to it with a message for the qadi of Fayyum?[61] On the other hand, he was sure the Fayyum witnesses were wrong. He never questioned the astronomical determination that the moon would not have been visible that night, either in Cairo or Fayyum,[62] and he even referred to it frankly as the thirtieth of Sha'ban (rather than the first of Ramadan or a more neutral term, like "the night of doubt," *laylat al-shakk*).[63]

In the shadow of this controversy, Rida reflected on the legitimacy of "acting according to those who calculate in devotional practice."[64] Now, he framed the issue rather differently than he had in the 1903 fatwa supporting *ru'ya*, and how he would continue to frame it in the later fatwas taking the same position. In the case of Egypt in 1904, Rida acknowledged that "one can prevent making the determination of the month the province of some calculators, which would constitute a form of religious power and leadership that is prohibited in Islam (*sulṭa wa-ri'āsa dīniyya mamnū'a fī al-islām*), if one acts on their opinion only in a land (*balad*) where there are many trusted calculators. And their credibility is established when their calendars agree."[65] Here, the possibility emerged that accessibility of knowledge and unity of practice did not mandate *ru'yat al-hilāl*. If calculation was actually the most broadly accepted kind of knowledge in a society, then calculation would produce the most unified practice, without the danger of a "religious leadership."[66] Importantly,

---

[59] "Ithbat Ramadanina hadha fi Misr," 697.

[60] "Ithbat Ramadanina hadha fi Misr," 698.

[61] "Ithbat Ramadanina hadha fi Misr," 697.

[62] See his report of the conversation with Muhtar Pasha, discussed presently, in "Ra'y mashayikh al-'asr fi dhalik," *al-Manar* 7, no. 18 (16 Ramadan 1322/24 November 1904), 700.

[63] "Ithbat Ramadanina hadha fi Misr," 697.

[64] "Al-'amal bi-hisab al-hasibin fi al-'ibadat," *al-Manar* 7, no. 18 (16 Ramadan 1322/24 November 1904), 698.

[65] "Al-'amal bi-hisab al-hasibin fi al-'ibadat," 699.

[66] Ahmad Shakir echoed this argument in his 1939 essay in favor of a mathematically determined calendar. Moosa, "Ahmad Shakir," 77.

however, the accessibility of knowledge was no longer defined as the ability of the "masses" (al-'āmma) or even "the ignorant" (al-jāhil) to participate equally in determining the new month themselves (through ru'yat al-hilāl), but rather the breadth of society among which the calculators – unlike the 'ulama' – enjoyed credibility. And they achieved that credibility when they – unlike the 'ulama' – agreed among themselves. The unity of their predictions would produce the broad unity in practice that Rida so desired.

Was Egypt – unlike the Muslim communities of Bangkok, or Russia, or China – a place in which calculators were sufficiently in agreement, and enjoyed sufficiently broad credibility, to assume responsibility for determining the calendar? In the same 1904 essay, Rida recounted a fascinating debate that had recently occurred at the house of Muhtar Pasha. The Ottoman officer and astronomical savant had invited Rida, along with some prominent scholars (mashāyikh), for a Ramadan visit (perhaps an iftar).[67] Presumably because of the recent controversy, the conversation turned to the issue of determining the start of Ramadan by calculation. The scholars held that calculation is not a form of Sharia evidence (al-bayyināt al-shar'iyya). Muhtar Pasha responded that God, knowing how "the leaders of prior religions" had used calculation to "mess around with people's devotional practices" (yatalā' ab bihi fi 'ibādāt al-nās), made Islamic practice dependent upon witnessing so that all would be equal, "for not every place has precise, trustworthy calculators."[68] The Pasha's language, in Rida's account, echoed Rida's own reasoning: the problem with calculation is not the essential notion of a knowledge elite, but rather the historical problem that knowledge elites tended to be unreliable and manipulative.

Unlike Rida, however, Muhtar Pasha took the next logical step: if a place does have precise, trustworthy calculators – "for example, if the government gives them an observatory that issues calendars and determines the times, to make it easy for people" – then what could be the problem?[69] This conversation was happening in 1904, just a year since the opening of the Helwan Observatory, and two years since the first Survey Department almanacs had come out. The space of the observatory, and the printed pages that bore its imprimatur, were quickly adopted by those who sought to question the knowledge of the 'ulama'. To Muhtar Pasha's challenge, however, the scholars offered two rejoinders. First, they pointed out that the calculators were not, in fact, trustworthy, since

---

[67] "Ra'y mashayikh al-'asr fi dhalik," 699–701.
[68] "Ra'y mashayikh al-'asr fi dhalik," 700.
[69] "Ra'y mashayikh al-'asr fi dhalik," 700.

even in Egypt they produced conflicting calendars. Second, the Shafi'i scholars offered their intermediary position: calculation should be followed by those who are able to practice it, and those who believe them. The Pasha began to answer their first objection, using examples to show that "calculation is definitive and infallible" (*qaṭʿī la yumkin an yukhṭiʾ*).[70]

At this point, intriguingly, Rida entered the debate in defense of Muhtar Pasha. He explained the disagreement among Egyptian calendars as a result not of error in calculation, but a misunderstanding among certain calendar-makers as to the Sharia definition of the new month. (Thus, Rida foreshadowed by almost twenty years the Ministry of Waqfs' call for a single, correct calendar.) In an argument reminiscent of 'Abd Allah Fikri Pasha's treatise, Rida further pointed out that no less a Muslim scholar than Ghazali had granted the certainty of calculating the motions of the sun and moon.[71] As for the Shafi'i position, Rida remarked that if one who believes the person able to calculate should follow their calculations, then essentially everyone in Egypt should follow the computed dates of Ramadan, because "we see everyone believe them regarding the times of prayer."

By linking the issue of Ramadan with the issue of prayer timing, Rida's fatwa suggests that the plausibility of standardizing the calendar in the early twentieth century rested on the prior emergence of astronomical calculation – specifically, the printed calculations of the observatory – as a standard in determining the times of prayer. Similarly, in his 1939 work advocating an astronomically determined calendar, Ahmad Shakir wrote:

Our country, Egypt, has an observatory considered to be one of the best observatories, staffed with experts in astronomy. ... They can, according to the scientists, issue a conclusive and authoritative ruling based on compelling knowledge. What harm would it do us if we accepted their word and knowledge and relied on their calculations in this matter [of the calendar], as we already do on their calculations for prayer times and other acts of devotion?[72]

Arguing for the adoption of an astronomically determined calendar, Shakir apparently took it for granted that such a calendar would be produced by the observatory, and that everyone already agreed on the authority of the observatory to regulate prayer time. In other words, arguments for basing the calendar on astronomical calculation rested not only on a new and explicitly articulated epistemology, but also on

[70] "Ra'y mashayikh al-'asr fi dhalik," 700.
[71] "Ra'y mashayikh al-'asr fi dhalik," 700. Compare with Fikri, *Risala*, 20.
[72] I use the translation of Ebrahim Moosa, "Ahmad Shakir," 77.

the already common *experience* of praying at the times determined by a specific institution.

At the end of his account of the debate at Muhtar Pasha's house, Rida challenged the 'ulama' to produce grounds for distinguishing between the calculation of prayer times and the calculation of the new month. Either both were legitimate, he asserted, or neither was.[73] In Rida's own articles and fatwas on these issues over a thirty-year period, however, it is not clear that he ultimately picked one side or the other. Most of the time, he sided with *ru'ya* over *hisāb*. In each case, however, he was addressing a specific set of circumstances. Under these circumstances, *ru'ya* was not the timeless practice of Muslims as opposed to modern astronomy. Rather, it was a way of undercutting a variety of practices that Rida associated with unreliable 'ulama' who only bred dissension and disunity. In a place like Egypt in 1904, where it was *ru'ya* that Rida associated with the incompetence of the 'ulama', and the observatory and its publications could be relied upon to issue correct, widely respected information, Rida took a much more favorable view of calculation.[74]

Conclusion

Given the ability of a leading activist to shift between supposedly opposite views on the question of the calendar, perhaps it is best not to think of *ru'ya* and *hisāb*, seeing and calculating, in contrast to each other, but to focus on what they had in common. Rida's reading of foundational texts, particularly hadith, was doubtless crucial to his preference for *ru'yat al-hilāl*. But Rida also thought that *ru'yat al-hilāl* was, under most circumstances he encountered, the method most likely to result in a unified observance of the fast – and the importance of this consideration for him should not be underestimated, either. Confronted by a plethora of traditional practices, calendars at variance with each other, and communities in doubt from China to the Sudan, Rida, not unlike the Ministry of Endowments, sought to create "unity in the order of practice" – not only in Cairo, crucially, but across Muslim communities. Observation and calculation were both debated as ways of bringing Muslims together,

---

[73] "Ra'y mashayikh al-'asr fi dhalik," 701.

[74] In most of his discussions of *ru'yat al-hilāl*, Rida attributed a positive role to the government. One of the reasons he accepted the validity of testimony by telegraph was that the government punished forgers severely. See "Ithbat Ramadanina hadha fi Misr," 697; and *al-Manar* 6, no. 18 (16 Ramadan 1321/5 December 1903), 707.

transnationally, in a unified, correct practice. For reasons not only rooted in a new valuation of certain texts, but also contingent upon technological and political circumstances, it was observation that seemed to offer a more certain path toward this objective.

Rashid Rida's concern that Muslims not "disagree" about the timing of Ramadan was distinctive in its assumption that the correct observance of the religious duty would be a synchronized observance. There are, after all, other ways of conceptualizing the unity or agreement of a religious community that do not require everyone to perform actions at the same time. Of course, in their highly developed legal and astronomical traditions of knowledge regarding the determination of prayer times and the calendar, Muslim scholars had long emphasized the value, indeed the necessity, of performing these duties at the *right* time. In *al-Manar*, Rida drew on centuries of Islamic legal discourses that had debated, for example, the significance of a lunar sighting in one place for the beginning of Ramadan in other locales.[75] Such debates, however, had always assumed some geographical limitation to the communication of the new moon's appearance. Consequently, what was at stake in these debates was not whether everyone would actually fast on the same days, but whether a community that did not see the new moon of Ramadan would be liable for a missed day of fasting if another community had seen it.

In one of his early ruminations on the importance of *ru'yat al-hilāl*, Rida envisioned something novel:

If it became easy for every region to communicate with every other through the telegraph, which is secured against forgery, and if Muslims had a Great Imam whose Sharia rulings were executed in all their lands, and it were easy for him to inform them of the sighting of the moon that was determined before him, and they fasted accordingly – there would be considerable good in that.[76]

If every Muslim community were connected by telegraph, Rida imagined, the Caliph ("Great Imam") could disseminate the announcement of the new moon's appearance instantaneously, so that Muslims the world over would begin the fast on the same day. This dream of a central determination of time, channeled through the veins of the telegraph to effect a synchronization of society, would surely have claimed the sympathy of

---

[75] See e.g., *Al-Manar* 6, no. 17 (1 Ramadan 1321/20 November 1903), 655.
[76] "Fasl fima yuthbat bihi al-sawm wa-l-fitr," *al-Manar* 6, no. 17 (1 Ramadan 1321/20 November 1903), 646.

Henry Lyons, who was busy moving the center of just such a system from 'Abbasiyya to the Helwan Observatory that very year.[77]

Rida had no observatory of his own, but he had his "Lighthouse." Its circulating pages were the beginnings of the pan-Islamic telegraph he imagined, connecting once-distant communities to a single, clear, and correct signal. By 1930, the seamlessly coordinated global *umma* that Rida envisioned in 1903 must have seemed within reach. In Singapore that year, although observers delegated by the Mohammedan Advisory Board failed to see the lunar crescent on the day that almanacs predicted it would become visible, a report of the crescent's appearance in Southern Morocco arrived by telegraph, and a twenty-one-gun salute announced the beginning of the fast based on testimony from halfway around the world.

In the years since, however, a unified, transnational observance of Ramadan has proved elusive. Today, in countries like Egypt with strong traditions of nationalized Islamic authority, *ru'yat al-hilāl* is practiced within the framework of the nation-state: observation is performed by a national observatory or other scientific institution, whose reports are then certified by a national religious authority.[78] Elsewhere, and perhaps especially where Muslims are a minority comprising multiple ethnic diasporas, a plurality of arrangements remains the rule: some rely upon local observation of the moon, while others depend upon the declaration of the fast in a designated Islamic "center," such as Mecca.[79] In other words, a unified Islamic calendar has failed to materialize not because of the triumph of *ru'yat al-hilāl* over *ḥisāb*, but because the building of a global, technological network around *ru'yat al-hilāl* depended upon the acknowledgment of a single, global Muslim authority – just as the spread of uniform mean time in the same period required its own, globally acknowledged center at Greenwich. While the relative success of these projects diverged over the course of the early twentieth century, the impulse behind them had much in common. Others have rightly pointed

---

[77] See Chapter 6. The time balls at Alexandria and Suez, in addition to the Citadel cannon, had been controlled since 1901 by the Observatory at 'Abbasiyya, the functions of which passed to the Helwan Observatory upon the latter's opening in 1904. By 1908, Lyons could claim that "the clocks of the Telegraph offices and Railway stations throughout the country are daily corrected by reference to the signal sent automatically by the Observatory clock." Wade, "Report of the Superintendent," 8; Lyons, "Civil Time in Egypt," 158.

[78] For a recent example, see Thanassis Cambanis, "Egyptians Look for a Heavenly Sign to Start Ramadan," *New York Times*, 12 August 2010, p. A14.

[79] See, for example, the discussion of debates in South Africa: Moosa, "Ahmad Shakir," 87–89.

to the model of the hegemonic nation-state as an important source for the modern, unitary reconception of Islamic law of which Rashid Rida's writing is illustrative.[80] But for those, like Rida, who believed they were reconstructing the original Islam, as for those who were constructing a new Egyptian state at the same time, the meaning of unity was inseparable from the work of scientific actors, institutions, and techniques that arose during the viceregal and colonial periods.

[80] Dallal, "Appropriating the Past," 357.

Conclusion: Astronomy, the State, and Islamic Authority at the End of the Day

The Egyptian Survey Department, now the "General Survey Authority," still publishes an almanac of prayer times for every day of the year. As of 2011, it included twenty-six locations, from Damietta to Aswan and from the Western Desert to Sinai.[1] Meanwhile, a popular website offers the Survey Authority's method as the default way to calculate prayer times for over 40,000 locations in Egypt, and one of six ways in which web users can choose to find prayer times for six million locations worldwide.[2] Thus, the Survey of Egypt, founded as an organ of imperial financial administration, has come to serve as an instrument for Muslims the world over to define the accuracy with which they fulfill their duty to pray.

One feature that distinguishes the Survey Authority's method from other ways of calculating prayer times online is its definition of *'ishā'*, the time of the evening prayer. According to the Survey Authority, evening begins with "the disappearance of the red twilight," defined as the moment when the center of the sun's disk has set to 17°.5 below the horizon.[3] This definition is interestingly composite. The cessation of "red twilight" (*al-shafaq al-aḥmar*) is a category drawn from Islamic law, in which it marks the beginning of *'ishā'* according to the Shafi'i and Maliki legal schools.[4] To define this phenomenon mathematically, however, the Survey Authority turns to another tradition. According to an explanatory

---

[1] The almanacs from 2005 to 2011 are available at www.esa.gov.eg/CalendarData.aspx. All websites cited in this conclusion were last accessed during April 2015.

[2] "Egypt Prayer Times," www.islamicfinder.org/cityPrayerNew.php?country=Egypt. All pages on islamicfinder.org are available in Arabic as well as English.

[3] Ahmad Isma'il Khalifa, "*Muqaddima falakiyya wa-judiziyya*," www.esa.gov.eg/flaky.pdf.

[4] The Hanafi tradition prefers the later, "white twilight" (*al-shafaq al-abyaḍ*), while the Hanbali prefers the "white twilight" when conditions make it difficult to observe the sunset (e.g., in cities), and the "red twilight" when the sunset is more easily visible. See Muhammad ibn Ahmad Ibn Rushd, *Bidayat al-mujtahid wa-nihayat al-muqtasid*, vol. 1 (Cairo: Maktabat al-Kulliyyat al-Azhariyya, 1970), 114; for a discussion that includes the Hanbali positions, see Muhammad ibn al-Hasan al-Tusi, *Kitab al-khilaf*, vol. 1 (Qom: Mu'assasat al-Nashr al-Islami, 1425 [2004]), 262–64.

supplement published by Dr. Ahmad Isma'il Khalifa, President of the Almanacs Committee of the Survey Authority (as of 2011), the value of 17°.5 was adopted from "the recommendation of two foreign experts, Melthe [sic] & Lehmann, who undertook research on the twilight in Aswan in the winter of 1908, as they were tasked by the Survey Authority."[5] Adolf Miethe (d. 1927) was a professor of photochemistry at the University of Berlin-Charlottenburg, where he was best known for developing new methods in three-color photography and astrophotography.[6] Along with E. Lehmann, he traveled to Aswan in 1908 to take advantage of the clear environment to observe various astronomical and meteorological phenomena with new photographic technology.[7] In other words, the work of early German photochemists contributed to a definition of Muslim prayer times that continues to circulate widely on the internet. Thus, websites like www.islamicfinder.org and www.islamicity.org point to the authority that new scientific actors, working in conjunction with the state, gained over Islamic practice at the beginning of the twentieth century – and to the durability, even expansion, of this authority in our supposedly fragmented age of digital media.

This book has sought to explain the origins of such articulations of science, Islam, and the state as a contingent product of late Ottoman debates, rather than a necessary consequence of the rise of modern sciences. After all, why did late Ottomans need German photochemists to tell them when sunset was? Within Adolf Miethe's lifetime, 'ulama' like Muhammad al-Khudari practiced a tradition of astronomical knowledge rooted in the spaces and norms of Islamic learning. The diminished authority of such figures cannot be explained simply by the increasing availability of new kinds of astronomical texts, tools, or models. Muslim scholars had been translating such novelties into their own idiom since the eighteenth century, whether integrating mechanical timekeeping into a new subfield of mīqāt, or rendering Lalande into the latest zīj. In fact, insofar as members of the old learned elite were able to act as its interpreters, new astronomy served to enhance their role as arbiters of knowledge, rather than to marginalize their position in society.

New kinds of scientific authorities did not fill a vacuum in late Ottoman Egypt. Rather, they were conceived in a new relationship between science and the state. It is no accident that Miethe and Lehmann, and the

[5] Khalifa, "Muqaddima falakiyya wa-judiziyya," 7–8.
[6] Klaus Hentschel, "Adolf Miethe's Autobiography," *Journal for the History of Astronomy* 44, no. 155 (2013): 223–25.
[7] A. Miethe and E. Lehmann, "Dämmerungsbeobachtunger in Assuan im Winter 1908," Separat-Abdruck aus der *Meteorologischen Zeitschrift* (1909): 97–114.

new mathematical definition of *'ishā'*, came to be promoted specifically by the Survey Department. Astronomy assumed a new importance to Ottoman-Egyptian politics precisely because of its ability to define territory, both cartographically and historiographically. Long before the advent of European empire in Egypt, the Ottoman-Egyptian viceroys erected new observatories, conscripted peasants into new technical academies, and sent promising students to train and work around the globe: in Paris, Madrid, and Washington, as well as the Sudan. These new astronomers owed their objectives, their personal sustenance, their technical resources, and a large degree of their prestige to service of the state. Meanwhile, by mapping Egypt in Arabic, and writing its history specifically through the pre-Islamic sites of Alexandria and the pyramids, they laid some of the first building blocks for conceiving of Egypt outside the Ottoman sphere – even as the political context for constructing Egyptian nationalism out of these blocks lay several decades in the future.

The viceregal astronomers also began a process that would ultimately extend the reach of the new, state science into realms previously governed by the knowledge of 'ulama'. When Mahmud al-Falaki and Isma'il al-Falaki published almanacs of prayer times and engineered a new, public signal of noon, their peculiar tools and resources both appropriated the knowledge of *mīqāt* and displaced its old practitioners from their previously privileged position. At the same time, however, their French training left Mahmud and Isma'il at a certain distance from Islamic discourses in the middle of the nineteenth century. Thus, when Mahmud al-Falaki used his astronomical skills to produce a new determination of the Prophet Muhammad's birth-date, thirty years passed before someone translated his argument into Arabic, and another twenty years passed before it made its way into a popular, Arabic biography of the Prophet.

Astronomers of Mahmud al-Falaki's training and accomplishments eventually came to speak to a broad audience, and from a position of authority, in part because of another space in which astronomy was remade in late Ottoman Egypt: the Arabic press in the late nineteenth and early twentieth centuries. From missionary-educated journalists to viceregal bureaucrats and Muslim reformists, the authors, publishers, and activists who shaped the Arabic press in this period shared a conception of modern science as the work of new actors, with new technologies, in new spaces. As part of a narrative of Ottoman-Islamic decline through which new cultural elites understood themselves, the practice of science among the old learned elite of 'ulama' was increasingly constructed as an object of historical memory. Islam certainly remained relevant to public debates about astronomy, but strictly through the interpretation of its "Sharia texts."

The public authority of new astronomers received a critical boost from two interrelated developments in the early twentieth century. Under the British Occupation, astronomy took shape around the priorities of administering a global empire: surveying, to be sure, but more broadly the reorganization of "almost all scientific work in Egypt connected with measurement," and particularly the promotion of uniform mean time.[8] Meanwhile, an increasing number of Muslims turned to the new authorities of measurement to define their religious practices: setting their watches by the Helwan Observatory's time service, finding their prayer times in the tables it produced, measuring the purity of the *kiswa*'s thread in the Survey Department laboratory, and arguing that the observatory could predict the beginning and end of Ramadan. They did so not only because of the qualities internal to these new institutions, but also for a variety of their own reasons, which included post-Ottoman sensitivity to the need for unity in the global Muslim community. Especially for the emergent Salafi movement, a modern notion of synchrony informed the vision of an Islamic community united around a single, clear, and correct practice.

Put another way, the promotion of the Survey Department's method for calculating prayer times – like the citation of German photochemical work as the authoritative determination of the end of twilight – reflects the convergence of a new politics of science with a new religious sensibility. As an Islamic authority, the Survey of Egypt has not simply taken the place that 'ulama' like Muhammad al-Khudari occupied in the early nineteenth century. In fact, Muhammad al-Khudari took relatively little interest in using astronomy to define prayer times. In genres such as the *zij* and its commentary, practitioners of scholarly astronomy enabled major areas of social activity, particularly related to the calendar and astrological judgment, but the specifically ritual applications of their knowledge were much less widely used. It took a culture of science rooted in imperial state-building, with its emphasis on unitary, portable standards determined in central institutions, to enable the popularization of what was once a rare and elite religious practice. The labors of viceregal and British administrators provided a powerful resource for religious reformists who aimed to promote a unitary understanding of Islam in the early twentieth century. The rise of new scientific actors and institutions in relation to the late Ottoman state, and the linkages of colonialism, standardization, and religious reform that emerged in this period, make Dr. Khalifa's explanation for the Survey Authority's definition of twilight

---

[8] Murray, "The Survey of Egypt," 1.

plausible. Thanks to a newly close relationship of Islam, science, and the state, it became possible in Egypt to combine a category of *fiqh* (the "red twilight") with the work of German photochemists (the value of 17°.5) in order to institutionalize and disseminate a single, precise interpretation of an Islamic practice.

And yet, it is worth pointing out that the value Miethe and Lehmann found for the angular depression of the sun at the end of twilight was not 17°.5, but slightly less than 15° (14°54′). They had gone to Aswan for their own purposes, moreover, and not at the behest of the Survey Department. It was suggested in 1909 that the department take their work into consideration for the computation of prayer times, but the idea never made it into the almanacs.[9] In fact, the value of 17°.5 originated with Mahmud Effendi Naji, the Survey Department official who was responsible for the calculation of the government almanacs for the first several decades of the twentieth century. In a 1907 article that Naji wrote for the fourth issue of the department's in-house journal, he explained the definition of *'ishā'* as follows:

There are various opinions about the angular depression of the sun's centre below the horizon necessary for the disappearance of the red twilight, some saying that the depression must be 17° and some 18°. For the almanacs computed in the Department the mean 17°30′ is taken...[10]

Naji did not name his sources for the "various opinions" that he averaged to compute the time of *'ishā'*. However, 18° was the value employed by some of the most prominent of the early Muslim astronomers, including Habash al-Hasib (d. 874) and Ibn Yunus (d. 1009), while 17° predominated in the timekeeping tables that were used in Cairo from the thirteenth century onward.[11] Or, as a contemporary of Naji's who taught astronomy at al-Azhar put it, 18° was the view of the "ancients," while 17° was used by most *muwaqqit*s in his time.[12] These sources were likely the basis for Naji's computation.

In making use of these older sources, however, Naji applied a distinctly modern technique. The notion that the best value for observed phenomena should be derived from an average of observations, rather than chosen at the discretion of the observer, was characteristic of a new

[9] "Observations of Twilight at Aswan," *Cairo Scientific Journal* 3 (1909), 92. Khalifa's attribution of the value to Miethe and Lehmann appears to be based on a misreading of this article.

[10] Nagi, "The Mohammedan Hours of Prayer," 152.

[11] The same value was also used in the Damascus corpus of tables. E. Wiedemann and D.A. King, "al-Shafak," *Encyclopaedia of Islam*, 2nd edition (Brill Online).

[12] Muhammad Abu al-'Ala' al-Banna, *al-Mudhakkirat fi 'ilmay al-hay'a wa-l-miqat* (Cairo: al-Matba'a al-Rahmaniyya, 1924), 72.

approach to error that developed over the eighteenth and nineteenth centuries.[13] With the rise of statistical thinking, the reliability of data came to inhere less in the personal reputation of individuals, and more in the extent to which unavoidable errors could be managed with new concepts like "personal equation" and Gauss's bell curve.[14] That was why Mahmud al-Falaki had averaged his measurements of the height of the pyramids, and then averaged his average with the measurements taken by Jomard and others. On a smaller scale, for Mahmud Naji, given contrasting data from the tradition of 'ilm al-mīqāt, the obvious thing to do was not to pick one authority over another, but to produce a new value by taking the mean.

In other words, the Survey Department's method for calculating the time of 'ishā' turns out to have a history almost precisely opposite to the official explanation. Rather than an Islamic appropriation of modern science – using the work of German photochemists to determine the timing of prayer – what Mahmud Naji performed was a modern appropriation of science that Muslim scholars had been doing for centuries. This reversal represents a larger problem in our understanding of the global history of modern science. It has become almost a truism that the movement of knowledge between Europe and other parts of the world in the nineteenth century entailed culturally specific adaptation: that translation from French into Arabic, for example, was a process of transformation as much as replication. We think far less, however, about the ways in which traditions like scholarly astronomy provided the technical knowledge from which distinctively modern practices emerged. In a sense, the problem is akin to the challenge of broadening our understanding of the history of science and empire to include the active role of a "colonized colonizer" like Ottoman Egypt. Just as modern disciplines were not the only sources of mathematical or natural knowledge available to a Survey Department bureaucrat like Mahmud Naji, European states were not the only empires that facilitated the global circulation of scientific knowledge and objects – as the French use of Isma'il al-Falaki's survey instrument to remeasure the Paris meridian reminds us.

But one can hardly blame Dr. Khalifa, of today's Survey Department, for not seeing that the value of 17°.5 derived ultimately from the work of Fatimid and Mamluk astronomers, who had never used this number themselves. In fact, Mahmud Naji's statistical manipulation is not the only reason the history of this number may be difficult for contemporary observers to glimpse. In an era when science in Egypt – even the

---

[13] Alder, *Measure of All Things*, 305–7.
[14] Alder, *Measure of All Things*, 307; Schaffer, "Astronomers Mark Time."

science of computing prayer times – has long been assumed to operate external to the pursuit of specifically Islamic learning, it is difficult even to imagine a Survey Department bureaucrat turning to the tradition of mosque timekeeping as a source of mathematical knowledge. Only a very specific set of Islamic disciplines are assumed to inform modern life. Thus, a Muslim blogger recently offered an explanation of the Survey Department method's history that differs from Dr. Khalifa's official account, relying instead on a legal manual that the Ministry of Islamic Endowments published in 1967.[15] Following this manual, the blogger understands the value of 17°.5 as a holdover from Egypt's Ottoman past, when state affairs followed the Hanafi legal school. The dominant Hanafi opinion defines evening not by the cessation of the red twilight, but by the end of the later, "white twilight." From this perspective, to observe ʿishāʾ by the Survey Authority tables is not to follow the opinion of Miethe and Lehmann, but to participate in an ongoing disagreement that dates to the early development of Islamic law. All of the premises of this explanation are accurate: the role of the Hanafi *madhhab* under Ottoman rule, the Hanafi definition of ʿishāʾ, even the intuition that the Survey Authority's method must bear some relation to premodern Islamic debates. Yet Naji had taken 17°.5 to be the measure of the red, not the white, twilight. Whereas Dr. Khalifa's error flows from (and perpetuates) the notion that only modern scientific disciplines can provide natural knowledge, the blogger's account represents a corollary, equally flawed assumption: to the extent that an Islamic tradition remains relevant, it can only be the tradition of law.

The history of astronomy in late Ottoman Egypt expands our view of the kinds of knowledge that mattered to the emergence of a modern state, science, and religion. But it also sheds light on the fate of knowledge traditions, such as law, which already receive considerable attention. The attempt of many states to codify Islamic jurisprudence over the last two centuries is another example in which Islamic discourses have supplied data, as it were, which modern epistemologies and methods have transformed – perhaps beyond recognition. With respect to law, one prominent historian of Islam has gone so far as to characterize this process as the "structural death" of the Sharia.[16] At least in the case of science, however, we should not confuse a phenomenon that became difficult to see with something that disappeared. The rise of the Ottoman-Egyptian

---

[15] "Shabakat al-multazim al-islamiyya." www.mltzm.com/vb/showthread.php?t=36289 #axzz2Fpn6ar00.

[16] Wael B. Hallaq, *Shariʿa: Theory, Practice, Transformations* (Cambridge: Cambridge University Press, 2009), 15.

and Anglo-Egyptian empires, the growth of the Arabic press, and the emergence of Islamic reformism as a mass-mediated movement altered the circumstances and structures within which Islamic traditions of knowledge were conceived and practiced. Mahmud Naji, like many of his contemporaries among Muslim intellectuals of the early twentieth century, participated in a new state institution, in the service of which he employed new methodologies. If the values derived from the history of Islam became subject to new processes, however, they remained a basis for action – perhaps more than ever.

Appendix: Muhammad al-Khudari al-Dimyati's Introduction to his *Commentary on the Brilliancy of the Solution of the Seven Planets*

In the name of God, the merciful, the compassionate: *Blessed is He Who made constellations in the skies, and placed therein a Lamp and a Moon giving light. And it is He Who made the Night and the Day to follow each other: for such as have the will to celebrate His praises or to show their gratitude. It is He Who made the sun to be a shining glory and the moon to be a light (of beauty), and measured out stages for her; that ye might know the number of years and the count (of time). Nowise did Allah create this but in truth and righteousness. (Thus) doth He explain His Signs in detail, for those who understand. Verily, in the alternation of the night and the day, and in all that Allah hath created, in the heavens and the earth, are signs for those who fear Him.*[1] Praised and exalted be He, the creator of beings with no model preexistent, the fashioner of creatures perfect in order and alignment. We praise Him, sublime and mighty, for what he has displayed of grace in abundance, and done for all mankind in varieties of benevolence. And we pray and ask peace for our lord Muhammad, pole of the circle of eternal fortune; his family and companions, planets of guidance to the rising points of everlasting dominion; and for those who walk in their way and follow their pleasing path, so long as a star rise and set, or a full moon emerge and be whole.

Says the servant in need of his Lord's mercy, Muhammad al-Khudari al-Shafi'i (God pardon his sins and overlook his faults in both worlds):

Whereas: the science (*fann*) of timekeeping is among the most noble occupations in which lives and time may be spent, and the scholar of it is a scholar of the most noble of the sciences after the religious sciences, for by means of it one arrives at the contemplation of the heavenly bodies, and by it one is aided in the consideration of the kingdom of the heavens and earth, and the varying courses of the travelling bodies in longitude and latitude, their arrangement in this marvelous fashion, and their precise ordering according to perfect system and variation, so that

[1] Qur'an 10:5–6.

the observer is baffled by the subtleties of wisdom and marvels of creation they comprise, and he submits to the greatness of their creator and
the majesty of their inventor, saying, "*Our Lord! not for naught hast Thou
created (all) this! Glory to Thee!*";[2]

And (whereas) the (most commonly) used of (this science's) books
in our time is *Kitab al-Lumʿa fi Hall al-Kawakib al-Sayyara al-Sabʿa*
because of the slackening of interest in other tomes, and people's contentment with the apparent phenomena (*ẓawāhir al-aʿmāl*) rather than
the examination of aspects of proofs;

And (whereas) it is so utterly obscure, and completely resistant to the
one who is occupied with it (*al-muʿānī*), and I desired a commentary
that would lift the veil from its face and remove doubt and confusion
from its puzzle, but all I found was a fragment of commentary by the
eminent scholar (*al-ʿallāma*) Abu al-Khayr al-Husayni, which is, at its
best, obscure,[3] and the extent of its scholarship is insufficient for one
who needs it to clarify ambiguities and obscurities;

(Therefore) it occurred to me, after failing to know of anyone who
unveiled this text's face, to urge (my) swift quills to a commentary that
would tame its unruly meanings, and be a guide to the one laboring
to realize (his) desires. But my shortcomings impeded me, along with
lack of information, and my preoccupation with many troubles and
controversies.

I was still beginning and collecting (material) when help came to me,
and I began (the work) full of hope for guidance from God. For I saw,
while asleep, that I drank of the water that gushed from between the
fingers of (the Prophet Muhammad) peace be upon him, so that I drew
extraordinary knowledge from it, after being quite enlivened by drinking
it. People were hastening towards it, crowding around it, and I rejoiced in
my vision, saying, "Oh, my heralds of good tidings!" So (the book) drew
strength from resolve, and earlier decisions helped me by the power of
God, and by the grace of God it became a commentary fit to open hearts,
and a decisive word upon whose grove sings the nightingale of correct
inquiry – although I do not offer it on condition that it be flawless.

---

[2] Qur'an 3:191.
[3] "At its best, obscure" (*ghāyatuhā al-khusūf*): *al-khusūf* is the term for a lunar eclipse.
It may be that Khudari is saying that the fragment he possessed of Abu al-Khayr al-
Husayni's commentary only addressed, or perhaps only reached, the section of the handbook that discusses lunar eclipses. In the line that follows, however, Khudari uses *kusūf*,
the term for a solar eclipse, to mean "obscurity." Such astronomical puns run throughout the text, and I think it most likely that, paired with the subsequent pun on *kusūf*,
ghāyatuhā al-khusūf is also a playful expression of Khudari's low opinion of the older
commentary, rather than a description of the amount of the text to which he had access.

However, much good cancels a little bad, and that which cannot be fully realized should not be left completely undone.

And who is it whose attributes are all pleasing? Sufficiently noble is one whose flaws may be counted. So may God have mercy on one who considers (this book) fairly, forgives where it has stumbled, and offers it assistance. I ask the gracious and almighty, the one of judgment and power, to account [this work] as offered sincerely to God, and a means toward the attainment of paradise; to place it at the apogee of acceptance and benefit, and to deem it far from the perigee of the ignorant and the riffraff. Verily He is capable of what He wills, and the one who by nature answers. Said the author (Ibn Ghulam Allah):

"In the name of God, the merciful, the compassionate..."

Bibliography

In alphabetizing last names, I have ignored initial articles such as al-, de, and von, but not Ibn. For authors who bore multiple Arabic *nisba* names, the alphabetization follows the name by which they are most commonly known (e.g., "al-Khudari al-Dimyati, Muhammad," rather than "al-Dimyati, Muhammad al-Khudari").

ARCHIVES

Bibliotheca Alexandrina
  Dhakirat Misr al-Mu'asira http://modernegypt.bibalex.org/collections
Dar al-Watha'iq al-Qawmiyya (Egyptian National Archives)
  Diwan al-Ashghal al-'Umumiyya
  Diwan al-Jihadiyya
  Al-Diwan al-Khidiwi
  Diwan al-Maliyya
  Majlis al-Nuzzar wa-l-Wuzara'
  Watha'iq 'Abidin
Jackson Library, University of North Carolina, Greensboro
  Ethel Mary Smyth letters
L'Observatoire de Paris
  Carnets d'observations d'astronomes, 1841–1908
    Ismail Effendi Moustapha, 1854–61
  Les éclipses
  Longitudes et latitudes astronomiques, 1862–64
  Manuscrits Yvon Villarceau
Royal Astronomical Society (London)
  Reynolds Papers
  Add Mss.
Science Museum (London and Wroughton)
  Henry Lyons Papers
Special Collections Library, Cambridge University
  Royal Greenwich Observatory Archives
    George Biddell Airy Papers
    William Christie Papers
    Frank Dyson Papers
    George Lyon Tupman Papers

JOURNALS AND NEWSPAPERS

Al-Ahram
Cairo Scientific Journal
Comptes rendus hebdomadaires des séances de l'Académie des sciences
Eastern Daily Mail and Straits Morning Advertiser
Fortnightly Review
Fun
Himarat Munyati
Al-Jinan
Al-Manar
Monthly Notices of the Royal Astronomical Society
Al-Mu'ayyad
Al-Muqattam
Al-Muqtataf
Quarterly Review
Rawdat al-Madaris
Review of Reviews
Ruznâme-i Vekâyi-i Mısriye
Singapore Free Press and Mercantile Advertiser
The Straits Times

MANUSCRIPTS

Al-'Abbani, Shaykh Sa'd ibn Ahmad. *Mu'arraba li-Sana Shamsiyya 1242*. MS 371, Jawhari 42103, Azhar.

(Anonymous.) *Tahdhib al-anam fi ta'rib lalandnama*. MS TR 182, ENL.

Al-'Azzazi, Khalil. *Al-Kawkab al-Azhar fi al-'aml bi-l-rub' al-muqantar*. MS K 3990, ENL.

Muqaddima fi 'amal mawaqi' 'aqarib al-sa'at 'ala qadr al-hisas al-shar'iyya li-kull 'ard. MS TR 204, ENL.

Batru, 'Isa. *Qatf azhar rawdat al-ka'inat, wa-ma qasahu Adam min al-imtihanat*. MS Or. 12.334, Leiden University.

Al-Biruni, Abu Rayhan. *Al-Isti'ab bi-l-wujuh al-mumkina fi sina'at al-asturlab*. MS K 8528, ENL.

Al-Dimashqi, 'Abd al-Latif. *Al-Manhaj al-Aqrab li-Tashih Mawqi' al-'Aqrab*. Isl Ms. 808, Michigan; MSS DM 812, DM 1104, TR 286, Z 822, ENL.

Al-Husayni, Ahmad. *Tabaqat al-Shafi'iyya*. MS Taymur: Tarikh 1411, ENL.

Hüsnü, Hüseyin. *Tercume-i zij-i Lalande*. Ms T 6553, Istanbul University.

Ibn Hasan, Ibrahim "Shihab al-Din." *Sharh al-Lum'a fi Hall al-Kawakib al-Sab'a*. MS DM 638, ENL.

Sharh Muqaddimat Mahmud Qutb al-Mahalli. MS TM 225, ENL.

Ibn Qasim, Ahmad. *Tuhfat al-Ikhwan*. MS DM 1016, ENL.

Wasilat al-mubtadi'in li-'ilm ghurrat al-shuhur wa-l-sinin. MS DM 1016, ENL.

Al-Jurjani, 'Ali b. Muhammad al-Sharif. *Sharh al-tadhkira*. MSS TH 32, TH 42, DH 103, DH 91, DH 86, ENL.

Al-Kawm al-Rishi, Abu'l-'Abbas Ahmad ibn Ghulam Allah ibn Ahmad. *Al-Lum'a fi Hall al-Kawakib al-Sab'a*. MS Arabe 2526, BnF; MS Arab 249, Houghton Library, Harvard University.

Al-Khashshab, 'Ali. *Risala mukhtasira fi ta'dil al-zaman wa-l-kusufat wa-l-khusuf*. MS DM 261, ENL.

Al-Khawaniki, Ramadan b. Salih. *Jadawil Mawqi' al-Sa'at 'ala Hasab Awqat al-Salawat*. Isl. MS 808, Michigan; MS DM 812 and Z 822, ENL.

Al-Khudari al-Dimyati, Muhammad. *Sharh al-Lum'a fi Hall al-Kawakib al-Sab'a*. Isl. Ms. 772, Michigan; MSS 397 Bakhit 45606, 379 'Arusi 42758, 374 'Arusi 42753, Azhar; MSS 523.4089927 and 520.89927, QNL.

Sharh Zad al-Musafir fi ma'rifat wad' fadl al-da'ir. MS ṬM 124, ENL.

De Lalande, J.J. *Al-Mukhtasar fi 'ilm al-falak*. Trans. Basili Fakhr et al. MSS Arabe 2554 and 2555, BnF.

Al-Nabuli, Muhammad ibn 'Abd al-Rahman. *Kashf al-Hijab 'an Murshid al-Tullab*, Ms 386 'Arusi 42765, Azhar.

Natijat mawqi' 'aqrab al-sa'at 'ala qadr hisas awa'il awqat al-salawat fi al-shuhur al-qibtiyya. MS 317 al-Saqqa 28898, Azhar.

Al-Nisaburi, Nizam al-Din. *Tawdih al-tadhkira*. MS DH 88, ENL.

Al-Sannar, Muhammad. *Kitab Rawdat al-Fasaha fi 'Ilmay al-Handasa wa-l-Misaha*. MS TR 95, ENL.

Al-Shafi'i, Mustafa Muhammad. *Risala fi dhawat al-adhnab wa-takhti'at man za'm fana' al-'alam bi'stidam al-ard bi-najm dhi dhanab*. MS TR 82, ENL.

Al-Sharbatli, Ahmad ibn Ibrahim. *Dustur Usul 'Ilm al-Miqat*. Isl. Ms. 760, Michigan.

Al-Tahhan, Ibn Abi al-Khayr al-Husayni al-Armayuni. *Al-Durra al-Mudi'a fi Sharh al-Lum'a al-Bahiyya*. MS K 4009, ENL.

Voulgaris, Eugenios. *Kitab fi al-mi'a sina al-ula min tajassud al-masih al-mukhallis*. Trans. 'Isa Batru. Ms Isl. Ms. 3rd Ser. no. 828, Princeton University.

Yahya, Husayn. *Jadwal al-tawqi'at al-yawmiyya wa-fihi samt al-qibla*. MS K 4011.2, ENL.

Zayid, Husayn. *Al-Matla' al-sa'id fi hisabat al-kawakib al-sab'a 'ala al-rasd al-jadid*. MS ṬM 193, ENL.

PUBLISHED SOURCES

Al-'Abbasi, Muhammad. *Al-Fatawa al-Mahdiyya fi al-Waqa'i' al-Misriyya*. Cairo: al-Matba'a al-Azhariyya, 1301 [1883–84].

Al-Alusi al-Baghdadi, Abu al-Thana' Shihab al-Din Mahmud al-Hasani al-Husayni. *Al-Fayd al-Warid 'ala Rawd Marthiyyat Mawlana Khalid*. Cairo: al-Matba'a al-Kastaliyya, 1278 [1861–62].

Amin, Ahmad. *My Life*. Trans. Issa J. Boullata. Leiden: Brill, 1978.

Al-'Azzazi, Khalil. *Nukhbat qawl al-sadat fi ma'rifat ma yata'allaq bi-l-munharifat*. Cairo, 1315 [1897–98].

Tashil al-raqa'iq fi hisab al-daraj wa-l-daqa'iq. N.p.: al-Matba'a al-Bahiyya, 1299 [1881–82].

Baedeker, Karl, ed. *Egypt: Handbook for Travellers*, 5th edition. Leipsic: Karl Baedeker, 1902.

Egypt and the Sudan: Handbook for Travellers, 6th edition Leipzig: Karl Baedeker, 1908.

Al-Banna, Muhammad Abu 'l-'Ala'. *Al-Mudhakkirat fi 'ilmay al-hay'a wa-l-miqat.* Cairo: al-Matba'a al-Rahmaniyya, 1924.

Barkley, H.C. *Bulgaria before the War: During Seven Years' Experience of European Turkey and Its Inhabitants.* London: John Murray, 1877.

Bartky, Ian. *Selling the True Time: Nineteenth-Century Timekeeping in America.* Stanford, CA: Stanford University Press, 2000.

Bayram, Mustafa. *Tarikh al-Azhar.* Cairo, 1903.

Beaujour, Louis-Auguste Félix de. *A View of the Commerce of Greece, Formed after an Annual Average, from 1787 to 1797.* London: H.L. Galabin, 1800.

Bertrand, J. "Éloge historique de M. Yvon Villarceau, membre de l'Institut." Paper read at the Académie des sciences, 24 December 1888.

Bon, Gustave Le. *La civilisation des Arabes.* Syracuse [Italy]: IMAG, 1969.

Al-Bujayrimi, Sulayman ibn 'Umar ibn Muhammad. *Hashiyat al-Bujayrimi 'ala al-manhaj al-musammat al-tajrid li-naf' al-'abid 'ala sharh manhaj al-tullab li-Shaykh al-Islam Abi Yahya Zakariyya al-Ansari,* vol. 2. Cairo: Mustafa al-Babi al-Halabi, 1369 [1950].

Cambanis, Thanassis. "Egyptians Look for a Heavenly Sign to Start Ramadan." *New York Times,* 12 August 2010, p. A14.

Çelebi, Katip. *Kashf al-Zunun 'an Asami al-Kutub wa-l-Funun.* Beirut: Dar Ihya' al-Turath al-'Arabi, 1995.

Chennells, Ellen. *Recollections of an Egyptian Princess, by Her English Governess, being a record of five years' residence at the court of Ismael Pasha, Khedive.* Edinburgh: William Blackwood and Sons, 1893.

Chrysostom, John. *Qatf maqalat al-qiddis Yuhanna famm al-dhahab 'an mutala'at al-kutub al-muqaddasa.* Trans. 'Isa Batru. Beirut, 1836.

Commission des monuments d'Égypte. *Description de l'Égypte: Antiquités,* vol. 2. Paris: Imprimerie Royale, 1818.

La Description de l'Égypte: État modern, vol. 1. Paris: L'Impremerie impériale, 1809.

Dağlı, Yücel and Hamit Pehlivanlı. "Introduction." *Takvimü's-Sinîn,* by Gâzî Ahmed Muhtar Paşa. Ankara: Genelkurmay Basımevi, 1993.

Dowson, E.M. *A Report on the Work of the Survey Department in 1909.* Cairo: al-Mokattam Printing Office, 1910.

A Report on the Work of the Survey Department in 1910. Cairo: al-Mokattam Printing Office, 1911.

Duff Gordon, Lucie. *Letters from Egypt.* London: Virago, 1983.

Fakhr, Basili. *Al-Jawahir al-Fakhriyya 'an al-'illa al-ibthaqiyya.* Jerusalem: Matba'at al-Qabr al-Muqaddas al-Batriyarkiyya, 1861.

Al-Falaki, Mahmud. *Khartat Mudiriyyat al-Gharbiyya.* Cairo: Kaufmann, 1289 [1872].

Khartat Mudiriyyat al-Qalyubiyya. Cairo: Kaufmann, 1289 [1872].

Nata'ij al-afham fi taqwim al-'arab qabla al-islam wa-fi tahqiq mawli-dihi 'alayhi al-salat wa-l-salam. Trans. Ahmad Zaki Afandi. Jidda: Dar al-Manara, 1992.

Risala 'an al-Iskandariyya al-qadima wa-dawahiha wa-l-jihat al-qariba minha allati uktushifat bi-l-hafriyyat wa-a'mal sabr al-ghawr wa-l-mash wa-turuq al-bahth al-ukhra. Trans. Mahmud Salih al-Falaki. Alexandria: Dar Nashr al-Thaqafa, 1966.

Al-Zawahir al-falakiyya al-murtabita bi-bina' al-ahram. Trans. Mahmud Salih al-Falaki. Cairo: Maktabat al-Anjlu al-Misriyya, n.d.

Faye, H. *Cours d'astronomie de l'école polytechnique.* 2 vols. Paris: Gauthier-Villars, 1881–83.

Fihrist al-kutub al-'arabiyya al-mahfuza bi-l-Kutubkhana al-Khidiwiyya, 7 vols. Cairo: Matbaʻat ʻUthman ʻAbd al-Razzaq, 1305–11 [1888–93].

Fikri Basha, ʻAbd Allah. *Al-Athar al-Fikriyya,* ed. Amin Fikri. Introd. Muhammad ʻAbduh. Bulaq: al-Matbaʻa al-Kubra al-Amiriyya, 1897.

"Risala fi muqaranat baʻd nusus al-hay'a bi-l-warid fi al-nusus al-sharʻiyya." *Rawdat al-Madaris* 7, no. 5 (15 Rabiʻ al-Awwal 1293 [1876]): 1–23.

Forté, Joseph and Tadrus Hana. *Marsad Hilwan.* Cairo: Al-Matbaʻa al-Amiriyya, 1927.

Gaillot, A. "Sur la direction de la verticale à l'Observatoire de Paris." *Comptes rendus* 87 (July–December 1878): 684–704.

Gellion-Danglar, Eugène. *Lettres sur l'Égypte contemporaine (1865–1875).* Paris: Sandoz et Fischbacher, n.d.

Ghayth, ʻAbd al-Hamid Mursi. *Kitab al-Manahij al-Hamidiyya fi Hisabat al-Nata'ij al-Sanawiyya.* Cairo, 1923.

Graves, Robert Perceval. *The Life of Sir William Rowan Hamilton,* vol. 2. Dublin: Hodges, Figgis & Co., 1885.

Hall, Trowbridge. *Egypt in Silhouette.* New York: Macmillan, 1928.

Halawa, Sulayman. *Al-Kawkab al-Zahir fi Fannn al-Bahr al-Zakhir.* Cairo, 1291 [1874–75].

Al-Hamawi, Salim. *Al-Barahin al-qatʻiyya ʻala ʻadam dawaran al-kura al-ardiyya.* Alexandria: Matbaʻat al-Kawkab al-Sharqi, 1293 [1876].

Ibn Abi al-Fadl, Muhammad ibn ʻAli. *Al-Riyad al-Zahirat fi al-'Aml bi-Rubʻ al-Muqantarat.* Cairo: al-Matbaʻa al-'Amira al-Sharqiyya, 1322 [1904–05].

Ibn Abi al-Hazm, ʻAli. *The Theologus Autodidactus of Ibn al-Nafis.* Trans. and ed. Max Meyerhof and Joseph Schacht. Oxford: Clarendon, 1968.

Ibn ʻAbidin, Muhammad Amin al-shahir bi-. *Hashiyat radd al-muhtar ʻala al-durr al-mukhtar,* vol. 1. Beirut: Dar al-Fikr, 1399 [1979].

Ibn Rushd, Muhammad ibn Ahmad. *Bidayat al-mujtahid wa-nihayat al-muqtasid,* vol. 1. Cairo: Maktabat al-Kulliya al-Azhariyya, 1970–74.

Al-Jabarti, ʻAbd al-Rahman. *'Aja'ib al-Athar fi al-Tarajim wa-l-Akhbar,* vol. 2, ed. Hasan Muhammad Jawhar et al. Cairo: Lajnat al-Bayan al-'Arabi, 1958–67.

'Aja'ib al-Athar fi al-Tarajim wa-l-Akhbar, ed. Ibrahim Shams al-Din. Cairo: Maktabat Madbuli, 1997.

Napoleon in Egypt: al-Jabarti's Chronicle of the French Occupation, 1798. Trans. and ed. Shmuel Moreh. Princeton, NJ: Markus Weiner Publishers, 2004.

Keeling, B.F.E. "Khedivial Observatory, Helwan." *MNRAS* 73 no. 4 (February 1913): 254–55.

Al-Khafaji, Ahmad b. Muhammad b. ʻUmar. *Shifa' al-Ghalil fi ma fi Kalam al-'Arab min al-Dakhil,* ed. Muhammad Kashshash. Beirut: Dar al-Kutub al-'Ilmiyya, 1998.

Khalifa, Ahmad Isma'il. *Muqaddima falakiyya wa-judiziyya.* www.esa.gov.eg/flaky.pdf. Accessed April 2015.

Al-Khudari Bey, Muhammad. *Nur al-yaqin fi sirat sayyid al-mursalin.* Cairo: Dar al-Basa'ir, 2008.

Al-Khudari al-Dimyati, Muhammad. *Hashiya 'ala Sharh ibn 'Aqil 'ala Matn al-Alfiyya.* Cairo: Matba'at al-Hujar al-Nayyira, 1272 [1856].

Knox-Shaw, H. "Khedivial Observatory, Helwan." *MNRAS* 74, no. 4 (February 1914): 311–12.

"Observations of Halley's Comet made at the Khedivial Observatory, Helwan." *Survey Department Paper no. 23.* Cairo: National Printing Department, 1911.

Kudsi, Bakülu. *Efkâr ül-Ceberût fi tercümet-i Esrar il-melekût.* Istanbul: Dar üt-Tıbaat ül-Amire, 1265 [1848].

Lane, E.W. *An Account of the Manners and Customs of the Modern Egyptians: The Definitive 1860 Edition.* Cairo: AUC Press, 2003.

Lalande, Jerome de. *Astronomie,* 3rd edition. Paris: Desaint, 1792.

Compendio di astronomia, ed. Vincenzo Chiminello. Padua: Tommaso Bettinelli, 1796.

Leverrier, Urbain. "Refutation de quelques critiques et allegations portées contre les travaux de l'Observatoire impérial de Paris, et denuées de toute espèce de fondement." *Comptes rendus* 65 (January–June 1865): 105–15.

Lucas, A. and B.F.E. Keeling. "The Manufacture of the Holy Carpet." *Cairo Scientific Journal* 7, no. 81 (June 1913): 129–30.

Lyons, H.G. *The Cadastral Survey of Egypt, 1892–1907.* Cairo: National Printing Department, 1908.

"Civil Time in Egypt." *Cairo Scientific Journal* 2 (1908): 50–59.

The History of Surveying and Land-Measurement in Egypt. Cairo: National Printing Department, 1907.

A Report on the Work of the Survey Department in 1905. Cairo: al-Mokattam Printing Office, 1906.

A Report on the Work of the Survey Department in 1906. Cairo: al-Mokattam Printing Office, 1907.

Mahmoud-Bey. *L'âge et le but des Pyramides, lus dans Sirius.* Extrait des Bulletins de l'Académie Royale de Belgique, 14, 2nd Ser. Alexandria: Imprimerie Francaise Mourès, Rey & Cᵉ, 1865.

Mémoire sur l'antique Alexandrie. Copenhagen: Imprimerie de Bianco Luno, 1872.

Rapport à son altesse Mohammed Saïd, vice-roi d'Égypte, sur l'Éclipse totale de soleil observé à Dongolah (Nubie), le 18 Juillet 1860, par Mahmoud-Bey, astronome de son altesse. Paris: Mallet-Bachelier, Imprimeur-Libraire du bureau des longitudes, de l'école polytechnique, 1861.

Mahmoud Effendi. "Mémoire sur le calendrier arabe avant l'islamisme, et sur la naissance et l'âge du prophète Mohammed." *Extrait du Journal Asiatique* no. 2 (1858).

"*Mémoire sur le calendrier arabe avant l'islamisme, et sur la naissance et l'âge du prophète Mohammed.*" Brussels: Hayez, 1858.

"Mémoire sur les Calendriers judaïque et musulman: du Calendrier judaïque." *Mémoire présenté à la classe des sciences de l'Académie royale de Belgique,* 8 May 1855. Brussels: Hayez, 1855.

288   Bibliography

"Identité du rôle de l'auxiliaire *avoir* et du verbe كان, lié avec un autre verbe." *Journal Asiatique*, 5th Ser., 13 (April–May 1859): 293–309.

Al-Mawardi, Abu 'l-Hasan 'Ali bin Muhammad ibn Habib al-Basri. *Adab al-Dunya wa-l-din*. Beirut: Dar al-Kutub al-'Ilmiyya, 1987.

May, William Page. *Helwân and the Egyptian Desert*. London: George Allen, 1901.

Montucla, J.F. [*sic*]. *Histoire des mathématiques: nouvelle edition*, vol. 1. Paris: Henri Agasse, Year 7 [1798–99].

Moustapha, Ismail-Bey. "Notice nécrologique de S.E. Mahmoud Pacha El Falaki." *Notices biographiques sur S.E. Mahmoud-Pacha El Falaki*. Cairo: Impremerie Nationale, 1886.

Moustapha, Ismaïl-Effendi. *Recherche des coefficients de dilatation et étalonage de l'appareil à mesurer les bases géodesiques appartenant au gouvernement Égyptien*. Paris: Impremerie de V. Goupy et C, 1864.

Mubarak, 'Ali. *Al-Khitat al-tawfiqiyya al-jadida*, 20 vols. Bulaq: al-Matba'a al-Kubra al-Amiriyya, 1304–06 [1886–89].

Muhtar Paşa, Gâzî Ahmed. *Islah-ü Takvim*. Cairo: Matba'at Muhammad Afandi Mustafa, 1307 [1889–90].

Mukhtar Basha, Ahmad. *Riyad al-Mukhtar*. Cairo: Matba'at Bulaq, 1885.

Mukhtar Bey, Muhammad. "Tarjamat hayat al-'alim al-fadil al-maghfur lahu Mahmud Basha al-Falaki." *Notices biographiques sur S.E. Mahmoud-Pacha El Falaki*. Cairo: Impremerie Nationale, 1886.

Al-Muradi, Muhammad Khalil ibn 'Ali. *Silk al-Durar fi A'yan al-Qarn al-Thani 'Ashar*. Beirut: Dar al-Kutub al-'Ilmiyya, 1997.

Al-Murati, 'Umar b. al-Murabah b. Hasan b. 'Abd al-Qadir, ed. *Muqaddima Sharifa Kashifa lima Ihtawat 'alayhi min Rasm al-Kalimat al-Qur'aniyya wa-Dabtiha wa-'Add al-Ayy al-Munifa*. Ismailia: Maktabat al-Imam al-Bukhari, 1427 [2006].

Musa, Salama. *The Education of Salama Musa*. Trans. L.O. Schuman. Leiden: Brill, 1961.

Al-Muti'i, Muhammad Bakhit. *Kitab Irshad ahl al-milla ila ithbat al-ahilla*. Beirut: Dar Ibn Hazm, 1421[2000].

Al-Muti'i, Muhammad ibn Bakhit. *Al-Fatawa lil-Imam al-'Allama Muhammad bin Bakhit al-Muti'i*, ed. Muhammad Salim Abu 'Asi. Lebanon: al-Siddiq li-l-'Ulum, Dar Nur al-Sabah, Dar al-'Ulum, 2012.

Al-Nabuli, Muhammad. *Fath al-Mannan 'ala al-Manzuma al-Musamma Tuhfat al-Ikhwan*. Cairo: Matba'at Mustafa al-Babi al-Halabi, 1325 [1907].

Al-Nabulusi, 'Abd al-Ghani ibn Isma'il. *Jawahir al-Nusus fi Hall Kalimat al-Fusus*, ed. 'Asim Ibrahim al-Kayyani al-Husayni al-Shadhili al-Darqawi. Beirut: Dar al-Kutub al-'Ilmiyya, 2008.

Nagi, Mahmud. "The Mohammedan Hours of Prayer." *Survey Notes* no. 4 (January 1907): 151–53.

Naji, Mahmud. *Natijat al-Dawla al-Misriyya li-Sanat 1354 Hijriyya*. Cairo: al-Matba'a al-Amiriyya, 1935.

Pellissier, Augustin. *Rapport addressé a M. Le Ministre d'instruction publique et des cultes, par M. Pellissier, Professeur de philosophie, chargé d'une mission en Orient, sur l'état d'instruction publique en Égypte*. Paris: Paul Dupont, n.d.

Perceval, Caussin de. "Mémoire sur le calendrier arabe avant l'islamisme." *Journal Asiatique* 4th Ser., 1 (1843): 342–79.

Pond, John. *Astronomical Observations Made at the Royal Observatory at Greenwich.* London: T. Bensley, 1830.

Public Works Ministry (Egypt). *Report upon the Administration of the Public Works Department for 1899 by Sir W. E. Garstin, K.C.M.G., under secretary of state for public works department, with reports by the officers in charge of the several branches of the administration.* Cairo: National Printing Department, 1900.

Al-Qazwini, Zakariyya. *'Aja'ib al-makhluqat wa-ghara'ib al-mawjudat,* ed. Faruq Sa'd. Beirut: Dar al-Afaq al-Haditha, 1973.

Raghib, Idris Bey. *Kitab Tib al-nafs bi-ma'rifat al-awqat al-khams.* Cairo: Matba'at al-Muqtataf, 1312 [1894–95].

Al-Razi, Fakhr al-Din Muhammad b. 'Umar. *Al-Tafsir al-Kabir li-l-Fakhr al-Razi,* vol. 22. Tehran: Shirkat Ṣaḥāfi Nawin, 198-.

Al-Sabban, Muhammad. *Hashiya 'ala Mullawi lil-sullam.* Cairo: Dar al-Kutub al-'Arabiyya al-Kubra, 1332 [1913–14].

Sacy, Sylvestre de. "Mémoire sur divers événemens de l'histoire des Arabes avant Mahomet." *Mémoires de littérature, tirés des registres de l'Académie royale des inscriptions et belles-lettres* 48 (1808): 484–762.

Al-Safadi, Khalil ibn Aybak. *Al-Ghayth al-musjam fi sharh lamiyyat al-'ajam.* Beirut: Dar al-Kutub al-'Ilmiyya, 1975.

Saint-Hilaire, J. Barthélemy. *Lettres sur l'Égypte.* Paris: Michel Lévy, 1857.

Sarhank, Isma'il. *Haqa'iq al-akhbar 'an duwal al-bihar.* Bulaq: al-Matba'a al-Miriyya, 1314 [1896–97].

"Shabakat al-multazim al-islamiyya." www.mltzm.com/vb/showthread.php?t= 36289 #axzz2Fpn6ar00. Accessed April 2015.

Al-Sharnubi al-Azhari, 'Abd al-Majid. *Sharh dala'il al-khayrat li-l-Juzuli.* Cairo: Maktabat al-Adab, 1994.

Sharubim, Mikha'il. *Al-Kaafi fi tarikh Misr al-qadim wa-al-hadith.* Bulaq: al-Matba'a al-Kubra al-Amiriyah, 1898.

Shirwani, 'Abd al-Hamid. *Hawashi 'Abd al-Hamid al-Shirwani wa-Ahmad ibn Qasim al-'Abbadi 'ala Tuhfat al-muhtaj bi-sharh al-Minhaj,* vol. 3. Beirut: Dar Sadir, n.d.

Sidqi, Muhammad Tawfiq. *Al-Din fi nazar al-'aql al-sahih.* Cairo: Matba'at al-Manar, 1346 [1927].

Smyth, Charles Piazzi. *Life and Work at the Great Pyramid, during the Months of January, February, April, and May, A.D. 1865.* Edinburgh: Edmonston & Douglas, 1867.

Survey Department (Cairo). *Meteorological Report for the Year 1903.* Cairo: National Printing Department, 1905.

A Report on the Meteorological Observations Made at the Abbassia Observatory, Cairo, during the Year 1900. Cairo: National Printing Office, 1902.

Survey Department (Egypt). *An Almanac for the Year 1902: Compiled at the Offices of the Survey Department, Public Works Ministry.* Cairo: National Printing Department, 1902.

"A Report on the Delimitation of the Turco-Egyptian Boundary, between the Vilayet of the Hejaz and the Peninsula of Sinai (June-September, 1906), by

E.B.H. Wade, together with additions by B.F. E. Keeling and J.I. Craig."
Survey Department Paper no. 4. Cairo: National Printing Department, 1907.

*A Report on the Meteorological Observations Made at the Abbassia Observatory,
Cairo, during the years 1898 and 1899.* Cairo: National Printing Department,
1900.

Report on the Meteorological and Physical Services for 1912. Cairo: Government
Press, 1913.

Survey Department and Physical Service (Egypt). *Helwan Observatory Bulletins,*
vol. 1. Cairo, 1923.

Survey of Egypt. "The Survey of Egypt, 1898–1948," ed. G.W. Murray. *Survey
Department Paper* no. 50 (1950).

Al-Tahtawi, Rifaʿa Rafiʿ. *Takhlis al-ibriz ila talkhis Bariz.* Bulaq: Dar al-Tibaʿa
al-Khidiwiyya, 1250 [1834–35].

Taylor, John. *The Great Pyramid: Why was it Built? & Who Built it?* London:
Longman, Green, Longman, and Roberts, 1859.

Tekeli, Sevim. *16'ıncı Asırda Osmanlılarda Saat ve Takiyüddin'in "Mekanik Saat
Konstrüksüyonuna Dair En Parlak Yıldızlar" Adlı Eseri.* Ankara: Ankara
Üniversitesi Basımevi, 1966.

Toderini, L'Abbé. *De la Littérature des turcs, traduit de l'Italien en François.* Trans.
l'Abbé de Cournand. Paris: Poinçot Libraire, 1789.

Turner, H.H. "From an Oxford Note-Book." *The Observatory* 32, no. 406
(1909): 110–14.

Al-Tusi, Muhammad ibn al-Hasan. *Kitab al-khilaf,* vol. 1. Qom: Muʾassasat
al-Nashr al-Islami, 1425 [2004].

ʿUrabi, Ahmad. *Mudhakkirat al-zaʿim Ahmad ʿUrabi,* ed. ʿAbd al-Munʿim Ibrahim
al-Jumayʿi. Cairo: Matbaʿat Dar al-Kutub, 2005.

Villarceau, A.-J.Yvon. "Ascensions droites et distances polaires apparentes de la
grande comète de 1861, conclues des observations equatoriales." *Comptes
rendus* 53 (July–December 1861): 1036.

"Comparison des déterminations astronomiques faites par l'Observatoire
impérial de Paris, avec les positions et azimuts géodésiques publiés par le
Dépôt de la Guerre." *Comptes rendus* 62 (January–December 1866): 805.

"Détermination astronomique de la longitude et de la latitude de
Dunkerque." *Annales de l'Observatoire impérial de Paris: mémoires* 8
(1866): 209–356.

"Longitudes, latitudes, et azimuts terrestres au moyen des observations faites
au cercle meridien no. II de Rigaud." *Annales de l'Observatoire impérial de
Paris: mémoires* 9 (1868): 1–124.

"Memoire sur les observations de l'éclipse totale de soleil du 18 Juillet 1860,
faites en Espagne par la Commission française (extrait)." *Comptes rendus* 67
(July–December 1868): 275–76, 278.

"Observations de la comète de Tempel, faites à l'equatorial de la tour de
l'ouest, à l'Observatoire imperial de Paris." *Comptes rendus* 49 (July–
December 1859): 484.

Al-Yaziji, Ibrahim. *As'ila ila Majallat "al-Bayan" wa-Ajwibat al-Shaykh Ibrahim
al-Yaziji ʿalayha,* ed.Yusuf Khuri. Beirut: Dar al-Hamraʾ, 1993.

Zaydan, Jirji. *Tarikh al-Tamaddun al-Islami.* Cairo: Matbaʿat al-Hilal, 1906.

Zayid, Husayn. *Al-Matla' al-sa'id fi hisabat al-kawakib al-sab'a 'ala al-rasd al-jadid.* Cairo: al-Matba'a al-Baruniyya, 1304 [1887].

Zéki Bey, Ahmed. "Notice biographique sur S.E. Ismail Pacha El-Falaki." *Bulletin de la société khédiviale de géographie,* 6th Ser. Cairo: Impremerie Nationale, 1908.

Zwemer, Samuel. "The Clock, the Calendar, and the Koran." *The Moslem World* 3 (1913): 262–74.

SECONDARY LITERATURE

'Abd al-Karim, Ahmad 'Izzat. *Tarikh al-ta'lim fi 'asr Muhammad 'Ali.* Cairo: Maktabat al-Nahda al-Misriyya, 1938.

Abu 'Ali, Ahmad, and Amin al-Watani. *Al-Maktaba al-Baladiyya: Faharis al-Tabi'iyyat wa-l-Riyadiyyat wa-l-Qawanin wa-l-shara'i'.* Alexandria: Sharikat al-Matbu'at al-Misriyya, 1347 [1928].

Abu-Lughod, Janet. "Tale of Two Cities: The Origins of Modern Cairo." *CSSH* 7 (1965): 429–57.

Adıvar, Adnan. *La science chez les Turcs Ottomans.* Paris: Librairie Orientale et Américaine, 1939.

Agrama, Hussein Ali. *Questioning Secularism: Islam, Sovereignty, and the Rule of Law in Modern Egypt.* Chicago: University of Chicago Press, 2012.

Ahmed, Asad Q. "Systematic Growth in Sustained Error: A Case Study in the Dynamism of Post-Classical Islamic Scholasticism." In *Islamic Scholarly Tradition,* ed. Asad Q. Ahmed, Behnam Sadeghi, and Michael Bonner. Leiden: Brill, 2011, 343–77.

Ahmed, Asad Q., Behnam Sadeghi, and Michael Bonner, eds. *The Islamic Scholarly Tradition: Studies in History, law, and Thought in Honor of Professor Michael Allan Cook.* Leiden: Brill, 2011.

Aksan, Virginia. *Ottoman Wars, 1700–1870: An Empire Besieged.* Harlow, UK: Longman/Pearson, 2007.

Alder, Ken. *The Measure of All Things.* New York: The Free Press, 2001.

Alleaume, Ghislaine. "L'école polytechnique du Caire et ses élèves: la formation d'une élite technique dans l'Égypte du XIXème siècle." Ph.D Diss., Université de Lyon II, 1993.

Alstadt, Audrey L. "Nasihatlar of Abbas Kulu Agha Bakikhanli." *Central Asian Monuments,* ed. H.B. Paksoy. Istanbul: ISIS Press, 1992. Accessed online at http://eurasia-research.com/erc/007cam.htm.

Anawati, G.C. "Fakhr al- Din al-Razi." In *Encyclopaedia of Islam,* 2nd edition, ed. P. Bearman et al. Brill Online.

Andrews, James. *Science for the Masses: The Bolshevik State, Public Science, and the Popular Imagination in Soviet Russia, 1917–1934.* College Station, TX: Texas A & M University Press, 2003.

Ansari, S.M. Razaullah. "European Astronomy in Indo-Persian Writings." In *History of Oriental Astronomy,* ed. S.M. Razaullah Ansari. Dordrecht: Kluwer, 2002. 134–44.

Arjomand, Kamran. "The Emergence of Scientific Modernity in Iran: Controversies Surrounding Astrology and Modern Astronomy in the

Mid-Nineteenth Century." *Iranian Studies* 30, nos. 1–2 (Winter/Spring, 1997): 5–24.

Aroian, Lois Armine. *The Nationalization of Arabic and Islamic Education in Egypt: Dar al-ʿUlum and al-Azhar*. Cairo Papers in Social Science 6, no. 4. Cairo: AUC, 1983.

Asad, Talal. *Formations of the Secular*. Stanford, CA: Stanford University Press, 2003.

"The Idea of an Anthropology of Islam." Georgetown University Center for Contemporary Arab Studies Occasional Paper Series. 1986.

Aubin, David, et al., eds. *The Heavens on Earth: Observatories and Astronomy in Nineteenth-Century Science and Culture*. Durham, NC: Duke University Press, 2010.

Ayalon, Ami. *The Press in the Arab Middle East*. New York: Oxford University Press, 1995.

ʿAzzawi, ʿAbbas. *Tarikh ʿilm al-falak fi al-ʿIraq wa-ʿalaqatihi bi-l-aqtar al-Islamiyya wa-l-ʿArabiyya fi al-ʿuhud al-taliyah li-ayyam al-ʿAbbasiyyin*. Baghdad: Matbaʿat al-Majmaʿal-ʿIlmi al-ʿIraqi, 1958.

Barak, On. *On Time: Technology and Temporality in Modern Egypt*. Berkeley, CA: University of California Press, 2013.

Baron, Beth. *The Women's Awakening in Egypt: Culture, Society, and the Press*. New Haven, CT: Yale University Press, 1994.

Barton, Ruth. "Just before 'Nature': The Purpose of Science and the Purpose of Popularization in Some English Popular Science Journals of the 1860s," *Annals of Science* 55 (1998): 1–33.

Basalla, George. "The Spread of Western Science." *Science* 156, no. 3775 (1967): 611–22.

Baskerville, Jr., John. "From Tahdhiib al-Amma to Tahmiish al-Ammiyya: In Search of Social and Literary Roles for Standard and Colloquial Arabic in late 19th Century Egypt." Ph.D. Diss., University of Texas, Austin, 2009.

Bayly, C.A. *The Birth of the Modern World*. Oxford: Blackwell, 2004.

Ben-Zaken, Avner. *Cross-Cultural Scientific Exchanges in the Eastern Mediterranean, 1560–1660*. Baltimore: Johns Hopkins University Press, 2010.

"The Heavens of the Sky and the Heavens of the Heart: The Ottoman Cultural Context for the Introduction of Post-Copernican Astronomy," *BJHS* 37 (2004): 1–28.

Reading Hayy Ibn Yaqzan. Baltimore: Johns Hopkins University Press, 2011.

Berlekamp, Persis. *Wonder, Image, and Cosmos in Medieval Islam*. New Haven, CT: Yale University Press, 2011.

Biger, Gideon. "The First Political Map of Egypt," *Cartographica* 19 (1982): 83–89.

Blake, Stephen. *Time in Early Modern Islam*. Cambridge: Cambridge University Press, 2013.

Boistel, Guy, et al. *Jérôme Lalande*. Rennes: Presses universitaires de Rennes, 2010.

Bourdieu, Pierre. "Time Perspectives of the Kabyle." In *The Sociology of Time*, ed. John Hassard. London: Macmillan, 1990. 219–37.

Brentjes, Sonja. "Astronomy a Temptation? On Early Modern Encounters across the Mediterranean Sea." In *Travellers from Europe in the Ottoman and Safavid*

Empires, 16th-17th centuries: Seeking, Transforming, Discarding Knowledge. Aldershot, UK: Ashgate, 2010.

"The Prison of Categories—'Decline' and Its Company." In *Islamic Philosophy, Science, Culture, and Religion: Studies in Honor of Dimitri Gutas,* ed. Felicitas Opwis and David Reisman. Leiden: Brill, 2012. 131–56.

Brentjes, Sonja and Robert G. Morrison. "The Sciences in Islamic Societies." In *The New Cambridge History of Islam,* vol. 4, ed. Robert Irwin with William Blair. Cambridge: Cambridge University Press, 2011. 564–639.

"Bringing Sound into Middle East Studies." *IJMES* 48 (2016): 113–55.

Brown, Daniel. *Rethinking Tradition in Modern Islamic Thought.* Cambridge: Cambridge University Press, 1996.

Burnett, D. Graham. *Masters of All They Surveyed: Exploration, Geography, and a British El Dorado.* Chicago: University of Chicago Press, 2000.

Cairo University. *Helwan Observatory.* Cairo: Cairo University Press, 1958.

Canales, Jimena. *A Tenth of a Second.* Chicago: University of Chicago Press, 2009.

Carlebach, Elisheva. *Palaces of Time: Jewish Calendar and Culture in Early Modern Europe.* Cambridge, MA: Belknap Press, 2011.

Casanova, José. *Public Religions in the Modern World.* Chicago: University of Chicago Press, 1994.

Çelik, Zeynep. *Empire, Architecture, and the City: French-Ottoman Encounters, 1830–1914.* Seattle, WA: University of Washington Press, 2008.

Chamberlain, Michael. *Knowledge and Social Practice in Medieval Damascus, 1190–1350.* Cambridge: Cambridge University Press, 1994.

Charette, François. *Mathematical Instrumentation in Fourteenth-Century Egypt and Syria: The Illustrated Treatise of Najm al-Din al-Misri.* Leiden: Brill, 2003.

Cioeta, Donald. "Ottoman Censorship in Lebanon and Syria." *IJMES* 10 (1979): 167–86.

Coen, Deborah. *The Earthquake Observers.* Chicago: University of Chicago Press, 2013.

Cohn, Bernard. *Colonialism and Its Forms of Knowledge.* New York: Oxford University Press, 2004.

Cole, Juan. *Colonialism and Revolution in the Middle East.* Princeton, NJ: Princeton University Press, 1993.

Collins, Harry. *Changing Order: Replication and Induction in Scientific Practice.* London: SAGE, 1985.

Commins, David. "Religious Reformers and Arabists in Damascus, 1885–1914." *IJMES* 18, no. 4 (November 1986): 405–25.

Crowe, Michael. *Theories of the World from Antiquity to the Copernican Revolution.* Mineola, NY: Dover Publications, 1990.

Crozet, Pascal. *Les sciences modernes en Égypte: transfert et appropriation, 1805–1902.* Paris: Geuthner, 2008.

"La Trajectoire d'un scientifique égyptien au XIXe siècle: Mahmud al-Falaki (1815–1885)." *Entre reforme sociale et mouvement national: Identité et modernization en Égypte.* Cairo: CEDEJ, 1995. 285–309.

Dallal, Ahmad. "Appropriating the Past: Twentieth-century Reconstruction of Pre-Modern Islamic Thought." *Islamic Law and Society* 7, no. 3 (2000): 325–58.

Islam, Science, and the Challenge of History. New Haven, CT: Yale University Press, 2010.

An Islamic Response to Greek Astronomy. Leiden: Brill, 1995.

Davis, Ralph. *Aleppo and Devonshire Square.* London: Macmillan, 1967.

Delanoue, Gilbert. *Moralistes et politiques musulmans dans l'Egypte du XIXe siècle,* vol. 2. Cairo: Institut Français d'archéologie orientale du Caire, 1982.

DeWeese, Devin. "Authority." In *Key Themes for the Study of Islam,* ed. Jamal Elias. Oxford: Oneworld, 2010. 26–52.

Di-Capua, Yoav. *Gatekeepers of the Arab Past: Historians and History-Writing in Twentieth-Century Egypt.* Berkeley, CA: University of California Press, 2009.

Dikötter, Frank, and Ian Brown, eds. *Cultures of Confinement: A History of the Prison in Africa, Asia, and Latin America.* Ithaca, NY: Cornell University Press, 2007.

Dimirdash, Ahmad. *Mahmud Hamdi al-Falaki.* Cairo: al-Dar al-Missriyya lil-ta'lif wa-l-tarjama, 1966.

Dixon, Thomas, et al., eds. *Science and Religion: New Historical Perspectives.* Cambridge: Cambridge University Press, 2010.

Dohrn-van Rossum, Gerhard. *History of the Hour: Clocks and Modern Temporal Orders.* Trans. Thomas Dunlap. Chicago: University of Chicago Press, 1996.

Dumont, Simone. *Un Astronome des lumières: Jérôme Lalande.* Paris: Vuibert/ Observatoire de Paris, 2007.

Ebert, Johannes. *Religion und Reform in der arabischen Provinz: Ḥusayn al-Ġisr al-Tarâbulusî (1845–1909) – Ein islamischer Gelehrter zwischen Tradition und Reform.* Frankfurt am Main: Peter Lang, 1991.

Eccel, Chris. *Egypt, Islam, and Social Change: al-Azhar in Conflict and Accommodation.* Berlin: K. Schwarz, 1984.

Eickelman, Dale F. and James Piscatori. *Muslim Politics.* Princeton, NJ: Princeton University Press, 1996.

El Hadidi, Hager. "Survivals and Surviving: Belonging to Zar in Cairo." Ph.D. Diss., University of North Carolina, 2006.

El-Rouayheb, Khaled. "Opening the Gate of Verification: The Forgotten Arab-Islamic Florescence of the 17th Century." *IJMES* 38 (2006): 263–81.

Relational Syllogisms and the History of Arabic Logic. Leiden: Brill, 2010.

El Shakry, Omnia. *The Great Social Laboratory: Subjects of Knowledge in Colonial and Postcolonial Egypt.* Stanford, CA: Stanford University Press, 2007.

Ellis, Matthew. "Between Empire and Nation: the Emergence of Egypt's Libyan Borderland, 1841–1911." Ph.D. Diss., Princeton University, 2012.

Elman, Benjamin. *On Their Own Terms: Science in China, 1550–1900.* Cambridge, MA: Harvard University Press, 2005.

Elshakry, Marwa. "The Gospel of Science and American Evangelism in Late Ottoman Beirut." *Past and Present* 196 (2007): 173–214.

"Knowledge in Motion: The Cultural Politics of Modern Science Translations in Arabic." *Isis* 99 (2008): 701–30.

Reading Darwin in Arabic. Chicago: University of Chicago Press, 2013.

"When Science Became Western: Historiographical Reflections." *Isis* 101 (2010): 98–109.

Escovitz, Joseph H. *The Office of Qadi al-Qudat in Cairo under the Bahri Mamluks.* Berlin: Klaus Schwarz, 1984.

Euben, Roxanne and Muhammad Qasim Zaman, eds. *Princeton Readings in Islamist Thought* Princeton, NJ: Princeton University Press, 2009.

Fahmy, Khaled. *All the Pasha's Men.* Cambridge: Cambridge University Press, 1997.

"Dissecting the Modern Egyptian State." *IJMES* 47 (2015): 559–62.

"Medicine and Power: Towards a Social History of Medicine in Nineteenth-Century Egypt." *Cairo Papers in Social Science* 23, no. 2, ed. Enid Hill. Cairo: AUC Press, 2000.

Mehmed Ali: From Ottoman Governor to Ruler of Egypt. Oxford: Oneworld, 2009.

Al-Fandi, Muhammad Jamal al-Din, et al. *Tarikh al-Haraka al-'Ilmiyya fi Misr al-Haditha: al-'ulum al-asasiyya: al-irsad al-jawiyya wa-l-falakiyya wa-l-jiyufiziqiyya.* 1990.

Farag, Nadia. "Al-Muqtataf 1876–1900: A Study of the Influence of Victorian Thought on Modern Arabic Thought." Ph.D. diss., Oxford University, 1969.

Fleck, Ludwik. *Genesis and Development of a Scientific Fact.* Trans. Fred Bradley and Thaddeus J. Trenn. Chicago: University of Chicago Press, 1979.

Flood, Finbarr Barry. "Lost Histories of a Licit Figural Art." *IJMES* 45 (2013): 566–69.

Floor, Willem and Hasan Javadi. *Introduction to The Heavenly Rose-Garden: A History of Shirvan & Daghestan,* ed. Abbas Qoli Aqa Bakikhanov. Washington, DC: Mage, 2009.

Fortna, Benjamin. *Imperial Classroom: Islam, the State, and Education in the Late Ottoman Empire.* Oxford: Oxford University Press, 2002.

Galison, Peter. *Einstein's Clocks, Poincaré's Maps: Empires of Time.* New York: Norton, 2003.

Image and Logic: A Material History of Microphysics. Chicago: University of Chicago Press, 1997.

Gavish, Dov. *A Survey of Palestine under the British Mandate.* London: RoutledgeCurzon, 2005.

Gelvin, James and Nile Green, eds. *Global Muslims in the Age of Steam and Print.* Berkeley, CA: University of California Press, 2014.

Genuth, Sara Schechner. *Comets, Popular Culture, and the Birth of Modern Cosmology.* Princeton, NJ: Princeton University Press, 1997.

Georgeon, François and Frédéric Hitzel, eds. *Les Ottomans et le temps.* Leiden: Brill, 2012.

Gerber, Haim. *Islamic Law and Culture 1600–1840.* Leiden: Brill, 1999.

Gershoni, Israel and James Jankowski. *Commemorating the Nation.* Chicago: Middle East Documentation Center, 2004.

Egypt, Islam, and the Arabs: the Search for Egyptian Nationhood, 1900–1930. New York: Oxford University Press, 1986.

Gesink, Indira Falk. *Islamic Reform and Conservatism: al-Azhar and the Evolution of Modern Sunni Islam.* London: I.B. Tauris, 2014.

Ghazal, Amal. *Islamic Reform and Arab Nationalism: Expanding the Crescent from the Mediterranean to the Indian Ocean.* London: Routledge, 2010.

Gieryn, Thomas. "Three Truth Spots." *Journal of the History of the Behavioural Sciences* 38 (2002): 113–32.

Gilbar, Gad G. "Muslim tujjar and the Middle East and Their Networks in the Long Nineteenth Century." *Studia Islamica* 100/101 (2005): 138–202.

Glass, Dagmar. *Der Muqtataf und seine Öffentlichkeit: Aufklaerung, Raesonnement und Meinungsstreit in der fruehen arabischen Zeitschriftenkommunikation.* Würzburg: Ergon Verlag, 2004.

Glennie, Paul and Nigel Thrift. "Reworking E.P. Thompson's 'Time, Work-Discipline, and Industrial Capitalism.'" *Time and Society* 5 (1996): 275–99.

Goichon, A.M. "Hikma." In *Encyclopaedia of Islam*, 2nd edition, ed. P. Bearman et al. Brill Online.

Goldschmidt, Jr., Arthur. *Biographical Dictionary of Modern Egypt.* Cairo: AUC Press, 2000.

Goldstein, Bernard. "Copernicus and the Origins of the Helicentric System," *JHA* 33 (2002): 219–35.

Gorman, Anthony. *Historians, State and Politics in Twentieth Century Egypt: Contesting the Nation.* London: RoutledgeCurzon, 2003.

Gould, Rebecca. "Cosmopolitical Genres and Geographies: Poetry and History in the Nineteenth Century Caucasus." *Comparative Literature* 70 (2018, forthcoming).

Grafton, Anthony. *The Footnote: A Curious History.* Cambridge, MA: Harvard University Press, 1999.

Gran, Peter. *Islamic Roots of Capitalism.* Cairo: AUC Press, 1999.

Von Grunebaum, G.E. "Muslim World View and Muslim Science." *Dialectica* 17 (1963): 353–67.

Gutas, Dimitri. "Aspects of Literary Form and Genre in Arabic Logical Works." In *Glosses and Commentaries on Aristotelian Logical Texts: The Syriac, Arabic and Medieval Latin Traditions*, ed. Charles Burnett. London: Warburg Institute, 1993. 28–76.

Greek Thought, Arabic Culture. New York: Routledge, 1998.

Haddad, Mahmoud. "Arab Religious Nationalism in the Colonial Era." *JAOS* 117 (1997): 253–77.

Hallaq, Wael. *Authority, Continuity, and Change in Islamic Law.* Cambridge: Cambridge University Press, 2001.

"Model *Shurut* Works and the Dialectic of Doctrine and Practice." *Islamic Law and Society* 2, no. 2 (1995): 109–34.

Shari'a: Theory, Practice, Transformations. Cambridge: Cambridge University Press, 2009.

Hamza, Dyala. "From 'Ilm to Sihafa or the politics of the public interest (maslaha): Muhammad Rashid Ridā and his journal al-Manār (1898–1935)." In *The Making of the Arab Intellectual*, ed. Dyala Hamza. New York: Routledge, 2013. 90–127.

Hanioğlu, M. Şükrü. "Blueprints for a Future Society: Late Ottoman Materialists on Science, Religion, and Art." In *Late Ottoman Society: The Intellectual Legacy*, ed. Elisabet Özdalga. London: RoutledgeCurzon, 2005. 28–116.

A Brief History of the Late Ottoman Empire. Princeton, NJ: Princeton University Press, 2008.

The Young Turks in Opposition. New York: Oxford University Press, 1995.

Hanna, Nelly. *In Praise of Books: A Cultural History of Cairo's Middle Class, Sixteenth to the Eighteenth Century*. Cairo: AUC Press, 2004.

Hanssen, Jens. *Fin de siècle Beirut: The Making of an Ottoman Provincial Capital*. New York: Oxford University Press, 2005.

Hasan, on Muhammad 'Abd al-Ghani. *'Abd Allah Fikri*. Cairo: al-Mu'assasa al-Misriyya al-'Amma lil-ta'lif wa-l-anba' wa-l-nashr, n.d.

Hathaway, Jane. *The Politics of Households in Ottoman Egypt: The Rise of the Qazdağlis*. Cambridge: Cambridge University Press, 1997.

Hatina, Meir. *'Ulama', Politics, and the Public Sphere: An Egyptian Perspective*. Salt Lake City: University of Utah Press, 2010.

Headrick, Daniel. *The Tools of Empire: Technology and European Imperialism in the Nineteenth Century*. New York: Oxford, 1981.

Heller, H.G. *Rudolf Falb: Eine Lebens- und Charakterskizze nach persönlichen Erinnerungen*. Berlin: Friedrich Gottheiner, 1903.

Hentschel, Klaus. "Adolf Miethe's Autobiography." *Journal for the History of Astronomy* 44, no. 155 (2013): 223–25.

Heyworth-Dunne, J. "Printing and Translations under Muhammad 'Ali of Egypt: The Foundation of Modern Arabic," *JRAS* 3 (1940): 325–49.

Hilgartner, Stephen. "The Dominant View of Popularization: Conceptual Problems, Political Uses," *Social Studies of Science* 20 (1990): 519–39.

Hill, Donald. *Arabic Water-Clocks*. Aleppo: University of Aleppo Institute for the History of Arabic Science, 1981.

Hill, Peter. "Early Translations of English Fiction into Arabic." *Journal of Semitic Studies* 60 (2015): 177–212.

"The First Arabic Translations of Enlightenment Literature: The Damietta Circle of the 1800s and 1810s." *Intellectual History Review* 25 (2015): 209–33.

Hill, Richard and Peter Hogg. *A Black corps d'élite: an Egyptian Sudanese conscript battalion with the French Army in Mexico, 1863–1867, and its survivors in subsequent African history*. East Lansing: Michigan State University Press, 1995.

Holt, P.M. and M.W. Daly. *A History of the Sudan*, 6th edition. Harlow, UK: Pearson, 2011.

Hostetler, Laura. *Qing Colonial Enterprise: Ethnography and Cartography in Early Modern China*. Chicago: University of Chicago Press, 2001.

Hourani, Albert. *Arabic Thought in the Liberal Age*. Cambridge: Cambridge University Press, 1962.

Huber, Valeska. *Channelling Mobilities: Migration and Globalisation in the Suez Canal Region and Beyond, 1869–1914*. Cambridge: Cambridge University Press, 2013.

Hunter, Robert. *Egypt Under the Khedives 1805–1879: From Household Government to Modern Bureaucracy*. Cairo: AUC Press, 1999.

Ibin, Michael W. "Printing of the Qur'an." In *Encyclopaedia of the Qur'an*, ed. Jane Dammen McAuliffe. Brill Online.

İhsanoğlu, Ekmeleddin. *Science, Technology and Learning in the Ottoman Empire: Western Influence, Local Institutions, and the Transfer of Knowledge*. Aldershot, UK: Ashgate, 2004.

İhsanoğlu, Ekmeleddin, et al., eds. *Osmanlı Astronomi Literatürü Tarihi.* Istanbul: IRCICA, 1997.

Jacob, Wilson Chacko. *Working Out Egypt: Effendi Masculinity and Subject Formation in Colonial Modernity, 1870–1940.* Durham: Duke University Press, 2011.

Jami' al-Azhar. *Fihris al-Kutub al-Mawjuda bi-l-Maktaba al-Azhariyya,* 6 vols. Cairo, 1946–52.

Jami'at Fu'ad al-Awwal. *Al-Marsad al-Maliki bi-Hilwan.* Cairo: Matba'at Jami'at Fu'ad al-Awwal, 1949.

Jami'at al-Qahira. *Ma'had al-Irsad bi-Hilwan: nabdha tarikhiyya.* Cairo: Matba'at Jami'at al-Qahira, 1958.

Jansen, J.J.G. *The Interpretation of the Koran in Modern Egypt.* Leiden: Brill, 1974.

Jomier, Jacques. *Le commentaire coranique du Manâr; tendances modernes de l'exégèse coranique en Egypte.* Paris: Maisonneuve, 1954.

Jung, Sandro. "Illustrated Pocket Diaries and the Commodification of Culture." *Eighteenth-Century Life* 37 (2013): 53–84.

Katz, Jonathan. *Dreams, Sufism and Sainthood: The Visionary Career of Muhammad al-Zawawi.* Leiden: Brill, 1996.

Kennedy, E.S. *Studies in the Islamic Exact Sciences,* ed. David King and Mary Helen Kennedy. Beirut: American University of Beirut Press, 1983.

"A Survey of Astronomical Tables." *Transactions of the American Philosophical Society* 46 (1956): 123–77.

Kerr, Malcolm. *Islamic Reform: The Political and Legal Theories of Muhammad 'Abduh and Rashid Rida.* Berkeley, CA: University of California Press, 1966.

Khafaji, Muhammad 'Abd al-Mun'im. *Al-Azhar fi alf 'am,* vol. 3. Cairo: al-Matba'a al-Muniriyya bi-l-Azhar, 1374 [1954–55].

Khalidi, Rashid. "'Abd al-Ghani al-'Uraisi and al-Mufid: The Press and Arab Nationalism before 1914." In *Intellectual life in the Arab East, 1890–1939,* ed. Marwan Buheiry. Beirut: AUB Press, 1983. 38–61.

Khuri, Ibrahim. *Fihris Makhtutat Dar al-Kutub al-Zahiriyya: 'Ilm al-Hay'a wa-Mulhaqatuhu.* Damascus: Majma' al-Lugha al-'Arabiyya, 1389 [1969].

Khuri-Makdisi, Ilham. *The Eastern Mediterranean and the Making of Global Radicalism, 1860–1914.* Berkeley, CA: University of California Press, 2010.

King, David A. "The Astronomy of the Mamluks." *Isis* 74 (1983): 531–55.

Astronomy in the Service of Islam. Aldershot, UK: Ashgate, 1993.

The Call of the Muezzin. Leiden: Brill, 2004. ProQuest ebrary, Web. Last accessed 11 January 2015.

Fihris al-Makhtutat al-'Ilmiyya al-Mahfuza bi-Dar al-Kutub al-Misriyya, 2 vols. Cairo: al-Hay'a al-'Amma al-Misriyya lil-Kitab, 1986.

A Survey of the Scientific Manuscripts in the Egyptian National Library. Winona Lake, IN: Eisenbrauns, 1986.

In Synchrony with the Heavens: Studies in Astronomical Timekeeping and Instrumentation in Medieval Islam. 2 vols. Leiden: Brill, 2004.

Koç, Gülçın Tunalı. "An Ottoman Astrologer at Work: Sadullah El-Ankaravi and the Everyday Practice of İlm-ı Nücum." In *Les Ottomans et le temps,* ed. François Georgeon and Frédéric Hitzel. Leiden: Brill, 2012. 39–59.

Kozma, Liat. *Policing Egyptian Women: Sex, Law, and Medicine in Khedival Egypt.* Syracuse, NY: Syracuse University Press, 2011.

Krämer, Gudrun and Sabine Schmidtke. "Introduction." In *Speaking for Islam: Religious Authorities in Muslim societies*, ed. Gudrun Krämer and Sabine Schmidtke. Leiden: Brill, 2006.

Kropf, Evyn. "Islamic Manuscripts Collection: About the Collection." University of Michigan Library Research Guides. http://guides.lib.umich .edu/islamicmss/about. Accessed 26 June 2015.

Kuhn, Thomas. *The Copernican Revolution*. Cambridge, MA: Harvard University Press, 1957.

Kurz, Otto. *European Clocks and Watches in the Near East*. Leiden: E.J. Brill, 1975.

Laffan, Michael. "An Indonesian Community in Cairo: Continuity and Change in a Cosmopolitan Islamic Milieu," *Indonesia* 77 (2004): 1–26.

Latour, Bruno. *Science in Action*. Cambridge, MA: Harvard University Press, 1987.

Lauzière, Henri. *The Making of Salafism*. New York: Columbia University Press, 2016.

"The Reconstruction of *Salafiyya*: Reconsidering Salafism from the Perspective of Conceptual History." *IJMES* 42 (2010): 369–89.

Leemhuis, Fred. "From Palm Leaves to the Internet." In *The Cambridge Companion to the Qur'an*, ed. Jane Dammen McAuliffe. Cambridge: Cambridge University Press, 2006. 146–61.

Lightman, Bernard. *Victorian Popularizers of Science*. Chicago: University of Chicago Press, 2007.

Livingston, John. "Western Science and Educational Reform in the Thought of Rifa'a al-Tahtawi." *IJMES* 28 (1996): 543–64.

Livingstone, David. *Putting Science in Its Place*. Chicago: University of Chicago Press, 2003.

Lory, Bernard and Hervé Georgelin. "Les temps entrelacés de deux villes pluricommunautaires: Smyrne et Monastir." In *Les Ottomans et le temps*, ed. François Georgeon and Frédéric Hitzel. Leiden: Brill, 2012.

Luqa, Anwar. *Voyageurs et écrivains égyptiens en France au XIXe siècle*. Paris: Didier, 1970.

Makdisi, George. *The Rise of Colleges: Institutions of Learning in Islam and the West*. Edinburgh: Edinburgh University Press, 1981.

Marsot, Afaf Lutfi al-Sayyid. "The Beginnings of Modernization among the Rectors of al-Azhar, 1798–1879." In *Beginnings of Modernization in the Middle East: The Nineteenth Century*, ed. William R. Polk and Richard L. Chambers. Chicago: University of Chicago Press, 1966. 267–80.

Egypt in the Reign of Muhammad Ali. Cambridge: Cambridge University Press, 1984.

Masud, Khalid, Brinkley Messick, and David Powers, eds. *Islamic Legal Interpretation: Muftis and Their Fatwas*. Cambridge, MA: Harvard University Press, 1996.

Meier, Olivier. *Al-Muqtataf et le débat sur le Darwinisme: Beyrouth, 1876–1885*. Cairo: CEDEJ, 1996.

Miethe, A. and E. Lehmann. "Dämmerungsbeobachtunger in Assuan im Winter 1908." Separat-Abdruck aus der *Meteorologischen Zeitschrift* (1909): 97–114.

Melnikova, Ekaterina. "Eschatological Expectations at the Turn of the Nineteenth-Twentieth Centuries: The End of the World is [Not] Nigh?" *Forum for Anthropology and Culture* 1 (2004): 253–70.

Melvin-Koushki, Matthew. "Powers of One: The Mathematicalization of the Occult Sciences in the High Persianate Tradition." *Intellectual History of the Islamicate World* 5 (2017): 127–99.

Messick, Brinkley. *The Calligraphic State: Textual Domination and History in a Muslim Society.* Berkeley, CA: University of California Press, 1993.

Mikhail, Alan. *Nature and Empire in Ottoman Egypt.* Cambridge: Cambridge University Press, 2011.

Mitchell, Timothy. *Colonising Egypt.* Berkeley, CA: University of California Press, 1988.

Rule of Experts. Berkeley, CA: University of California Press, 2002.

Moosa, Ebrahim. "Shaykh Ahmad Shakir and the Adoption of a Scientifically-Based Lunar Calendar." *Islamic Law and Society* 5, no. 1 (1998): 57–89.

Morley, William H. "Description of an Arabic Quadrant." *Journal of the Royal Asiatic Society of Great Britain and Ireland* 17 (1860): 322–30.

Morrison, Robert. *Islam and Science: The Intellectual Career of Nizam al-Din al-Nisaburi.* Abingdon: Routledge, 2007.

"The Islamic Aspects of Cosmology, Astronomy and Astrology." In *The New Cambridge History of Islam*, vol. 4, ed. Robert Irwin and William Blair. Cambridge: Cambridge University Press, 2010. 589–613.

Mukharji, Projit. *Nationalizing the Body: The Medical Market, Print, and Daktari Medicine.* London: Anthem Press, 2009.

Murphy, Jane Holt. "Ahmad al-Damanhuri and the Utility of Expertise in Early Modern Ottoman Egypt." *Osiris* 25 (2010): 85–104.

"Improving the Mind and Delighting the Spirit: Jabarti and the Sciences in Eighteenth-Century Ottoman Cairo." Ph.D. Diss., Princeton University, 2006.

Nappi, Carla. *The Monkey and the Inkpot: Natural History and Its Transformations in Early Modern China.* Cambridge, MA: Harvard University Press, 2009.

Nasser Rabbat, "Al-Azhar Mosque: An Architectural Chronicle of Cairo's History," *Muqarnas* 13 (1996): 45–67.

Nautical Astronomy. Scranton, PA: International Textbook Company, 1902.

Netz, Reviel. *The Transformation of Mathematics in the Early Mediterranean World: From Problems to Equations.* Cambridge: Cambridge University Press, 2004.

Neuwirth, Angelika. "Cosmology." In *Encyclopaedia of the Qur'an*, ed. Jane Dammen McAullife. Brill Online.

Ogle, Vanessa. *The Global Transformation of Time: 1870–1950.* Cambridge, MA: Harvard University Press, 2015.

"Whose Time Is It? The Pluralization of Time and the Global Condition, 1870s–1940s." *American Historical Review* 118 (2013): 1376–402.

Ophir, Adi and Steven Shapin. "The Place of Knowledge: A Methodological Survey." *Science in Context* 4 (1991): 3–21.

Outram, Dourinda. *The Enlightenment*, 2nd edition. New York: Cambridge University Press, 2005.

Ouyang, Wen-Chin. "Fictive Mode, 'Journey to the West', and Transformation of Space: 'Ali Mubarak's Discourses of Modernization." *Comparative Critical Studies* 4, no. 3 (2007): 331–58.

Owen, E. Roger. *Lord Cromer: Victorian Imperialist, Edwardian Proconsul.* Oxford: Oxford University Press, 2004.

The Middle East in the World Economy. London: I.B. Tauris, 1993.

Özdemir, Kemal. *Ottoman Clocks and Watches.* Istanbul: Creative Yayıncılık, 1993.

Perceval, Caussin de. "Mémoire sur le calendrier arabe avant l'islamisme." *Journal Asiatique* 4th Ser., 1 (1843): 342–79.

Philipp, Thomas and Guido Schwald. *A Guide to 'Abd al-Rahman al-Jabarti's History of Egypt.* Stuttgart: Franz Steiner Verlag, 1994.

Pingree, David. "An Astronomer's Progress." *Proceedings of the American Philosophical Society* 143, no. 1 (1999): 73–85.

Pollock, Sheldon. *The Language of the Gods in the World of Men.* Berkeley, CA: University of California Press, 2006.

Powell, Eve Troutt. *A Different Shade of Colonialism.* Berkeley, CA: University of California Press, 2003.

Prakash, Gyan. *Another Reason: Science and the Imagination of Modern India.* Princeton, NJ: Princeton University Press, 1999.

Proctor, Robert N. and Londa Schiebinger, eds. *Agnotology: The Making and Unmaking of Ignorance.* Stanford, CA: Stanford University Press, 2008.

Al-Qadi, Wadad. "East and West in 'Ali Mubarak's '*Alamuddin.*" In *Intellectual life in the Arab East, 1890–1939*, ed. Marwan Buheiry. Beirut: AUB Press, 1983. 21–37.

Qasim, 'Abd al-Hakim 'Abd al-Ghani Muhammad. *Tarikh al-ba'that al-Misriyya ila Urubba.* Cairo: Maktabat Madbuli, 2010.

Quadri, Syed Junaid. "Transformations of Tradition: Modernity in the Thought of Muhammad Bakhit al-Muti'i." Ph.D. Diss., McGill University, 2014.

Quataert, Donald. *The Ottoman Empire: 1700–1922*, 2nd edition. Cambridge: Cambridge University Press, 2005.

Ragep, F. J. "Freeing Astronomy from Philosophy: An Aspect of Islamic Influence on Science." *Osiris*, 2nd Ser., 16 (2001): 49–71.

Nasir al-Din al-Tusi's Memoir on Astronomy, vol. 1. New York: Springer-Verlag, 1993.

"When Did Islamic Science Die (and Who Cares)?" *Viewpoint: Newsletter of the BSHS* 85 (2008): 1–3.

Ragep, F.J., et al., eds. *Tradition, Transmission, Transformation.* Leiden: Brill, 1996.

Raj, Kapil. "The Historical Anatomy of a Contact Zone: Calcutta in the Eighteenth Century." *Indian Economic and Social History Review* 18 (2011): 55–82.

Relocating Modern Science. Basingstoke: Palgrave Macmillan, 2007.

Ratcliff, Jessica. *The Transit of Venus Enterprise in Victorian Britain.* London: Pickering & Chatto, 2008.

Raymond, André. *Artisans et commerçants au Caire au XVIIIe siècle.* Damascus: Institut français de Damas, 1974.

Reimer, Michael. *Colonial Bridgehead: Government and Society in Alexandria, 1807–1882.* Boulder, CO: Westview Press, 1997.

Reid, Donald Malcolm. *Whose Pharaohs? Archaeology, Museums, and Egyptian National Identity from Napoleon to World War I.* Berkeley, CA: University of California Press, 2002.

Riexinger, Martin. *Sana'ullah Amritsari (1868–1948) und die Ahl-i-Hadis im Punjab under britches Herrschaft*. Würzburg: Ergon, 2004.

Richards, E.G. *Mapping Time*. Oxford: Oxford University Press, 1998.

Riskin, Jessica. *Science in the Age of Sensibility: The Sentimental Empiricists of the French Enlightenment*. Chicago: University of Chicago Press, 2002.

Robson, J. "Al-Baydawi." *Encyclopaedia of Islam*, 2nd edition. Brill Online, 2012.

Rogaski, Ruth. *Hygienic Modernity*. Berkeley, CA: University of California Press, 2004.

Rosenfeld, B.A. and Ekmeleddin İhsanoğlu. *Mathematicians, Astronomers, and Other Scholars of Islamic Civilization and Their Works (7th-19th c.)*. Istanbul: Research Centre for Islamic History, Art and Culture, 2003.

Rosenthal, Franz. "'Blurbs' (*taqriz*) from fourteenth-century Egypt." *Oriens* 27/28 (1981): 177–96.

Ryad, Umar. *Islamic Reformism and Christianity*. Leiden: Brill, 2009.

"A Printed Muslim 'Lighthouse' in Cairo." *Arabica* 56 (2009): 27–60.

Ryzova, Lucie. *The Age of the Efendiyya: Passages to Modernity in National-Colonial Egypt*. Oxford: Oxford University Press, 2014.

Sabra, A.I. "The Appropriation and Subsequent Naturalization of Greek Science in Medieval Islam: A Preliminary Statement." *History of Science* 25 (1987): 223–43.

"Situating Arabic Science: Locality versus Essence." *Isis* 87 (1996): 654–70.

Sajdi, Dana, ed. *Ottoman Tulips, Ottoman Coffee*. London: I.B. Tauris, 2007.

Sajdi, Dana. *The Barber of Damascus: Nouveau Literacy in the Eighteenth-Century Ottoman Levant*. Stanford, CA: Stanford University Press, 2013.

Saliba, George. "Copernican Astronomy in the Arab East: Theories of the Earth's Motion in the Nineteenth Century." In *Transfer of Modern Science and Technology to the Islamic World*, ed. Ekmeleddin İhsanoğlu. Turkey: IRCICA, 1992. 145–55.

"The Development of Astronomy in Medieval Islamic Society" *Arab Studies Quarterly* 4 (1982): 211–25.

A History of Arabic Astronomy: Planetary Theories during the Golden Age of Islam. New York: New York University Press, 1994.

Islamic Science and the Making of the European Renaissance. Cambridge, MA: MIT Press, 2007.

"An Observational Notebook of a Thirteenth-Century Astronomer." *Isis* 74 (1983): 388–401.

Samaha, A.M. *The 50th Anniversary of Helwan Observatory*. Cairo: Cairo University Press, 1954.

Sami, Amin. *Misr wa-l-Nil*. Cairo: Matbaʿat Dar al-Kutub al-Misriyya, 1938.

Sarkis, Yusuf. *Muʿjam al-Matbuʿat al-ʿArabiyya wa-l-muʿarraba*. Cairo: Maktabat Yusuf Ilyan Sarkis, 1928.

Sawaie, Mohammed. "Rifaʿa Rafiʿ al-Tahtawi and His Contribution to the Lexical Development of Modern Literary Arabic." *IJMES* 32 (2000): 395–410.

Scalenghe, Sara. "Being Different: Intersexuality, Blindness, Deafness, and Madness in Ottoman Syria." Ph.D. Diss., Georgetown University, 2006.

Schacht, Joseph. "Abu 'l-Su'ud." *Encyclopaedia of Islam*, 2nd edition. Brill Online, 2012.

"Max Meyerhof." *Osiris* 9 (1950): 7–32.

Schaefer, Dagmar. *The Crafting of the 10,000 Things*. Chicago: University of Chicago Press, 2011.

Schaffer, Simon. "The Asiatic Enlightenments of British Astronomy." In *The Brokered World: Go-Betweens and Global Intelligence, 1770–1820*, ed. Simon Schaffer et al. Sagamore Beach: Science History Publications, 2009. 49–104.

"Astronomers Mark Time: Discipline and the Personal Equation." *Science in Context* 2 (1988): 115–45.

"Keeping the Books at Paramatta Observatory." In *Heavens on Earth*, ed. David Aubin et al. Durham, NC: Duke University Press, 2010. 118–47.

"Late Victorian Metrology and Its Instrumentation: A Manufactory of Ohms." In *Invisible Connections: Instruments, Institutions, and Science*, ed. Robert Bud and Susan Cozzens. Bellingham, WA: SPIE, 1992. 23–56.

"Metrology, Metrication, and Victorian Values." In *Victorian Science in Context*, ed. Bernard Lightman. Chicago: University of Chicago Press, 1997. 438–74.

"Oriental Metrology and the Politics of Antiquity in Nineteenth Century Survey Sciences." *Science in Context* 30 (2017): 173–212.

"When the Stars Threw Down Their Spears: Histories of Astronomy and Empire." Tarner Lectures delivered at Cambridge University, February–March 2010. http://sms.cam.ac.uk/media/741069;jsessionid=2C144475E4 4DBF24CAF626035BBE4967. Accessed 2 July 2013.

Schayegh, Cyrus. *Who is Knowledgeable, is Strong: Science, Class, and the Formation of Modern Iranian Society, 1900–1950*. Berkeley, CA: University of California Press, 2009.

Schivaon, Martina. "Geodesy and Mapmaking in France and Algeria: Between Army Officers and Observatory Scientists." Trans. Charlotte Bigg and David Aubin. In *Heavens on Earth*, ed. David Aubin et al. Durham, NC: Duke University Press, 2010. 199–224.

Schivelbusch, Wolfgang. *The Railway Journey: The Industrialization of Time and Space in the 19th Century*. Berkeley, CA: University of California Press, 1977.

Schölch, Alexander. *Egypt for the Egyptians!: The Socio-Political Crisis in Egypt 1878–1882*. London: Ithaca Press, 1981.

Schwartz, Kathryn A. "Meaningful Mediums: A Material and Intellectual History of Manuscript and Print Production in Nineteenth-Century Ottoman Cairo." Ph.D. Diss., Harvard University, 2015.

"The Political Economy of Private Printing in Cairo as Told from a Commissioning Deal Turned Sour, 1871." *IJMES* 49 (2017): 25–45.

Schwarz, Paul. *Escorial-Studien zur Arabischen Literatur- und Sprachkunde*. Stuttgart: Kohlhammer, 1922.

Secord, James. *Victorian Sensation*. Chicago: University of Chicago Press, 2000.

Shaker, A.A. "The Scientific Work Done by the 30 Inch Reynolds Reflector at Helwan." *NRIAG Journal of Astronomy and Astrophysics*, Special Issue, 2008.

Shank, Michael. "Made to Order." *Isis* 105 (2014): 167–76.

Al-Shayyal, Jamal al-Din. *Mujmal Tarikh Dimyat*. Alexandria: Matba'at Don Bosco, 1949.

Tarikh al-tarjama wa-l-haraka al-thaqafiyya fi 'asr Muhammad 'Ali. Cairo: Dar al-Fikr al-'Arabi, 1951.

Shears, Jeremy and Ashraf Ahmed Shaker. "Harold Knox-Shaw and the Helwan Observatory." *Journal of the British Astronomical Association* 125 (2015): 80–93.

Sheets-Pyeson, Susan. "Popular Science Periodicals in Paris and London: The Emergence of a Low Scientific Culture, 1820–1875," *Annals of Science* 42 (1985): 549–72.

Silvera, Alain. "Edmé-François Jomard and Egyptian Reforms in 1839," *Middle Eastern Studies* 7, no. 3 (October 1971), 301–16.

Sivasundaram, Sujit. "Focus: Global Histories of Science." *Isis* 101 (2010): 95–97.

"Science." In *Pacific Histories*, ed. David Armitage and Alison Bashford. Basingstoke: Palgrave Macmillan, 2014. 237–60.

Skovgaard-Petersen, Jakob. *Defining Islam for the Egyptian State.* Leiden: Brill, 1997.

Stanley, Matthew. "Predicting the Past: Ancient Eclipses and Airy, Newcomb, and Huxley on the Authority of Science." *Isis* 103 (2012): 254–77.

Stanton, Andrea and Carole Woodall, et al. "Bringing Stound into Middle East Studies." *IJMES* 48 (2016): 113–55.

Starrett, Gregory. *Putting Islam to Work: Education, Politics, and Religious Transformation in Egypt.* Berkeley, CA: University of California Press, 1998.

Stolz, Daniel. "By Virtue of Your Knowledge: Scientific Materialism and the *Fatwa*s of Rashid Rida." *BSOAS* 72 (2012): 223–47.

"The Lighthouse and the Observatory: Islam, Authority, and Cultures of Astronomy in Late Ottoman Egypt." Ph.D. Diss., Princeton University, 2013.

"Positioning the Watch Hand: 'Ulama' and the Practice of Mechanical Timekeeping in Cairo, 1737–1874." *IJMES* 47 (2015): 489–510.

Swerdlow, Noel. "Copernicus and Astrology." *Perspectives on Science* 20 (2012): 353–78.

Tarrazi, Filib Di. *Tarikh al-Sihafa al-'Arabiyya*, vol. 3. Beirut: Al-Matba'a al-Adabiyya, 1914.

Taton, René and Curtis Wilson, eds. *Planetary Astronomy from the Renaissance to the Rise of Astrophysics.* Cambridge: Cambridge University Press, 1989.

Taub, Liba. *Ptolemy's Universe: The Natural Philosophical and Ethical Foundations.* Chicago: Open Court, 1993.

Taylor, Charles. *Philosophical Arguments.* Cambridge, MA: Harvard University Press, 1995.

Thompson, E.P. "Time, Work-Discipline, and Industrial Capitalism." *Past and Present* 38 (1967): 56–97.

Tignor, Robert L. *Modernization and British Colonial Rule, 1882–1914.* Princeton, NJ: Princeton University Press, 1966.

Tilley, Helen. "Global Histories, Vernacular Science, and African Genealogies." *Isis* 101 (2010): 110–19.

Todes, Daniel. *Pavlov's Physiology Factory: Experiment, Interpretation, Laboratory Enterprise.* Baltimore: Johns Hopkins University Press, 2002.

Toledano, Ehud. "Forgetting Egypt's Ottoman Past." In *Cultural Horizons: A Festschrift in Honor of Talat S. Halman*, vol. 1, ed. Jayne Warner. Syracuse, NY: Syracuse University Press, 2001. 150–67.

"Social and Economic Change in 'The Long Nineteenth Century.'" In *Cambridge History of Egypt*, vol. 2. Ed. M.W. Daly. Cambridge: Cambridge University Press, 1998. 252–84.

State and Society in Mid-Nineteenth-Century Egypt. Cambridge: Cambridge University Press, 1990.

Tollefson, Jr., Harold H. "The 1894 British takeover of the Egyptian Ministry of Interior." *Middle East Studies* 26, no. 4 (October 1990): 547–60.

Tresch, John. *The Romantic Machine: Utopian Science and Technology after Napoleon*. Chicago: University of Chicago Press, 2012.

Tresch, John and Emily I. Dolan. "Toward a New Organology: Instruments of Music and Science." *Osiris* 28 (2013): 278–98.

Warwick, Andrew. *Masters of Theory: Cambridge and the Rise of Mathematical Physics*. Chicago: University of Chicago Press, 2003.

Watenpaugh, Keith. *Being Modern in the Middle East*. Princeton, NJ: Princeton University Press, 2006.

Weber, Stefan. *Damascus: Ottoman Modernity and Urban Transformation, 1808–1918*. Aarhus: Aarhus University Press, 2009.

Wiedemann, E. and D.A. King. "Al-Shafak." In *Encyclopaedia of Islam*, 2nd edition, ed. P. Bearman et al. Brill Online.

Weiss, Bernard. *The Spirit of Islamic Law*. Athens, GA: University of Georgia Press, 1998.

Wensinck A.J. and D.A. King. "Mikat." *Encyclopaedia of Islam*, 2nd edition, ed. P. Bearman et al. Brill Online.

Westman, Robert. *The Copernican Question: Prognostication, Skepticism, and Celestial Order*. Berkeley, CA: University of California Pres, 2011.

White, Ian. *English Clocks for the Eastern Markets*. Sussex: Antiquarian Horological Society, 2012.

Wise, M. Norton, ed. *The Values of Precision*. Princeton, NJ: Princeton University Press, 1995.

Wishnitzer, Avner. "'Our time': On the Durability of the Alaturka Hour System in the Late Ottoman Empire," *International Journal of Turkish Studies* 16 (2010): 47–69.

Reading Clocks, Alla Turca: Ottoman Temporality and Its Transformation during the Long Nineteenth Century. Chicago: University of Chicago Press, 2015.

Yahya, Nurfadzilah. "Craving Bureaucracy: Marriage, Islamic Law, and Arab Petitioners in the Straits Settlements." *The Muslim World* 105 (2015): 496–515.

"The Question of Animal Slaughter in the British Straits Settlements During the Early Twentieth Century." *Indonesia and the Malay World* 43 (2015): 173–90.

Yalçınkaya, M. Alper. *Learned Patriots: Debating Science, State, and Society in the Nineteenth-Century Ottoman Empire*. Chicago: University of Chicago Press, 2015.

Yousef, Hoda. *Composing Egypt: Reading, Writing, and the Emergence of a Modern Nation, 1870–1930*. Stanford, CA: Stanford University Press, 2016.

"Reassessing Egypt's System of Dual Education under Isma'il: Growing 'Ilm and Shifting Ground in Egypt's First Educational Journal, *Rawdat al-Madaris*." *IJMES* 40 (2008): 109–30.

Zaman, Muhammad Qasim. "Commentaries, Print, and Patronage: 'Hadith' and the Madrasas in Modern South Asia." *BSOAS* 62 (1999): 60–81.

Modern Islamic Thought in a Radical Age: Religious Authority and Internal Criticism. Cambridge: Cambridge University Press, 2012.

The Ulama in Contemporary Islam: Custodians of Change. Princeton, NJ: Princeton University Press, 2002.

Zeghal, Malika. *Gardiens de l'Islam: les oulémas d'Al Azhar dans l'égypte contemporain.* Paris: Presses de la Fondation nationale des sciences politiques, 1996.

"Religion and Politics in Egypt: The Ulema of al-Azhar, Radical Islam, and the State (1952–1994)." *IJMES* 31, no. 3 (1999): 371–99.

Index

Printed in the United States
By Bookmasters